Studies in Logic
Volume 3

Foundations of the Formal Sciences
The History of the Concept of the Formal Sciences

Volume 1
Proof Theoretical Coherence
Kosta Dosen and Zoran Petric

Volume 2
Model Based Reasoning in Science and Engineering
Lorenzo Magnani, editor

Volume 3
Foundations of the Formal Sciences IV: The History of the Concept of the Form Sciences
Benedikt Löwe, Volker Peckhaus and Thoralf Räsch, editors

Volume 4
Algebra, Logic, Set Theory. Festschrift für Ulrich Felgner zum 65. Geburtstag
Benedikt Löwe, editor

Studies in Logic Series Editor
Dov Gabbay   dov.gabbay@kcl.ac.

Foundations of the Formal Sciences
The History of the Concept
of the Formal Sciences

edited by
Benedikt Löwe
Volker Peckhaus
Thoralf Räsch

© Individual author and King's College 2006. All rights reserved.

ISBN 1-904987-29-X
College Publications
Scientific Director: Dov Gabbay
Managing Director: Jane Spurr
Department of Computer Science
Strand, London WC2R 2LS, UK

Original cover design by Richard Fraser
Cover produced by orchid creative www.orchidcreative.co.uk
Printed by Lightning Source, Milton Keynes, UK

All rights reserved. No part of this publication may be reproduced, stored in a retrieval syst
or transmitted, in any form, or by any means, electronic, mechanical, photocopying, recording
otherwise, without prior permission, in writing, from the publisher.

CONTENTS

Preface	vii
Schedule	xiii
Conference Photo	xiv
KEVIN DE LAPLANTE Sources of Domain-Independence in the Formal Sciences	1
JOHANNES EMRICH The infinite in mathematics	17
NORMA B. GOETHE Frege on understanding mathematical truth and the science of logic	27
IVOR GRATTAN-GUINNESS Classical mechanics as a formal(ised) science	51
LEILA HAAPARANTA Husserl's argument against naturalism and his own foundation of pure philosophy	69
JENS HØYRUP Bronze age formal science?	81
CHRISTOPH KANN Medieval logic as a formal science	103
DARYN LEHOUX Logic, Physics, and Prediction in Hellenistic Philosophy: x happens, but y?	125
JUSTUS LENTSCH The Logical Background of Pragmatism	143
JAAP MAAT The Status of Logic in the Seventeenth Century	157

MARTIN NEUMANN
A formal bridge between epistemic cultures 169

RAINER OSSWALD
On Formal Objects 183

MICHAEL OTTE
The Equals Sign: a Peircean View 199

SUSANNE PREDIGER
Mathematics — Cultural Product or Epistemic Exception? 217

DIRK SCHLIMM
Axiomatics and Progress in the Light of 20th Century Philosophy
of Science and Mathematics 233

RISTO VILKKO
Existence, Identity, and the Algebra of Logic 255

STEPHANIE WEBER-SCHROTH
The Formal Aspect of the Fourteenth Century Concept of
Consequence 267

Preface

The conference *"Foundations of the Formal Sciences IV: The History of the Concept of the Formal Sciences"* (FotFS IV) was held from February 14th to 17th, 2004 as an exceptional meeting in the sequence of FotFS conferences covering interdisciplinary aspects of logic and the formal sciences in general, as it allowed us to assume a meta-perspective from which we were able to survey the background of the whole conference series.

The conference was the fourth conference in the series after FotFS I in Berlin (May 1999), FotFS II in Bonn (November 2000), and FotFS III in Vienna (September 2001). The proceedings volumes of the first three conferences appeared in the journal *Synthese* and the book series *Trends in Logic*, respectively. After FotFS IV, we held another very successful meeting in the series, FotFS V, on the topic of "Infinite Games" in Bonn (November 2004). The next conference, FotFS VI: *Reasoning about Probabilities and Probabilistic Reasoning* is planned for May 2007 in Amsterdam.

'What are the *Formal Sciences*?'

The title "Foundations of the Formal Sciences" had been attracting many questions about the scope of the series for several years, in particular the most general of them all: 'What are the *Formal Sciences*?'

This question was discussed in the introductory paper *"The Formal Sciences: Their Scope, their Foundations, and their Unity"* (Synthese 133 (2002), p. 5-11) in which we tried to specify the scope of the conference series.

For the first FotFS conference in Berlin, we had understood "Foundations of the Formal Sciences" as a vague term encompassing foundations of mathematics and computer science, computational, mathematical and formal linguistics, and analytic philosophy. *Logic* was identified as the glue between these vastly different areas, where we took "logic" in its broadest and most liberal sense. We were hardly able to give precise criteria according to which the mentioned research areas were supposed to be classified as "Foundations of the Formal Sciences". And yet, there was a strong feeling that the areas were akin to each other. This proximity is partly explained by overlaps in subject matter and methodology, and partly by many centuries of a common historical background. A substantial factor for the common historical background is the Aristotelean *form–matter* distinction and its long history as one of the cornerstones of metaphysics, built in preeminently into the Western educational system.

The central question 'What are the *Formal Sciences*?' can be split up in many more concrete questions that we planned to tackle during the conference FotFS IV:

- Can we develop a theoretical classification of the sciences that juxtaposes a group of *formal sciences* including the listed areas to the well-known and fairly well delineated *natural sciences, social sciences,* and *humanities*?

- Can we give this theoretical classification solely by identifying common methodological features that define the *formal sciences*?

- If not, can we identify changes of the notion of *formal sciences* over time? Can we identify different concepts of *formal sciences* (not necessarily under this name) throughout the centuries?

- How were the areas that we listed as "Foundations of the Formal Sciences" classified throughout history? If they were not perceived as a separate class of sciences, how were they classified?

- Did the classification of these research areas influence the research done in the area?

If you want to find answers to an array of questions with this wide scope, you need researchers from many areas that can address the philosophical, historical and logical issues at hand. Investigating the *History of the Concept of the Formal Sciences* can only be done by a group of researchers interested in going beyond the traditional boundaries of their subjects. We were pleasantly surprised by the enthusiasm with which the various involved communities welcomed our conference and received a high number of high-quality submissions.

With so much expertise present at FotFS IV, we were able to shed some light at the historical, philosophical and methodological issues involved in approaching our question 'What are the *Formal Sciences*?'. At this conference, we opened the gigantic folio of the formal sciences, and we have just begun to read in it by opening some of its seals. Many more questions remain, and many more connections –both synchronic and diachronic– wait to be unearthed. The presentations at the conference and the papers in this volume stand witness to our efforts, and awake the hope that much more work will be done along these lines in the future.

The conference.

The conference FotFS IV was held at the *Mathematisches Institut* at the *Rheinische Friedrich-Wilhelms-Universität Bonn*, generously funded by the

the *Bonn International Graduate School for Mathematics, Physics, and Astronomy* (BIGS-MPA).

As it befits an interdisciplinary conference, the participants were welcomed by representatives of the medical sciences, the humanities, the natural sciences and mathematics: In the name of the rectorate of the university, *Prorektor* Andreas Hirner (by trade a surgeon at the *Klinik für Allgemein-, Viszeral-, Thorax- und Gefäßchirurgie*), greeted the participants while stressing the methodological differences between the formal sciences and his own area of expertise. Standing for the humanities and the natural sciences, *Dekan* Georg Rudinger of the *Philosophische Fakultät* and *Prodekan* Ingo Lieb of the *Mathematisch-Naturwissenschaftliche Fakultät* extended their welcomes to the participants, and last but not least, the managing director of the BIGS-MPA, Carl-Friedrich Bödigheimer from the *Mathematisches Institut* represented the host institute.

The first evening of the conference was at the same time part of FotFS IV and a special event for the graduate students at the BIGS-MPA, the "BIGS Student Afternoon in History and Philosophy of Mathematics".

During the preparatory phase of the conference, we had the unreserved support of our host institute for which we would like to extend our thanks, in particular to Peter Koepke. For the conference itself, we had a reliable staff of helpers consisting of Miriam Blum (Münster), Andreas Bösel (Potsdam), Stefan Bold (Bonn), Patrick Braselmann (Bonn), Frederik Herzberg (Bonn), Jorrit Kirsten (Bonn), Torsten Langer (Bonn), Oliver Lorscheid (Bonn), and Michael Möllerfeld (Bonn).

The following is a list of all presentations scheduled for FotFS IV (some were cancelled due to health reasons, but the abstracts were printed in our abstract booklet):

- Stefan Artmann (Jena): *"From the Forms of Life to the Life of Forms: Neo-Darwinism, Artificial Life, and Biosemiotics"*

- Jean-Yves Béziau (Neuchâtel): *"The Formal Character of Logic"*

- Jessica Carter (Odense): *"Ontology and Mathematical Practice"*

- Yury Chernoskutov (St.Petersburg): *"Frege as formal scientist: pro et contra"*

- Kevin de Laplante (Ames IA): *"Understanding the Formal Sciences"*

- Friedrich Dudda (Bochum): *"Mathematics as a General Science of Structure"*

- Johannes Emrich (Erlangen): *"The infinite in mathematics — a regulative idea?"*

- Catarina Dutilh Novaes (Leiden): *"Medieval Logic and the Modern Notion of 'Formal'"*

- Anthony Gardiner (Birmingham): *"The Platonic universe of ideal objects in mathematics"*

- Norma B. Goethe (Cordoba): *"Reasoning styles in modern mathematics: the ways of discovery, learning and the epistemic virtues of justification"*

- Ivor Grattan-Guinness (London): *"History or heritage? Historians and mathematicians on the history of mathematics"* (TALK AT THE BIGS PHD STUDENT AFTERNOON)

- Ivor Grattan-Guinness (London): *"How it means: What do mathematical theories say when they are used in physical theories?"*

- Leila Haaparanta (Tampere): *"On the Concepts of Formal, Pure and Naturalistic Philosophy"*

- Stefan Heßbrüggen-Walter (Marburg): *"Between Ambrose and the Arians: Augustine and the Value of Logic as a Formal Science"*

- Jens Høyrup (Roskilde): *"Bronze Age Formal Sciences?"*

- Franz Huber (Konstanz): *"Being Formal For the Sole Reason of Being Formal: Conditions of Adequacy for Formal Theories"*

- Christoph Kann (Düsseldorf): *"Dialectica est ars artium. Aspects of medieval logic as a formal science"*

- Ladislav Kvasz (Bratislava): *"On the Foundations of Formal Epistemology"*

- Javier Legris (Buenos Aires): *"Symbolic Knowledge and Formal Science in Frege and Schröder"* (cancelled)

- Daryn Lehoux (Halifax NS): *"The Formalization of Prediction in Antiquity"*

- Justus Lentsch (Bielefeld): *"C. S. Peirce's Lattice Theory as a Formal Framework for his Pragmatic Account of Meaning"*

- Dieter Lohmar (Köln): *"Die Formalwissenschaften aus dem Gesichtspunkt der Phänomenologie Husserls"*

- Jaap Maat (Amsterdam): *"The Status of Logic in the Seventeenth Century"*

- Ángel Nepomuceno-Fernández (Sevilla), Fernando Soler-Toscano (Sevilla): *"Concept Script: From Logic of Language to Language of Logic"*

- Martin Neumann (Osnabrück): *"Transforming concepts of Science into the Humanities: objective possibility in the times of 2nd German empire"*

- Olaf Neumann (Jena): *"What is Divine in the Divine Proportion?"*

- Terese M. O. Nielsen (Roskilde): *"Platonisms: Gödel, Quine and the Image of Mathematics"*

- Rainer Osswald (Hagen): *"On the Notion of a Formal Object"*

- Michael Otte (Bielefeld): *"A = B: Darstellen und Erklären in den formalen Wissenschaften"*

- Volker Peckhaus (Paderborn): *"The Historiography of Mathematics and Logic"* (TALK AT THE BIGS PHD STUDENT AFTERNOON)

- Tommaso Perrone (München): *"On the formalistic approach to the dynamic of scientific theories. The structural point of view"* (cancelled)

- Susanne Prediger (Bremen): *"Mathematics — cultural product or epistemic exception?"*

- Günter Schenk (Halle): *"Formalwissenschaftliches Denken bei Johann Heinrich Lambert"* (cancelled)

- Dirk Schlimm (Pittsburgh PA): *"The axiomatic method in the light of 20th century philosophy"*

- Charles A. Stewart (Dresden): *"Conceptual harmony and the semantics of programming languages"*

- Michael Stöltzner (Bielefeld): *"Hilbert's Axiomatic Method and Its Defective Reception by Logical Empiricists"*

- Christian Thiel (Erlangen): *"What is it like to be formal?"*

- Rüdiger Thiele (Leipzig): *"On the roots of Hilbert's proof theory; Some remarks in consideration of the canceled 24th problem"* (cancelled)

- Joanne Twining (Denver CO): *"Nitecki's Metaphorical Geometry and Unity of Knowledge"* (cancelled)

- Risto Vilkko (Helsinki): *"Existence, Identity, and the Algebra of Logic"*

- Stephanie Weber (Göttingen): *"The formal aspect of fourteenth century's theory of consequences"*

- Markus Werning (Düsseldorf): *"A neglect of semantics: What Frege might have objected to Gödel"*

This volume.

There are many interesting features of producing an interdisciplinary volume for a wide audience. This volume covers many different aspects connected to the topic of the conference, and none of our readers will be knowledgeable in all of them. Therefore papers in such a volume need to go beyond specialized research papers and try to cater for the experts and the non-experts at the same time.

Producing such a paper requires some extra effort and a lot of help. Forty anonymous referees helped to ensure the scientific quality of this volume while keeping in mind its intended wide audience and watching the quality of exposition in the submitted papers. Some of the referee reports were exceptionally detailed and came with many helpful suggestions to improve the presentation.

This volume was typeset with a LATEX class originally written by Philipp Rohde and later amended by Benedikt Löwe, Michael Möllerfeld and Thoralf Räsch. We had the help of Wioletta Ruszel and Stefan Bold for typesetting and bibliographic work and would like to gratefully acknowledge the support of Jane Spurr of College Publications. The final typesetting meetings in Amsterdam were partially supported by NWO grant DN 61-532.

Amsterdam/Paderborn/Potsdam, July 2005   B. L. V. P. Th. R.

Schedule

Time	Friday, February 14th	Saturday, February 15th	Sunday, February 16th	Monday, February 17th
09:00–10:00		DE LAPLANTE	THIEL	HØYRUP
10:00–10:15		*Break*	*Break*	*Break*
10:15–10:45		Haaparanta	Maat	Heßbrüggen-Walter
10:45–11:15		Vilkko	Lentsch	Lehoux
11:15–11:30		*Break*	*Break*	*Break*
11:30–12:00		M.Neumann	Steward	Schlimm
12:00–12:30		Béziau	Stölzner	O.Neumann
12:30–15:00		*Lunch Break*	*Lunch Break*	*Lunch Break*
15:00–15:30	**Opening**	PREDIGER	Carter	Weber
15:30–15:45	**Opening**	PREDIGER	Carter	Weber
15:45–16:00		PREDIGER	Lohmar	Dutilh Novaes
16:00–16:15	Peckhaus	Goethe	*Break*	*Break*
16:15–16:30	Peckhaus	Goethe	*Break*	*Break*
16:30–17:00		*Break*	Osswald	Kann
17:00–17:15	*Break*	Dudda	Emrich	**Closing**
17:15–17:30		Dudda	Emrich	**Closing**
17:30–17:45	Grattan-Guinness	Werning	*Break*	
17:45–18:00	Grattan-Guinness	Werning	Artmann	Chernoskutov
18:00–18:15		*Break*	Artmann	Chernoskutov
18:15–18:30		Nepumuceno-Soler	Perrone	Nielsen	Huber
18:30–18:45		Nepumuceno-Soler	Perrone	Nielsen	Huber
18:45–19:00		*Break*	*Break*	
19:00–19:30		GRATTAN-GUINNESS	Kvasz	
19:30–20:00		GRATTAN-GUINNESS	Gardiner	

xiv

Conference Photo

Benedikt **Löwe**, Volker **Peckhaus**, Thoralf **Räsch** (*eds.*)
Foundations of the Formal Sciences IV
The History of the Concept of the Formal Sciences

Sources of Domain-Independence in the Formal Sciences

KEVIN DE LAPLANTE

Iowa State University
Department of Philosophy and Religious Studies
402 Catt
Ames IA 50011-1306, United States of America
E-mail: kdelapla@iastate.edu

> ABSTRACT. The formal sciences are often contrasted with orthodox natural and social science, but it is not obvious how to draw this contrast. On the one hand, the formal sciences are closely related to mathematics; on the other hand, they are often used to describe and explain observable phenomena in the natural world. I propose a conception of formal science that situates the formal sciences between the world of pure mathematics and the world of general physical theory. The proposed conception is grounded in the notion of "domain-independence", and highlights the role of both formal and physical constraints in the description and explanation of domain-independent behaviors.

1 Introduction

Any discussion of the concept of *formal science* must acknowledge that the term is used in different ways, for different purposes, by different people. For some, the formal sciences are defined by the exclusive use of deductive methods for discovering, or reasoning about, the properties of formal, abstract systems. On this view, the formal sciences are synonymous with mathematics, formal logic, and certain branches of linguistics and computer science that emphasize the study of formal languages. For others, "formal science" means something like "exact science", or "formalized science". On this view, any scientific discipline that places heavy emphasis on mathematical or logical formalization of key theoretical concepts and theories, could be described as a formal science. This latter conception of formal science is much more liberal than the former, and would include all of physics, much of chemistry, and some parts of biology, ecology, psychology and economics, as well as newer computation-oriented disciplines like artificial life and artificial intelligence that do not fit easily within the traditional classification of the sciences.

In this paper I discuss a conception of formal science that is motivated by my work as a philosopher of science studying the so-called "complex systems sciences" — *i.e.*, sciences that purport to describe and explain emergent macro-level behaviors of systems composed of many interacting micro-level components. The central concept in the proposed conception of formal science is the notion of *domain-independence*: a science is "formal" to the extent that it describes and explains behavioral phenomena that may be observed in systems of differing natural kind-types (physical, chemical, biological, ecological, *etc.*) and at varying spatial and temporal scales. In terms of scope, the resulting conception lies somewhere in between the more restrictive and more liberal conceptions of formal science described above.

The most common way of understanding domain-independence is in "structuralist" terms — *i.e.*, a behavior is domain-independent in virtue of resulting from purely formal or structural properties of the natural system in question, properties that may be instantiated by systems of different natural kind-types. One of the aims of this paper is to challenge this interpretation of domain-independence. I argue that this interpretation fails to appreciate the distinctive role that physical (as opposed to formal) constraints play in the generation and explanation of domain-independent behaviors.

2 Domain-Independence and the Concept of a Formal Science

One interesting feature of the macro-level behaviors of complex systems is that they can often be realized in systems of different natural kind-types (physical, chemical, biological, ecological, *etc.*). To give just two examples:

i) the same critical point phenomena observed in phase transitions in gases and fluids can be observed in the transition from ferromagnetic to paramagnetic state in magnetic materials; and

ii) the same "period-doubling route" to chaotic dynamics has been observed in systems as diverse as fluids, chemical clocks, electrical circuits, lasers and acoustic systems [Les98].

These and other complex systems behaviors have the following generic features (I borrow this formulation from Robert Batterman's description of "universal" behaviors in thermodynamics and statistical physics [Bat02, p.13]):

1. The details of the system (those details that would feature in a complete causal-mechanical explanation of the system's behavior) are largely irrelevant for describing the behavior of interest.

2. Many different systems with completely different "micro" details will exhibit the identical behavior.

I call the behavior of a natural system *domain-independent* if it satisfies the above two criteria ("domain" refers to natural kind domains — physical systems, chemical systems, biological systems, ecological systems, *etc.*).

The concept of domain-independence suggests one way of characterizing a science as "formal". What do fluids, chemical systems, electrical circuits, lasers and acoustic systems have in common that would explain their common period-doubling route to chaotic dynamics? Whatever it is, it cannot have much (if anything) to do with the specific material properties of the components that make up these systems. Any explanation must refer to the relational or structural features that the systems have in common — in short, it must abstract away from the *matter* to identify the underlying *form* that is common to all the systems in question. In this sense, one can justifiably call a science that attempts to describe and explain the domain-independent features of natural systems –in this case, chaos theory and nonlinear dynamics– as a "formal science".

There are many disciplines that are formal in the sense that their primary focus is to describe, explain, predict (and often, construct and control) domain-independent phenomena. The paradigm examples are drawn from the "systems sciences": automata theory, linear and nonlinear systems theory, dynamical systems theory, control theory, network theory, information theory, *etc.* Other disciplines that qualify as formal in our sense have closer associations to the natural and social sciences. The domain-independent phenomenon of self-organization, for example, is often studied using computer models, but there is a thermodynamic school of self-organization theory that seeks to ground self-organization in the (suitably revised and understood) laws of thermodynamics. Game theory is a formal science by our definition, but its disciplinary home is typically in departments of economics, psychology or evolutionary biology.

Indeed, a conspicuous feature of the formal sciences is their trans-disciplinary character. On the one hand, the formal sciences are mathematical disciplines that can often be studied as formal systems without regard to applications. On the other hand, they are generally not regarded as mere bits of pure or applied mathematics; their intended domain of application is almost always the empirical sphere, the world of natural and social phenomena, and their practitioners typically have disciplinary ties to departments of natural, social or engineering science, rather than to mathematics (though this is much more true of North American than of the rest of the world).

3 A Structuralist Interpretation of Domain-Independence

Given the conception of formal science described above, it is natural to characterize the formal sciences in the following way:

> The formal sciences are sciences that attempt to describe and explain the formal, structural or relational properties –the "system-hood" properties– of real-world systems.

This view, or something very close to it, is held by the majority of systems scientists with respect to how they view the relationship between the systems sciences and traditional science. We can expand on this interpretation.

In the systems science literature, domain-independence is interpreted as a simple consequence of paying attention to the distinctly relational properties of systems. In general systems theories, a *system* is defined much as a logician would define a *model* or a mathematician a *structure* [Mar75], [Ros86], [Kli93]. Let a general system $\langle T, R \rangle$ be represented by a set of objects or "things" T, and a set of relations R defined on the set T. Formally, T may stand for a single set of elements, but it also may stand for a power set, or a power set of a power set, *etc.*, or any arbitrary subset of these. The things in T may have special properties by which systems are distinguished from one another. Robert Rosen refers to these as the *thinghood* properties of the system [Ros86]. The properties that distinguish a physical from a biological from a social system would be properties of the elements in T.

In most formal presentations of general systems theory, the relation R is represented by a subset of some Cartesian product of sets in T. For example, if T is a only single set, then R stands for a family of distinct types of relations: $R \subseteq T \times T$ (binary relations), $R \subseteq T \times T \times T$ (ternary relations), *etc.* Following Rosen again, we can refer to all the properties of the relation R as the *systemhood* properties of the system $\langle T, R \rangle$. Thus, any collection of elements –physical, chemical, biological, sociological, abstract– will "model" or "instantiate" the system $\langle T, R \rangle$ just in case the elements stand to each other in the relation specified by R. One can thus state the distinction between traditional and systems science as follows: traditional science is primarily concerned with the "thinghood" properties of natural systems, while systems science is primarily concerned with their "system-hood" properties [Kli93].

Now, if the domain-independence of the formal sciences is understood in this way, it is difficult to escape the conclusion that, with respect to the source of domain-independence, the formal sciences are not different in kind from applied mathematics. James Franklin agrees (note: Franklin's conception of "formal science" is essentially the same as the one developed above — *i.e.*, it focuses on the phenomenon of domain-independence, though he does not use this language):

> "... it is obvious that the formal sciences are either applied mathematics, or something very close to it. ... It may in fact be a historical accident that the formal sciences are not actually called applied mathematics and housed in departments of applied mathematics. In the mid-century, mathematics was going through a particularly pure phase, obsessed with rigour and generality, and was not receptive to new applied disciplines. Of the leading mathematicians, only von Neumann and Norbert Weiner took any serious notice of the new directions. By default, the formal sciences had to find academic homes in corners of departments of engineering, economics and business, psychology or whoever else would have them." [Fra94, p.523].

Franklin asserts that what makes the formal sciences distinctive as sciences is precisely what they share with mathematics — a concern with formal structure independent of the ontological makeup of the elements that instantiate that structure. The connection between formal science, applied mathematics and pure mathematics is, for Franklin, best understood in terms of an over-arching structuralist philosophy of mathematics. Mathematical structuralism comes in various forms, but all varieties assert that (a) mathematics is fundamentally a science of structure, and (b) structural properties can be instantiated by natural systems, and hence can be legitimate objects of empirical study.

> "A structuralist account of the formal sciences is, then, already available in structuralist philosophies of mathematics in general. The only thing to be added is an explanation of what structures exactly are studied by the particular disciplines. But this is a mathematical question, and the answer is found (if not always clearly expressed) in the axioms and definitions of each discipline. Topology studies one kind of structure, whose nature is captured by the definition of a topological space; information theory studies another. Conversely, a structuralist account of the formal sciences is an advantage for the philosophy of structuralism in mathematics. Since we recognize the similarity between the formal sciences, traditional applied mathematics, and pure mathematics, we should prefer a philosophy of mathematics that demonstrates their unity." [Fra94, p.526].

Whether a structuralist interpretation of the domain-independence of the formal sciences adds support to structuralist philosophies of mathematics is an interesting question, but I will not pursue it here. The point I wish to emphasize is that the two approaches to understanding the source of domain-independence in the formal sciences described above –*viz.* the orthodox system-theoretic approach of the systems scientists, and the structuralist approach defended by Franklin– agree that domain-independence has its source in relations of structural similarity (technically, isomorphisms or homomorphisms) that obtain between systems of different natural-kind types. The formal sciences are, fundamentally, sciences of structure. Let us call this the *structuralist* interpretation of domain-independence.

4 An Objection to the Structuralist Interpretation: Physical versus Formal Sources of Domain-Independence

As Franklin points out, the structuralist interpretation of domain-independence has the consequence that there is no principled difference between the way the formal sciences relate to the natural world, and the way that mathematical theories in general relate to the natural world. A formal science defines a mathematical structure of a certain kind, and this structure is applicable to the description and explanation of real-world systems if and only if the elements of the real-world system instantiate (partially or completely) this mathematical structure.

In this section I will suggest that a purely structural approach to domain-independence is inadequate as an account of the domain-independence observed in many of the formal sciences. In addition to the purely "formal" constraints imposed on the behavior of natural systems in virtue of them instantiating a particular mathematical structure, there are also "physical" constraints, imposed by contingent laws of nature, that function to generate the observed domain-independent behaviors. The formal sciences can be distinguished from pure and applied mathematics to the extent that such physical constraints play a role in generating domain-independent behaviors. I develop the case via two examples drawn from network theory. The first example will help clarify the notion of a formal constraint on the behavior of a natural system; the second will help clarify the notion of a physical constraint, and the relationship of formal and physical constraints in the generation of domain-independent behaviors.

4.1 The Königsberg Bridges

The river Pregel runs through the city of Königsberg. In the eighteenth century there were two islands in the river, and a network of bridges connecting the two islands to each other and to the banks of the river. One island (call it Island 1) had one bridge to each bank, and one to the other island (call it Island 2). Island 2 had two bridges to each bank, and one to Island 1. The citizens of Königsberg noticed that it seemed impossible to walk across all seven bridges over the river without walking across at least one of them twice. In 1736, Leonhard Euler proved this conjecture correct: there is no such path. His approach to the problem involved realizing that the size and shape of the land masses bounded by the water, as well as the length of the bridges, play no role whatsoever in the question; all that matters is the abstract network topology of the land (node) and bridge (edge) connections.

Euler reasoned that a node could be one of three types: (a) the beginning of a path, (b) the end of a path, or (c) an intermediate node. If the latter, then the path must both enter and exit the node; hence, such a node must have as many edges coming into it as going out. Therefore, the total number of edges incident on the node must be an even number. On the other hand, if a node is either the beginning or end of a path, there must be an odd number of edges incident upon it (namely, one). Since any path crossing each node exactly once must have a beginning and an end, it follows that any such path will have exactly two nodes with an odd number of edges. But for the Königsberg bridges we see that all four nodes have an odd number of edges (the two banks and Island 1 have three bridges, Island 2 has five). Ergo, there can be no path that that crosses all nodes without crossing one twice (in contemporary mathematical language, there are no *Euler paths* through this network).

The fact that there are no Euler paths through a given network is a formal property of the abstract network structure of the network, and can be known without any understanding of the material constitution of the network. This abstract structure imposes a "formal constraint" on the possible behaviors of natural systems that instantiate this structure — in this case, the citizens of Königsberg are constrained to follow paths that cross at least one bridge twice.

A formal constraint, then, is one that restricts the possible states or behaviors of a system simply in virtue of instantiating certain relational, organizational or structural properties of the system. Given an abstract structural description of the system, we can use deductive mathematical reasoning to discover these properties. We can be confident that a given natural system will exhibit these properties to the extent that we are confident that the natural system in fact instantiates the abstract structural description.

It follows that, on the structuralist interpretation of domain-independence, the only constraints that generate the domain-independent behaviors of natural systems are formal constraints.

4.2 Dominance of Indirect Effects

Now consider a second example, this one drawn from the network ecology literature. Here, the networks are not mere static configurations of nodes and edges, but digraphs representing flows of substance from one node to another. The nodes represent stocks of matter or energy resident in some functionally defined ecological entity (predators, microbiota, detritus, *etc.*), and the flows are transfers of matter or energy between nodes. Network ecologists have developed a variety of techniques for analyzing the structure and functioning of ecological networks. They can, for example, calculate total system through puts, numbers and magnitudes of flow cycles of varying path lengths, effects of one node on another, effects of one node on the network as whole, effects of the network as whole on one node, various indices of connectivity, feedback effects, *etc.* They can also derive theorems that describe very general features of network behavior. When data from ecosystem field studies are used to estimate stocks and flows for a particular ecosystem, domain-specific models can be developed to help explain and predict real-world ecosystem behaviors.

Let's take a look at one of these general theorems of systems ecology, called the *Dominance of Indirect Effects* theorem:

> Dominance of Indirect Effects: The ratio of the magnitude of indirect to direct flows in a network increases with increasing (a) system size (number of components), (b) system connectivity (density of interactions), (c) compartment storage (flow impedance), (d) feedback and non-feedback cycling, and (e) strength of direct flows. As a network becomes larger and more complex, the contribution of the indirect flows tends to exceed the contribution of the direct flows.

This theorem is developed by Higashi and Patten [HigPat86], [HigPat89] and elaborated in [Pat91]. To get a feel for it, consider a simple direct flow, from node A to node B, or $A \rightarrow B$. Now consider a flow chain, $A \rightarrow B \rightarrow C$. The node C receives a direct flow from B, but B is the recipient of a direct flow from A. If A were not present, the activity of B would be different, and hence the flow from B to C would be different. Thus, the output of A has an indirect effect on the input of C, through its direct effect on B. Higashi and Patten developed a formalism for calculating ratios of direct to indirect flow contributions for any given node, and for arbitrary path lengths. The theorem basically asserts that as a network grows in complexity, indirect feedback effects will come to dominate the activity of any given node in the network. Indeed, the theorem is one of several that Patten and his co-workers have developed that attempt to lay a foundation for a holistic conception of ecosystem structure and development.

Now, because the theorem is derived for general network flows, it may be applicable to systems as diverse as computer networks, neural networks, cellular metabolic pathways and economic systems, in addition to ecological systems. The Dominance of Indirect Effects appears to be a domain-independent network property of the sort characteristic of many of the formal sciences.

However, the domain-independence exhibited here, I claim, is not the product of purely formal constraints. The networks that are described by this theorem are representations of physical network flows of energy or material substance. In order to derive this result, one needs to assume that every flow transfer is subject to mass-balance, energy conservation and energy dissipation constraints. That is, the network needs to conform to basic conservation laws and respect the second law of thermodynamics, which requires that all physical processes involve the dissipation of a certain quantity of energy that can no longer be used to do work on the system. Without these assumptions, the Dominance of Indirect Effects theorem cannot be derived.

This is the key distinction between "physical" and "formal" constraints on the behavior of natural systems. The laws of thermodynamics are not reducible to or derivable from the relational, organizational or structural properties of natural systems, hence they are not formal constraints. They may be represented mathematically, but that alone does not warrant treating them as purely formal or relational in character (for more on this, see the exchange between [dLa99a], [dLa99b] and [Fra99]). The laws of thermodynamics are contingently true natural laws that constrain the behavior of all physical processes. Natural systems of varying natural kind-types (physical, chemical, biological, sociological) and of varying spatio-temporal scales are all subject to these laws. In this respect the laws of thermodynamics share the generality of formal constraints on natural systems, but they are not strictly speaking domain-independent — they have a physical domain (the domain of physical processes involving the transfer of energy into work and heat), but this domain is so broad that it cuts across traditional boundaries between the sciences. Formal constraints contribute to domain-independence in virtue of being genuinely formal; physical constraints, like the laws of thermodynamics, contribute to domain-independence in virtue of having physical domains of applicability that are nearly universal in scope.

This example from network ecology illustrates a further point. Just as we cannot interpret the domain-independence of the Dominance of Indirect Effects theorem as a purely formal property of a natural system instantiating a given mathematical structure, nor can we interpret it as a result of purely physical constraints. The Dominance of Indirect Effects theorem does not follow from thermodynamic considerations alone. It follows, rather, from the coupling of physical constraints (thermodynamic laws) and formal constraints (network topology).

5 Formal Science and the Physical-Formal Spectrum

The domain-independence of the Dominance of Indirect Effects theorem in systems ecology is a product of the coupling of physical and formal properties of natural systems. This is true not just of this particular theorem, but, due to the ubiquity of network formalisms and appeals to thermodynamic constraints, of the whole discipline of systems ecology.

The distinction between physical and formal constraints suggests a way of situating different sciences along a spectrum representing varying degrees of influence of formal versus physical constraints in generating domain-independent behaviors. At one end of the spectrum are theories that give purely formal, structural descriptions of natural systems. These would correspond to the various branches of pure and applied mathematics. At the other end of the spectrum are theories that state very general physical constraints on the behavior of natural systems. These would correspond to sets of natural laws described by physical theories, such as the laws of thermodynamics, or the general laws and principles of classical, relativistic and quantum mechanics (*e.g.*, the law of inertia, the invariance of the speed of light, the principle of relativity, *etc.*).

The domain-independence exhibited by Euler's theorem concerning the non-existence of an Euler path over the Königsberg bridges corresponds to the formal end of the spectrum — the theorem describes a purely formal constraint on the behavior of natural systems. The domain-independence exhibited by, say, general results concerning the efficiency of heat engines, or the non-existence of perpetual motion machines, corresponds to the physical end of the spectrum — these are purely physical constraints on the behavior of natural systems. The domain-independence exhibited by the theorems of systems ecology is a product of physical-formal coupling; I would place it somewhere in the middle of the spectrum.

My view is that particular instances of domain-independence in the sciences are usually the result of some form of physical-formal coupling, *i.e.*, that this is the rule rather than the exception. When the influence of physical constraints dominates, we drift toward more traditional areas of physics; when the influence of formal constraints dominates, we drift toward mathematics.

To give an example of the latter, consider the domain-independent period-doubling route to chaos mentioned earlier. A period-doubling system is one whose macroscopic behavior exhibits different kinds of periodicity for different values of a certain control parameter. For a certain value of the control parameter (say, temperature), the value of the behavioral variable is regular (say, $2, 2, 2, 2, \ldots$). At a certain critical parameter value, two alternating values appear (say, $1, 3, 1, 3, 1, 3, \ldots$) — this is the period-2 regime. As the control parameter is further increased, it hits another critical value, and a period-4 regime emerges (say, 4, 1, 3, 2, 4, 1, 3, 2, ...). Beyond this point, the system may enter a chaotic regime, and beyond that, return to another periodic regime. What is so striking is that for systems of very different physical types (fluids, electronic circuits, acoustic systems, *etc.*) the key structural features of the observed pattern of bifurcations are identical; hence, there must be an underlying structural dynamics common to all these systems.

Is this common structural dynamics a purely formal structure? Does it follow simply from the arrangement of the parts of the system? Is it a deductive consequence of the relational properties of the system? There is no room for a detailed analysis here, but the short answer is "no". This behavior is only possible for highly damped, dissipative systems. Thus, volumes in phase space, representing possible initial conditions of the system, contract as the system evolves. This results in a dramatic reduction in the degrees of freedom for the evolution of the system, converging on either a fixed point, a limit cycle, or a "strange" or chaotic attractor. The point is this: if there were no processes occurring within the system that irreversibly transform useful work energy into non-reusable heat energy, then the resulting dynamics would not be observed. Thus, the presence of contingent physical constraints is a necessary condition for the specific domain-independent behavior observed in these systems.

Now, the details of this domain-independent behavior are not explained by appeal to purely physical constraints. We do have a good understanding of why the period-doubling route to chaos has the particular structural features it has, but the explanation invokes the mathematical technique known as *renormalization,* and is cast at a very abstract level [Les98]. My intuition in this case is to say that formal constraints play a significantly greater role in the explanation of the period-doubling route to chaos than do physical constraints, but that physical constraints do play a necessary role (they impose, for example, constraints on the possible solutions of the dynamical equations describing the system).

Hence, I'm inclined to place period-doubling phenomena, and the theories that explain them, toward the formal end of the physical-formal spectrum. There are other examples that I would place closer to the physical end of the spectrum, though precisely because of this, fewer people will be inclined to regard the domain-independent phenomena in question as belonging to "formal science" (though they still qualify as "formal" on our definition, since systems of different natural-kind types can exhibit the same phenomena, *e.g.*, Prigogine's minimum entropy production principle for near-equilibrium systems [dGrMaz84]).

To sum up: The structuralist interpretation of domain-independence is flawed, because it fails to recognize the distinctive nature and contribution of physical constraints in the generation and explanation of domain-independent behaviors. The explanation for any given domain-independent behavior, however, or may be dominated by one or the other type of constraint. As the formal contribution increases and the physical contribution decreases, we move toward a purely structural description of a system, and the structuralist interpretation of domain-independence becomes more and more appropriate. As the formal contribution decreases and the physical contribution increases, we have cases of domain-independence that are more readily interpreted as arising from simple subsumption under very general physical laws. I suggest, however, that very few interesting cases of domain-independence arise from purely formal or purely physical constraints. The general case is one in which both formal and physical constraints are present; the resulting domain-independent behavior is a consequence of the coupling of these two distinct types of constraint.

6   A Structuralist Objection, and a Response

The argument above rests on there being a principled distinction between formal (structural) and physical constraints. But what if there is no principled distinction? What if physical constraints really are, at some deeper level of analysis, just another kind of structural constraint? Such a view may well be a consequence of the various structuralist philosophies that have been proposed over the years and that are having something of a renaissance today in the philosophy of science. There are two distinct varieties: *epistemological structuralism* states that all we can know in science, all that science ever delivers, is knowledge of structure; *ontological structuralism*, by contrast, challenges traditional object-property ontologies by stating that physical objects are, ultimately, structural in nature (it's structure "all the way down"). It is unclear to me at this stage how the truth or falsity of epistemological structuralism bears on the formal/physical distinction I draw here. However, the following claim has some initial plausibility: if ontological structuralism is true, then all constraints are, fundamentally, structural, and consequently, the structuralist interpretation of domain-independence is *a fortiori* true.

I shall withhold judgment on whether ontological structuralism is true, and consider only whether the above conditional claim is warranted. To do this, we need to be more precise about the sort of structuralism that is being considered. For present purposes, the relevant issue is whether the sorts of general law-like constraints described by the principles and postulates of general physical theories –what I choose to interpret as physical constraints– can be interpreted as purely structural constraints. Now, one motivation for such a view is the observation that conservation laws in physics are always correlated with some set of structure-preserving transformations over a suitably-defined mathematical space, which defines an equivalence class of spaces within which the particular conservation laws hold (this is known as "Noether's Theorem"). For example, the laws of Newtonian mechanics define an abstract space-time with certain structural properties. The invariance of Newton's laws under spatial translation is equivalent to the law of conservation of momentum; invariance under spatial rotations is equivalent to the law of conservation of angular momentum; invariance under time translations is equivalent to the law of conservation of energy. These invariance relations are defined by structure-preserving transformations within the mathematical structure of a Newtonian space-time. Physics is filled with these sorts of equivalences. All the fundamental forces are correlated with certain (local gauge) symmetries of quantum fields. Conservation of charge is correlated with symmetry properties of the quantum wave function for single particles. All of these symmetries can be represented as invariance structures within a suitably defined mathematical space.

Thus, the argument might go, the general physical laws that figure so strongly in my conception of physical constraints are really just structural properties of mathematical models, models which are amenable to a purely structural description. Recall our earlier definition:

> A formal constraint is one that restricts the possible states or behaviors of a system simply in virtue of instantiating certain relational, organizational or structural properties of the system. Given an abstract structural description of the system, we can use deductive mathematical reasoning to discover these properties. We can be confident that a given natural system will exhibit these properties to the extent that we are confident that the natural system in fact instantiates the abstract structural description.

In modern physics, we construct mathematical models defined entirely in terms of structural properties of an abstract system, explore these properties using deductive methods, and then make claims to the effect that the elements of a particular natural system realize or instantiate the structure of the model. The symmetries associated with the conservation laws described above are merely structural properties of mathematical models. Thus, there is no principled distinction between the formal (structural) constraints described by mathematical theories, and the physical constraints imposed by general physical laws.

In response, I am prepared to admit that it may be impossible to give a set of necessary and sufficient conditions for distinguishing formal from physical constraints, and that in many cases, the constraints on the behavior of natural systems described by general physical laws may also be described in terms of natural systems instantiating certain structures defined on an appropriate mathematical space. But I don't see this as a threat to the legitimacy or importance of the distinction. If we abandon the physical/formal distinction, we will still need a way to distinguish the constraint imposed on the behavior of the citizens of Königsberg by, say, the non-existence of Euler paths across a given network of bridges, from the constraint on attempts to cross this network of bridges imposed by, say, the upper limit on velocities set by the speed of light. The former is what we would normally call a "mathematical" property of the bridge network; the latter is clearly not. The fact that we can represent the finiteness of the speed of light within a mathematical model, and use deductive methods to draw inferences on the possible behaviors of natural systems from this model, does not warrant the conclusion that the speed of light is a "mathematical" property of natural systems that instantiate the model; this would trivialize the distinction between mathematical and physical properties. No, a universal structuralism that abandons the physical/formal distinction will simply have to find some other way of expressing what is intuitively different in these two cases, and in doing so, I believe, will necessarily recover what matters in the distinction that I am trying to draw.

7 Conclusion

In this paper I offered a conception of formal science centered around the notion of domain-independence. The most common way of understanding domain-independence is in structural terms, but I've tried to show that interesting cases of domain-independence in the sciences typically involve the influence of both formal (structural) and physical constraints on the behavior of natural systems. Domain-independent behaviors, and the theories that attempt to describe and explain them, can be arrayed along a physical-formal spectrum, with general physical theories and mathematical theories at the respective poles. I considered a structuralist objection to the physical-formal distinction, but argued that, while some form of philosophical structuralism may indeed turn out to be true, any such structuralism will have to find some way of recovering the key intuitive features of the physical-formal distinction that I emphasize here.

References.

[Bat02] Robert **Batterman**, The devil in the details: asymptotic reasoning in explanation, reduction, and emergence, Oxford University Press 2002

[dGrMaz84] Sybren R. **de Groot** and Peter **Mazur**, Non-equilibrium thermodynamics, Dover Publications 1984

[Fra94]	James **Franklin**, The formal sciences discover the philosopher's stone, **Studies in History and Philosophy of Science** 25 (1994), p.513-533
[Fra99]	James **Franklin**, Structure and domain-independence in the formal sciences, **Studies in History and Philosophy of Science** 30 (1999), p.721-723
[HigBur91]	Masahiko **Higashi** and Thomas P. **Burns** (*eds.*), Theoretical studies of ecosystems: the network perspective, Cambridge University Press 1991
[HigPat86]	Masahiko **Higashi**, Bernard C. **Patten**, Further aspects of the analysis of indirect effects in ecosystems, **Ecological Modelling** 31 (1986), p.69-77
[HigPat89]	Masahiko **Higashi**, Bernard C. **Patten**, Dominance of indirect causality in ecosystems, **American Naturalist** 133 (1989), p.288-302
[Kli93]	George **Klir**, Systems science: a guided tour, **Journal of Biological Systems** 1 (1993), p.27-58
[dLa99a]	Kevin **de Laplante**, Certainty and domain-independence in the sciences of complexity: a critique of James Franklin's account of formal science, **Studies in History and Philosophy of Science** 30 (1999), p.699-720
[dLa99b]	Kevin **de Laplante**, Response to Franklin's comments on 'Certainty and domain-independence in the sciences of complexity', **Studies in History and Philosophy of Science** 30 (1999), p.725-728
[Les98]	Annick **Lesne**, Renormalization methods: critical phenomena, chaos, fractal structures, John Wiley & Sons 1998
[Mar75]	Joseph H. **Marchal**, The concept of a system, **Philosophy of Science** 42 (1975), p.448-467
[Pat91]	Bernard C. **Patten**, Network ecology: indirect determination of the life-environment relationship in ecosystems, *in:* [HigBur91, p.288-351]
[Ros86]	Robert **Rosen**, Some comments on systems and systems theory, **International Journal of General Systems** 13 (1986), p.1-3

Received: May 17th, 2003;
In revised version: November 6th, 2003;
Accepted by the editors: December 17th, 2003.

Benedikt **Löwe**, Volker **Peckhaus**, Thoralf **Räsch** (*eds.*)
Foundations of the Formal Sciences IV
The History of the Concept of the Formal Sciences

The infinite in mathematics
A regulative idea?

JOHANNES EMRICH

Institut für Philosophie der Universität Erlangen-Nürnberg
Bismarckstraße 1
91054 Erlangen, Germany
E-mail: jsemrich@phil.uni-erlangen.de

> ABSTRACT. In this paper a mediating conception of the infinite is suggested between platonistic and anti-platonistic views. The Kantian distinction of constitutive and regulative ideas is applied to different kinds of infinity in mathematics, especially to the infinity of the set of all real numbers. Comfronting this distinction with the notions of completeness and determination yields a formalistic concept of the real numbers which is open to but does not depend on metaphysical (*i.e.*, platonistic) extensions.

"*Die Mathematik ist die Wissenschaft vom Unendlichen.*" [Wey25, p.1]

With these words Hermann Weyl introduced a lecture held in 1925. Indeed every philosophical reflection on mathematics will sooner or later be confronted with the problem of infinity.[1]

So it is not surprising that the infinite was one of the major topics of the "*Grundlagenstreit*" of the 1920s, in which Weyl was one of the main actors, standing somewhere between the opponents Hilbert (who advocated classical mathematics) and Brouwer (the constructivist challenger). Up to now

---

[1] Some of these problems are discussed in my PhD dissertation [Emr04] which deals with attempts to justify the law of excluded middle against the critique of the intuitionists. In my own attempt I try to take a view aiming to mediate between constructivists and platonists. The resulting considerations lead to a position which I call "construction-realism." This position allows to justify the law of excluded middle and principles from classical mathematics such as the axiom of choice and impredicative definitions with the exception of some very special cases, *e.g,* the application of the axiom of choice in the proof of the Banach-Tarski-Paradox remains unjustified. These cases have as far as I see no relevance for mathematical practice.

the question which idea, which concept of the infinite the mathematicians have (or should have), has not been settled. Still some constructivists are challenging classical mathematics by stigmatizing its "platonistic" conception of the infinite as metaphysical and thus unjustified, but not only the constructivists which may nowadays not be as influential as they were in the 1920s do so. A large variety of positions in the philosophy of mathematics agrees in being "anti-platonistic", *i.e.*, in rejecting the platonistic ontology of a realm of mathematical entities existing independently of any human access to it. In this paper I will suggest a mediating conception of the infinite between platonism and anti-platonism(s). To develop it, I will mainly take into account constructivism, especially intuitionism, as a paradigmatic opponent of platonism. The aim is not to give an account which satisfies all ontological and epistemological demands of both sides (this could hardly ever be done), but to lay aside most of these demands and look only for a mediating concept of the infinite (especially the uncountable infinite) concerning the question, in which way we are allowed to apply it in mathematical practice. For this it will prove helpful to reconstruct the controversy on the basis of a formalistic view of mathematics, which says that mathematics is what can be derived within formal systems or, more precisely, in *recursively axiomatizable* systems. This basis is adequate because formalism behaves neutral in all questions concerned here.[2]

In 1925 the notion of the infinite was apparently booming. In the paper "On the infinite" [Hil25], Weyl's teacher David Hilbert gave a comprehensive exposition of his proof-theoretical program. At the outset he stresses the value of the new theories of the infinite, especially the set theoretic approach by Georg Cantor.[3]

But then, facing the paradoxes of set theory, Hilbert calls into question the certainty of mathematical reasoning in the case of its application to infinite objects. To obtain certainty, he says, we first have to get full clarity about the essence of the infinite. According to Hilbert this clarification and *a fortiori* justification is to be obtained by proof-theoretic methods, *i.e.*, by showing that infinite objects can be seen as *ideal* elements, and statements about them as *ideal* propositions, which can never yield any contradiction when added to finite mathematics. What does "ideal" mean here? Hilbert

---

[2] Of course, there are ontological and epistemological problems in the philosophy of mathematics where formalism is not neutral. Such problems will, however, not be touched here. We choose the formalistic view only for pragmatic reasons in the context of the aim of mediation: The clearest and thus least problematic way to talk about mathematics, which should be taken to be able to mediate between the arguing parties, is to talk about formal systems.

[3] Here one finds Hilbert's well-known words: "*Aus dem Paradies, das Cantor uns geschaffen, soll uns niemand vertreiben können*" [Hil25, p.170].

introduces this manner of speech *per analogiam* by parallelizing the adjunction of (propositions about) the infinite with the adjunction of the complex numbers to the real numbers. Except for this *mathematical* explanation he gives one small hint at a *philosophical* way of understanding "ideal":

> The role that remains to the infinite is, rather, merely that of an idea — if, in accordance with Kant's words, we understand by an idea a concept of reason that transcends all experience and through which the concrete is completed so as to form a totality.[4]

If we try to understand the infinite as an idea in the sense of Kant, a first distinction is forced on us: Kant's distinction of a regulative and a constitutive understanding of ideas of reason. It may not be clear if Kant would agree to an application of it to mathematics, but there is a quite obvious way of understanding the idea of the infinite in a constitutive respectively a regulative manner: If such an idea determines uniquely the totality of the elements of an infinite set (or more generally determines the extension of an infinite mathematical object), thus "constitutes" its extension, then we may speak of a constitutive idea. If, however, the idea only guides us somehow in finding more and more elements of the set, but provides no (at least no obvious) way of exhausting it, then we may speak of a regulative idea. An example of an infinity that can be understood in a constitutive manner in the above sense is the set of all natural numbers. We start with the number 0 and move on step by step from the current number to its successor. This process generates the whole set uniquely. The underlying constitutive idea is that of *iteration*: we can *in principle* repeat on and on the application of the successor-operation. Since there is not much disagreement on this point between constructivists and platonists, there is nothing to do in mediating.

The situation changes, however, if we turn to the question whether together with the set itself all *properties* of the set are determined. With regard to the set of all properties of the set of the natural numbers there is a special case which is, in all interesting respects, equivalent to the general case: the case of the subset containing all those properties defining a real number. This subset will now be taken into account.

The essential idea that takes us from the rational numbers to the real numbers is the idea of *completeness*. If we want to have a number-set that can represent all points on a continuous line then the set of all rational numbers does not suffice. There are "gaps" between the rational numbers which

---

[4]Quoted from [vHe67, p.392]. The German original reads: "*Die Rolle, die dem Unendlichen bleibt, ist vielmehr lediglich die einer Idee — wenn man, nach den Worten Kants, unter einer Idee einen Vernunftbegriff versteht, der alle Erfahrung übersteigt und durch den das Konkrete im Sinne der Totalität ergänzt wird*" [Hil25, p.190].

have to be filled to get a satisfying representation of the continuum. This completization can be done, *e.g.*, by postulating that for every bounded set of rationals there exists a real number which is the least upper bound of the set. Now the question is: Does such a postulate, does the idea of completeness constitute the set of all real numbers uniquely? The platonist may say "yes", the constructivist "no". With this negative answer the constructivist is on the defensive since almost every mathematician agrees with the platonistic position. So a "natural" first move from a position trying to mediate between the opponents will consist in examining the arguments of the constructivist to understand his point.

A well-known attempt to explain the differences uses the distinction between the views of the infinite as potential on the side of the constructivists and as actual on the other. This distinction does not provide a good explanation of the differences, however, when applied to the set of the real numbers as a whole. Even Brouwer, the founder of the intuitionistic school, considered the continuum as an actual infinite object of mathematical intuition.[5]

And he was able to model this actual totality by means of his choice sequences. Why then do the constructivists deny the unique constitution of the set of all real numbers? — Because the single element, the single real number or choice sequence is not determined! According to the intuitionist the single choice sequence exists only as a potential infinite object. At every moment only a finite initial-segment of the sequence is determined, the other elements of the sequence still have to be chosen, and in general no law is known which would determine these choices. Now two questions arise:

(1) Is it true that there are sequences (real numbers) not determined by a law?

(2) If there are sequences not determined by a law, does it follow that they are not determined at all?

---

[5] *Cf.*, *e.g.*, [Bro29, p.154f], where he speaks of the primordial mathematical intuition which introduces "*das Unendliche als gedankliche Realität* [...] *und zwar* [...] *zunächst die Gesamtheit der natürlichen Zahlen, sodann diejenige der reellen Zahlen und schlie"slich die ganze reine Mathematik.*" Furthermore Brouwer introduces in [Bro30, p. 433], the notion of a set as a form, "*welche in der intuitionistischen Mathematik als primäre Schöpfung fertiger überabzählbarer Spezies auftritt.*" On Brouwer's view of the infinite as actual, *cf.* [Emr04, § 1.2]. Of course it would not be fair to identify constructivism with Brouwerian intuitionism. There are lots of other constructivisms, many of which criticize the concepts of intuitionism, especially the concept of choice sequences. (*Cf.*, *e.g.*, [Bis67, p.6]). On my point of view, however, these criticisms are at least to their greater extent due to misunderstandings concerning the ontological basis of choice sequences. For a perhaps more clarifying understanding of choice sequences see [Emr04, § 5.5].

Question (2) seems to be of a metaphysical kind. To say, as the platonist does, that the extension of every sequence is determined within some kind of mathematical reality existing independently of us, involves strong ontological commitments which can never be accepted by the constructivist. A position which wants to mediate has to take into account the constructivist point of view here. So on question (2) we reply: There is no persuasive argument to say "No, that does not follow; surely all sequences are determined."

Question (1) leads to the next distinction concerning the infinite in mathematics: the distinction between the countable and the uncountable infinite. In every language, especially in every formal language of a mathematical theory, which permits only finite expressions, there are only countably many formulas, and thus there can be formulated only countably many laws.

On the other hand it can be proved by Cantor's diagonal method that for every sequence of real numbers between 0 and 1 there exists a real number between 0 and 1, which does not appear in the sequence. This proof is accepted by almost all constructivists and even so the contraposition of the statement, which says that the set of all real numbers between 0 and 1 cannot be brought into a sequence, *i.e.*, cannot be countable. But if there are uncountably many real numbers and only countably many laws, there can be no one-to-one-correspondence between them. There must be real numbers without a law.

According to the intuitionist the idea of the completeness of the real numbers does constitute the set as a whole, but it cannot constitute every single element of the set. Surely there are single real numbers which are determined by their extension. So which numbers are, then, determined by a law, and which not? — In the formalistic picture adopted at the outset that obviously depends on the underlying language, *i.e.*, on the underlying formal-linguistic framework. Which elements are constituted does not depend on the idea of completeness, but on the *choice* of the linguistic framework. Stating a "choice" in this metamathematical context while just having talked about "choice" sequences as mathematical objects leads to the question:

Is there a structural similarity between choosing numbers when determining a choice sequence and choosing linguistic frameworks for the determination of more and more real numbers (and other more sophisticated mathematical objects)?

For answering this question it is helpful to draw a further distinction concerning the countable infinite, the distinction between the *determined* infinite and the *indetermined* infinite. A determined infinite object is a countable set for which we know a law given by a linguistic expression, which determines its extension. An indetermined infinite object consists of countably many elements, which have to be chosen successively, and for which we do not know a linguistic expression determining these choices. The paradigmatic cases of these types of infinite objects are lawlike sequences (*i.e.*, sequences determined by a law) respectively "lawless" sequences of natural numbers.[6]

But there is no necessity that the elements of the sequence are natural numbers (or other standard mathematical objects). We can also choose as elements metamathematical objects, *e.g.*, formal systems. So the question comes up: Can we interpret the development of mathematics as an indetermined infinite sequence of formal-linguistic deductive frameworks? On the basis of the formalistic view taken above, I think we can.[7]

This sequence is developing in time, just as Brouwer's choice sequences do. The elements of the sequence are the formal frameworks which are accepted at each particular moment by the members of the mathematical community as the basis of their scientific practice. Each time this basis changes, this is to be reconstructed within our formalistic picture of mathematics as the choice of a new framework by the community, *i.e.*, of a new element of the sequence. This process will never reach a definite end, since no formal framework will ever decide all mathematical problem as we know from Gödel's incompleteness theorem. So the sequence is (at least potentially) infinite. Furthermore the sequence is an indetermined object since nobody knows how it is going to develop in the future.[8]

Nobody can know if the current framework of classical mathematics, Zermelo-Fraenkel-set-theory ZFC, will ever be supplemented by a large-cardinal-axiom accepted by the community or by the axiom of constructibility or something else or if in the future the set-theoretic basis will be replaced by, *e.g.*, a category-theoretic one.

---

[6] The notion of a lawless sequence is fundamental in axiomatic approches to the theory of choice sequences; *cf.*, *e.g.*, [TrovDa88, §§ 12.2-12.4.]. I use this notion here in an informal way.

[7] Of course the biggest part of the development of mathematics takes place within fixed frameworks and consists of defining within these frameworks new concepts, of formulating new conjectures and of proving new theorems, so perhaps I should say: The development *of the foundations* of mathematics can be interpreted as an indetermined infinite sequence of formal-linguistic frameworks. In many cases the framework within which a mathematician works may not be made explicit by showing a formal system. But it is part of the self-image of nowadays mathematicians to be able in principle to give such an explication.

[8] For this reason of indeterminateness the countable union of all frameworks need not be itself a recursively axiomatizable framework, and consequently Gödel's theorem cannot be used to show the incompleteness of this union.

According to this formal picture of mathematic as a sequence of frameworks there is a development in the foundations of mathematics. But is there *progress*? We hope so. There is progress if we manage to develop the foundations of mathematics in a *cumulative* way, *i.e.*, if (all in all) every theorem once proven remains valid and new theorems become provable. This leads us back to the characterization of ideas as regulative. Surely a regulative idea should guide our acting and thinking *to make progress*.[9]

As a special case the idea of completeness of the real numbers as the idea of a higher infinity can be seen as telling us to extend the formal framework of mathematics bit by bit to determine more and more real numbers.

Let's summarize: We asked ourselves if the idea of completeness constitutes the set of real numbers in the sense that it determines all its elements extensionally. We found that from a mediating point of view we can not give a definite answer to the affirmative since there is no formal framework providing a linguistic expression for every real number. So can we give a definite answer to the negative, then, as the constructivists do? — Likewise no. According to the regulative idea underlying mathematics as a whole (and in principle all other sciences as well) the foundations of mathematics should develop in time in a cumulative way. If the scientific community manages to follow this idea without exception, then from a realistic point

---

[9] The formalistic picture aims at clarity. It may be objected, however, that the picture aims at *too much* clarity, *i.e.*, more clarity than mathematics actually provides. For what does the formal system currently accepted as its basis by the mathematical community exactly look like? It is not always clear which formal principles are currently accepted and which not, there is some uncertainty in the community concerning, *e.g.*, the axiom of choice. We are not warranted to say that it was accepted by the community from the year 1904 on, when Ernst Zermelo formulated it explicitly in his proof of the well-ordering-theorem, or from the year 1963 on, when Paul Cohen proved its independence of the other set-theoretic axioms, or to say that the axiom of choice is even nowadays not yet accepted. Indeed there are different opinions on the question: Should the axiom of choice be accepted or not? And that causes uncertainty. Now this uncertainty infects the above question if the foundations of mathematics always develop in a cumulative way. Suppose future mathematics will not accept the axiom of choice in its full extent. Then the answer to the question of cumulativity depends on whether the axiom is regarded as being fully accepted in the early 21st century or not. This could give rise to the question: Can a mathematical community accept a principle while being uncertain to be allowed to accept it, or while not knowing that it accepts it? And in the end cumulativity may even depend on whether one gives a realist or an anti-realist answer to the question: Is at every moment every mathematical principle either accepted by the community or not accepted? So we're back at highly controversial philosophical questions. Are these questions just produced by the formalistic picture? — No, they are special cases of more general philosophical questions which already exist. Our concern was just the problem of extensional determination of all real numbers. By using the formalistic picture we will surely not be able to answer those questions, but we will be able to avoid them in our context as the proposal at the end of this paper will show. To maintain clarity where clarity is not perceptible needn't be a bad thing. It's an idealization.

of view one can say: In the resulting sequence of formal frameworks every real number either will be determined in one of the frameworks (and thus by cumulativity in all following frameworks) or in none of them. I see no reason why there should necessarily exist any real number that will never be determined.

It may be objected that the totality of all real numbers determined in any framework is a countable union of countable sets, thus the totality of all numbers itself is countable and so the diagonal method can be used to show that there are real numbers not contained in this totality. Indeed the first claim can be accepted since we should consider only countably many frameworks and in every framework only countably many numbers are determined. But note that the expressions "countable", "union" and "set" are used here in a metamathematical sense, reflecting on a sequence of frameworks. So we have to be careful in applying theorems from standard set theory to them, especially if their proofs are making use of controversial principles. Now the classical theorem that every countable union of countable sets is itself countable, makes use of an instance of the axiom of choice.[10]

In this metamathematical case, an application of the theorem would be highly questionable. Since the above union ranges over an indetermined infinite sequence of objects (the frameworks), the existence of the needed choice-function will hardly be justifiable.[11]

So my proposal for a mediating point of view concerning the totality of the real numbers on the basis of a formalistic view of mathematics (which was taken for solely pragmatic reasons) has the form of a conditional statement:

> *If* one adopts the position that the mathematical process *indeed* will continue to extend the formal framework of mathematics bit by bit following certain regulative ideas such as that of the completeness of the real numbers, *then* one is allowed to claim that every real number will be determined sooner or later.

If the anti-platonist is willing to agree to this conditional, then the conflict with the platonist who holds that the idea of completeness *constitutes* the totality of the real numbers uniquely dissolves almost completely. In the sense of the mediating reconstruction above using a formalistic picture of mathematics it looks as follows: The platonist takes the adoption in the antecedent for granted and allows himself the succedent claim of the

---

[10] *Cf.* [Jec73, Theorem 10.6]. There it is proved that there is a model of ZF (without the axiom of choice!) in which the (uncountable) set of all real numbers is a countable union of countable sets. The proof goes back to [FefLév63].

[11] Note that it is consistent to use the theorem that every countable union of countable sets is countable in *mathematical* applications and to reject it in *metamathematical* applications such as the one above. For more detailed argumentation see [Emr04, § 5.3.1].

determination of all real numbers by interpreting the continuing extension of the frameworks as a process of "convergence"[12] against the timelessly given realm of mathematical entities and mathematical truth, whereas the anti-platonist is at least hesitant in doing so if not rejecting the adoption as unjustified or even false. The difference in interpreting the uncountable infinity of the real numbers in a constitutive respectively regulative way, *i.e.*, in holding that the mathematical process follows its guiding ideas in some unique and thus somehow determined way respectively denying this as unjustified and metaphysical, concerns, however, only real numbers not definable in current frameworks. And so this difference will hardly have any consequences for mathematical practice (except for pathological examples such as the non-measurable Vitali set, in the existence-proof of which the axiom of choice is applied to classes of currently undefinable reals).

References.

[Bis67]   Errett **Bishop**, Foundations of Constructive Analysis, McGraw-Hill 1967

[Bro29]   Luitzen E.J. **Brouwer**, Mathematik, Wissenschaft und Sprache, **Monatshefte fur Mathematik und Physik** 36 (1929), p.153-164; *reprint in:* [Hey75, p.417-428]

[Bro30]   Luitzen E.J. **Brouwer**, Die Struktur des Kontinuums, Wien 1930; *reprint in:* [Hey75, p.429-440]

[Emr04]   Johannes **Emrich**, Die Logik des Unendlichen, Rechtfertigungsversuche des tertium non datur in der Theorie des mathematischen Kontinuums, Logos Verlag Berlin 2004 [Logische Philosophie 14]

[FefLév63]   Solomon **Feferman** and Azriel **Lévy**, Independence results in set theory by Cohen's method II, **Notices of the American Mathematical Society** 10 (1963), p.593

[vHe67]   Jean **van Heijenoort** (*ed.*), From Frege to Gödel, A Source Book in Mathematical Logic 1879-1931, Harvard University Press 1967

[Hey75]   Arend **Heyting** (*ed.*), L. E. J. Brouwer, Collected Works I, Philosophy and Foundations of Mathematics, North-Holland 1975

[Hil25]   David **Hilbert**, Über das Unendliche, **Mathematische Annalen** 95 (1925), p.161-190

[Jec73]   Thomas J. **Jech**, The Axiom of Choice, North-Holland 1973 [Studies in Logic and the Foundations of Mathematics 75]

[TrovDa88]   Anne S. **Troelstra** and Dirk **van Dalen**, Constructivism in Mathematics, 2 volumes, North-Holland 1988 [Studies in Logic and the Foundations of Mathematics 121/123]

---

[12] Mathematically understood it would be more accurate to speak of "proper divergence" here, since the "limit", *i.e.*, the platonic heaven, is certainly an infinite object.

[Wey25]   Hermann **Weyl**, Die heutige Erkenntnislage in der Mathematik, **Symposion** 1 (1925), p.1-32

Received: May 17th, 2003;
In revised version: March 1st, 2004, April 21st, 2004;
Accepted by the editors: April 21st, 2004.

Benedikt **Löwe**, Volker **Peckhaus**, Thoralf **Räsch** (*eds.*)
Foundations of the Formal Sciences IV
The History of the Concept of the Formal Sciences

Frege on understanding mathematical truth and the science of logic

NORMA B. GOETHE[*]

School of Philosophy
Universidad Nacional de Córdoba
Centro de Investigaciones de la Facultad de Filosofía y Humanidades
Pabellón A. Tosco
Ciudad Universitaria
5000 Cordoba, Argentina
E-mail: `ngoethe@ffyh.unc.edu.ar`

The goal of the paper is to explore Frege's idea that mathematics has closer ties with logic then any other discipline. This idea engages Frege in a view of the science of logic with specific demands. On one hand, there is the demand to construct a system. On the other hand and underlying this requirement, there is the more fundamental demand to offer a deeper explanation of mathematical knowledge by providing 'insight' into the reasons ('grounds') for acknowledging the truths of arithmetic.

These demands are related, as Frege relates the ideal of system to the essence of explanation. This can be made plausible by re-establishing contact with the classical tradition's understanding of scientific demonstration. I argue that Frege's epistemic requirements are based upon specific assumptions: (a) there is a natural order of truth, which is always the same and (b) order and perspicuity can only be created by a system, which can be both traced to Leibniz.

---

[*]The author should like to thank Danielle Macbeth and Erich Reck for detailed comments. She should also like to thank Danielle Macbeth and Michele Friend for proofreading the paper.

1 Introduction: The bond between mathematics and philosophy

Towards the end of his career, Frege lectured at the University of Jena on "Logic in Mathematics".[1] As we read in his lecture notes dated about 1914, he argues for the view that "Mathematics has closer ties to logic then does any other discipline" and that "if one counts logic as part of philosophy, there will be a specially close bond between mathematics and philosophy" [Nachlaß, p.203]. This is confirmed, Frege thinks, by the history of these sciences, as the names of Plato, Descartes, Leibniz, Newton and Kant, remind us of times at which mathematics had great impact upon philosophy. However, Frege's starting point was pure mathematics, not philosophy. As a philosophically minded mathematician of his time, he felt compelled to point out time and again that the discipline of mathematics does not take its materials from history or psychology. Needless to say, this fact does not preclude the field of mathematics of having its own history.

Frege is faithful to the classical tradition in mathematics which his critics call "the Euclidean Myth".[2] Frege believes that "science comes to fruition only in a system" [Nachlaß, p.242]. That is, we must always distinguish between the progress of the history of sciences and system of science itself. In history, Frege argues, we have development, given that every alteration –the evolution of human knowledge included– takes place in time; "(t)he laws of number, however, are timeless and eternal. Time does not enter arithmetic or Analysis" [Nachlaß, p.237]. Moreover, it is only through the stability of a system that we can achieve complete perspicuity and order. In this regard, pure mathematics occupies a privileged position among the sciences:

> No science is in such command of its subject matter as mathematics and can work it up into such a perspicuous form, but perhaps also no science can be so enveloped in obscurity as mathematics, if it fails to construct a system. [Nachlaß, p.242]

Frege thus seems to be warning us that the failure to construct a system, to "work it up into such a perspicuous form", can lead mathematics not only to fragmentation but even to obscurity. As we shall see, what we are being faced with here is a striking ambiguity between actual mathematical practice and the demands of science guided by a philosophical vision.

---

[1] In 1914, Frege gave a course with the title "Logic in Mathematics", a course which Rudolf Carnap attended. (Cf. also the introductory essay of [RecAwo04]). According to the editors of Frege's Posthumous Writings the piece "Logic in Mathematics" [Nachlaß, p.203-50], presumably Frege's lecture notes, was written in the Spring of 1914.

[2] For a polemical attack of this idea, cf. [Her$_1$97]. For a different line of contemporary criticism of the logico-axiomatic tradition initiated by Frege, cf. [Gra92].

On the one hand, there is the demand to construct a system. The aim of systematic theory construction will lead to the requirement of perspicuity in the form of expression. This then motivates the invention of Frege's notation, as a "more sophisticated instrument" for the expression of pure thought.

On the other hand, and underlying this requirement, there is the more fundamental demand to offer a deeper explanation of mathematical knowledge by providing "insight" into the reasons ("grounds") for acknowledging the truths of arithmetic. Frege himself relates the issue of explanation with the ideal of "system of science": "the essence of explanation lies precisely in the fact that a wide, possibly unsurveyable, manifold is governed by one or a few sentences" [Nachlaß, p.36]. It goes without saying that such explanation must be sharply distinguished from one that simply traces the historical development of mathematics. The latter would simply run counter to the demand to construct a system. Moreover, Frege saw the "historic mode" of investigation often entangled with psychological considerations. And, he insisted, mathematics must refuse any assistance from psychology and accept instead "its close connection with logic" [Grundlagen, chapter 9]. Only in this way will the mathematician be in a position to appreciate the bond between mathematics and philosophy which is brought out by logic.

From early on, the mathematician-philosopher Frege was concerned about the fact that contemporary mathematicians mostly ignored the subject "logic". For instance, at about the time he publishes the *Begriffsschrift* (1879), the subject "logic" had been given almost no space in histories of mathematics, in spite of the fact that classical geometry made substantial use of deduction.[3]

How did the discipline get into this predicament? This seems all the more remarkable if we consider that up until the 17th century the textbook presentation of Euclidean geometry had been seen as the very paradigm of the deductive mode of presentation.[4] I shall return to this issue in the next section.

Underlying Frege's idea of the bond between mathematics and philosophy are two assumptions which are inter-connected: firstly, as already stated, "mathematics has closer ties with logic than does any other discipline", and secondly, logic is "part of philosophy" [Nachlaß, p.203]. Towards the end

---

[3] For instance, the first edition of Cajori's book on the history of mathematics from 1893 contains only a few incidental references to logic. According to [Gra92, p.104], however, the separation between logic and mathematics had long been in place.

[4] It should be noted, however, that this form of textbook presentation faced substantial criticism throughout the 17th century. For a detailed consideration of the debate concerning specific requirements for mathematical proofs and its impact on 17th century mathematical practice, *cf.* [Man96].

of the 19th century, this view was by no means obvious. On one hand, logic seemed to be mostly ignored by mathematicians. On the other hand, in philosophy, questions of logic were often mixed with psychological issues and thus regarded as pertaining to psychology, one of the empirical sciences.[5] This would seem to explain why most mathematicians did not agree with Frege on the great importance of logic to mathematics, or of philosophy to mathematics. Accordingly, it should not come as a surprise to find him devoting much energy to opposing the incursion of psychological considerations in current treatments of logic. In the *Grundgesetze*, Frege also complains about the lack of co-operation between mathematics and philosophy claiming that it was precisely the prevalence of the "psychological mode" of investigation in 19th century logic that made "a fruitful collaboration between mathematicians and logicians impossible" [*Grundgesetze*, p.25].[6]

2 "Deductive reasoning" and the remains of the seventeenth century

How exactly this "prevailing" confusion concerning the "deductive mode" of inference came about is a long story relating to issues of great complexity in the history of science. At least part of the problem goes back to the 17th century and is connected with the reception of methodological issues of relevance to the development of analytic geometry, and the emergence of the new science.[7] Like Bacon before him, although along different lines, Descartes argued against the Aristotelian logic taught at the Schools, whose

---

[5] Arguably, this does not apply to works in logic in the algebraic tradition, such as George Boole's work in England, Peirce's in the States, and Schröder's in Germany. For Frege's arguments against the Booleans, *cf.* [Slu87].

[6] It may be argued that by 1914 Frege was not completely on his own in his defence of the view that the link between mathematics and philosophy is logic. By then all volumes of Whitehead and Russell's *Principia Mathematica* (1910-13) had been published. Also, Hilbert and Bernays had started to do work on related issues in the 1900's and were quite actively engaged with logic and the foundations of mathematics in the 1910's in Hilbert's classes, as recent research has revealed. *Cf.* [SieSomTal02]. However, according to Norbert Wiener, a student at Göttingen in the Summer Term of 1914, things did not look very different from the way Frege described them. Wiener who was taking a course on Group Theory with Landau, a course on the Theory of Differential Equations with Hilbert, and three courses with Husserl, writes in a letter to Russell (from June or July 1914): "Symbolic logic stands in little favour in Göttingen. As usual, the mathematicians will have nothing to do with anything so philosophical as logic, while the philosophers will have nothing to do with anything so mathematical as symbols" [*Russell*, p.263-264].

[7] For a discussion of Descartes's conception of inference which takes into account the epistemological and mathematical reasons for his criticism of Aristotelian logic, see [Gau89]. For a discussion of the status of logic in the 17th century and the different lines of attack on the logic taught at the schools, *cf.* Jaap Maat's paper [Maa06] in this volume.

practitioners he called "dialecticians". His goal was to establish methodological rules for the direction of the intellect in the acquisition of knowledge; and according to Descartes, syllogistic logic taught only formal rules that were unable to offer any guide in the discovery of new truths. According to Descartes, such rules only serve to "mechanize" reasoning allowing our reason "to take a holiday" instead of engaging it in the resolution of problems in the search for truth. In other words, traditional formalisation was seen by Descartes not as advantage but as impediment to clear and attentive thinking.[8] Moreover, like Wallis and other 17th century mathematicians, Descartes complained that the synthetic mode of presentation of classical textbooks of geometry did not show the true way of discovery, the path by which a new truth was reached. Here what was being called into question is the "epistemic value" of the deductive mode in mathematical practice and scientific exposition.[9] In general terms, the different lines of attack launched by the 17th century critics were united by the perception that the traditional rules of deductive inference were useless from the point of view of the advancement of learning and instruction. Needless to say, this critical opposition was highly misleading. For instance, how could Descartes reject the "deductive mode" without contradicting the very principles underlying the epistemology advertised in his methodological writings?[10] Last but not least, the ambiguity of the Latin expression for inference "*deducere*" in the text of the *Regulae* may also have contributed to the blurring of the distinction between two different modes of inference, deduction and induction, which was once sharply drawn by Aristotle [Cla82]. The interpretation of this complex 17th century issue is a matter open to much debate, but according to contemporary commentators, it is as of this moment that the door was opened for the "psychological" views of logic and epistemology,

---

[8] *Cf.* [*Regulae*, Rule X]: "[...] Our principal concern here is thus to guard against our reason's taking a holiday while we are investigating the truth about some issue; so we reject the forms of reasoning just described as being inimical to our project. Instead we search carefully for everything which may help our mind to stay alert."

[9] As Mancosu [Man96] points out, there are mathematical but also logical and epistemological reasons for this criticism, which can be traced to earlier debates concerning the Aristotelian conception of scientific demonstration. Logical arguments must be truth-preserving, but while all valid forms of logical reasoning are certain, not all of them provide the evidence required by scientific demonstration. In other words, some mathematical proofs may be convincing without providing insight into the reasons why the theorem holds. This is the case with indirect proofs such as proofs by contradiction. The terms of the debate that occupies much of the philosophy of mathematics of the 17th century can be traced back to Proclus's comments on Euclid's text. *Cf.* also [Man00, p.108-12].

[10] For instance, [Her$_1$97, p.114] argues that in his methodological writings Descartes embraced the Euclidean ideal, while in his actual mathematical practice he flatly contradicted the ideas professed in those writings.

and it is these which Frege found to be offensively prevalent in logic textbooks towards the end of the 19th century.[11]

Now, we need to keep in mind this historical background in order to shed some light on Frege's efforts to clear the ground for the defence of his view that mathematics has closer ties with logic than any other science. Firstly, given that "the progress of the history of the sciences runs counter to the demands of (the science of) logic" [Nachlaß, p.242], the "historical mode" of investigation must be driven out of mathematics.[12] Secondly, the more fundamental issue whether there are modes of inference peculiar to mathematics, which for that very reason do not belong to logic, becomes important. However, the very notion of "inference" in logic textbooks at Frege's time appeared entangled with much of the 17th century ambiguity surrounding the term "deducere"; it thus becomes vital to separate sharply between the "deductive mode" and what Frege vaguely calls the "psychological mode" of investigation.

3 The distinction between discovery and the ground of truth

Frege argues for the view that the science of "logic" is of greater importance to mathematics than any other discipline. He emphasizes that "almost the entire activity of the mathematician consists in drawing inferences". Also, part of the job of the mathematician, besides drawing inferences, consists in giving definitions, and "inferring and defining are subject to logical laws" [Nachlaß, p.203]. Now this point of view will motivate Frege from early on to pursue a fundamental question, whether "there are perhaps modes of inference peculiar to pure mathematics, which for that very reason do

---

[11] Like Descartes, Kant critized syllogistic reasoning, which he, too, called "dialectics". Both rejected syllogism as "illusory" forms of knowledge. For Kant's criticism of Aristotle's axiomatic treatment of logic, cf. his 1762 essay "Über die Falsche Spitzfindigkeit der Vier Syllogistischen Figuren" ("On the Mistaken Subtlety of the Four Syllogistic Figures"). This essay is considered a crucial text for the development of Kant's critical ideas. Rejecting the tripartite scheme of traditional logic (with its division in concepts, judgments and syllogism), Kant argues for the centrality of judgment. The Kantian identification of understanding with judgment goes back to this work. What would be taught in a logic course in Germany and England towards the end of the 19th century? According to Kemp Smith [Kem18, xxxviii]: "Modern Logic, as developed by Lotze, Sigwart, Bradley and Bosanquet, is, in large part, the recasting of general logic in terms of the results reached by Kant's transcendental enquiry". Lotze was one of Frege's teachers at Jena.

[12] The text quoted above continues thus "We must always distinguish between history and system". This clearly indicates that for Frege "the demands of logic" include "the demand to construct a system". Cf. [Nachlaß, p.242]

not belong to logic" [*Nachlaß*, p.203].[13] It goes without saying that in order to decide about this issue we must take a closer look at mathematical practice. This is what Frege does. For, in his attempt to uncover the structure of mathematical reasoning, he does depend on mathematical work, mathematical thinking and the mathematician's sense of validity. To begin with, as Frege remarks in the preface to his first published book, arithmetic "was the starting point" [*Begriffsschrift*, p.vii] of the train of thoughts that led to *Begriffsschrift* (1879), and as the subtitle indicates his notation is "a formula language of pure thought modelled on that of arithmetic."[14]

However, Frege is not interested in accounting for mathematics as it takes place in actual research. As a matter of fact, he is rather critical of the way his fellow mathematicians present their research often showing a lack of precision in the way they use concepts and conduct proofs.[15] It is here that the characteristic ambivalence between the demands of science and actual mathematical practice comes into play. Frege's logicist aim is to uncover the deductive structure of derivation underlying mathematical reasoning whose true character, he thinks, is often obscured in ordinary mathematical practice.[16] The latter must be seen in the context of his project to prove

---

[13] *Cf.* [*Nachlaß*, p.203-4]. Here Frege discusses "the arithmetical mode of inference", *i.e.*, the inference by mathematical induction. The logical "reconstruction" of inductive reasoning is part of the task of bringing to light the "purely logical nature" of the arithmetical mode of inference. *Cf.* [*Begriffsschrift*, §§ 24-26].

[14] Macbeth [Mac05, p.8] argues that Frege's understanding of logic is deeply marked by his understanding of pure mathematics and of the formula language of arithmetic. The aim of Macbeth's book is to show that the latter informs Frege's formula language "at every level and at every stage in its development".

[15] For instance, focusing on a set of lecture notes he just is discussing [*Nachlaß*, p.217-221], Frege critizes Weierstrass for being extremely careless in the way he defines the concept "number". "How, we may ask, is it possible for so distinguished a mathematician to go so badly astray over this issue?" To this question Frege replies that Weierstrass "did not possess the ideal of a system of mathematics. We do not come across any proofs; no axioms are laid down [...] If he had made the attempt to construct a system from the foundation upwards, he could not have failed straightaway to see the uselessness of his definition" [*Nachlaß*, p.221]. This criticism relates to Frege's idea that when engaged in work in arithmetic and logic we ordinarily tend to work with concepts we understand incompletely. Frege also critizes Euclid who "had an inkling of this idea of a system" [*Nachlaß*, p.205], but failed to realize this ideal. In his proofs Euclid "often makes tacit use of presuppositions" [*Kleine Schriften*, p.85] instead of revealing all steps required for legitimating inferences. Another form of criticism against the lack of rigor among mathematicians is the one Frege addresses against Schröder. In [*Grundlagen*, p.viii], he argues that even mathematicians like Schröder confuse the physical or psychological conditions for proof with questions pertaining to logic.

[16] This idea relates to Frege's warning that "no science can be so enveloped in obscurity as mathematics, if it fails to construct a system" [*Nachlaß*, p.242]. Concerning the value and purpose of formal proofs, *cf.* Frege's comments on the *Begriffsschrift* [*Grundlagen*, §§ 90-91]. *Cf.* also Footnote 20.

that the laws of arithmetic are based upon the laws of logic, which, for Frege, are the most general laws. From the logicist perspective this is what it means to say that "arithmetic is part of logic".[17] In particular, it is in the context of his project where he intends to provide a deeper grounding of the theorems of arithmetic that Frege's notion of "logical proof" becomes of central importance. For, a deeper grounding of the truths of arithmetic amounts to "ultimate logical grounding" by proof.[18]

The first results of his investigation into "ultimate logical grounding" by proof appear in *Begriffsschrift* (1879).[19] Frege opens the Preface with a distinction between several stages in the recognition of a truth and grounding by proof. This concerns the issue of how to establish most firmly the truth of a proposition by finding its true ground. Frege's guiding idea here seems to coincide with the view he discusses much later in his lecture notes of 1914. In these notes, he focuses on the use of inference in mathematical practice and is more explicit about the significance of grounding a proposition by means of rigorous proof. Indeed, while Frege's results published in 1879 aim to satisfy the need for perspicuous treatment of proof structures by devising a bidimensional notation for the perspicuous expression of the "laws that govern inference", his goal in his lecture notes of 1914 is to motivate the need for rigorous deployment of proofs in pure mathematics.

According to the classical conception of scientific demonstration, there is a distinction to be drawn between proofs that convince and proofs that explain by providing the "grounds" for accepting a mathematical truth. Only the latter type of proof provides insight into the reasons why a theorem holds. For example, Bolzano relies on the distinction between proofs that aim at certainty and proofs that give the ultimate ground [*Wissenschaftslehre*, p.525]. Frege too seems to adhere to just this distinction when he claims that

> (i)n mathematics we must never rest content with the fact that something is

---

[17]This is of course a metaphor. In [*Grundlagen*, §87] the logicist thesis is stated to be that "arithmetical laws are analytic judgments". In [*Grundgesetze*, I, 1] it appears as the stronger thesis also maintained already in 1884: that "arithmetic is a branch of logic". However, as Dummett [Dum91, p.127] emphasizes, the expressions "analytic" and "synthetic *a priori*" do not play a central role after 1884. *Cf.* Footnote 23.

[18]I use the expression "deeper grounding" instead of "deeper foundation", as the German words *Begründung* (*begründen*) and *Grund* can be translated as 'offering reasons' and "reason" respectively. This terminology points to the link with classical tradition in scientific demonstration.

[19]Frege's logicist understanding of arithmetic was the starting point that motivated his *Begriffsschrift*. The core of his results in logic is intended to serve to test in the most reliable way the validity of the chain of inferences connecting a true conclusion with its premise(s). In order to assess his logicism he intends to apply his formal results to this science first. But Frege would also like to see this logical theory as a contribution to the general methodology of science [*Begriffsschrift*, p.viii].

obvious or that we are convinced of something, but we must strive to obtain
a clear insight into the network of inferences that support our conviction.
[Nachlaß, p.205]

It is worth noting, however, that for Frege the requirement to "strive to obtain a clear insight into the network of inferences" goes hand in hand with the specific demand to construct a system. This is what Frege means by the "the demands of (the science of) logic" [Nachlaß, p.242]: there is no proof without a method of proof and "the most perfect method of proof" is the deductive mode of presentation, which requires axiomatization. Grounding by proof thus appears as a holistic property to be assessed relative to a system.[20]

The construction of a system is made possible by regressive analysis. Frege indicates this when requiring that "in order to construct a system it is necessary that in any step forward we take we should be aware of the logical inferences involved" [Nachlaß, p.205]. For no matter how intricate the process of recognition of new truths may be, whenever we make advances in mathematics by finally stepping forward towards a true conclusion, "one can also trace the chains of inference backwards", *i.e.*, one can retrace step by step the inferences involved "by asking from what truths each theorem has been inferred" [Nachlaß, p.204]. Frege demands that "(w)hen an inference is being drawn, we must know what its premises are" because "if we have no clear recognition of what the premises are, we can have no certainty of arriving at the primitive truths, and failing that we cannot construct a system" [Nachlaß, p.205]. By thus constructing a formal proof that makes all steps explicit we will eventually come up against the primitive truths (to be taken as the axioms in the system). This is of the greatest importance, Frege tells us, for only in this way can we discover what the primitive truths are and, only if we succeeded in discovering the primitive truths can

---

[20]Frege is of course aware of the classical distinction between proofs that convince and proofs that provide insight into the grounds of truth. Proofs by contradiction, for instance, are proofs that aim at certainty, but being indirect, such proofs do not establish the ground of a truth; they only show convincingly that any attempt to negate it leads to contradiction, *cf.* Footnote 9. But Frege does not seem to be concerned about this issue. For a discussion of Frege's view that indirect proofs can be recast into direct proofs without difficulty, *cf.* [Man96, 4.3]. Instead, Frege's main concern here is with emphasizing the value of the "deductive mode of presentation". This relates to the ideal of system and "the essence of explanation" [Nachlaß, p.36]. The contrast he thus wants to point to is the one between formal and informal proofs. So, what he is saying is that "we must never rest content" with the use of informal proofs as they appear in ordinary mathematical practice. Commenting on the requirements that motivated his *Begriffsschrift* he discusses this issue in *Grundlagen*. While for Frege, the transformations which are of value in scientific usage are "step by step", unfortunately, "in proofs as we know them, progress is by jumps, which is why the variety of types of inference in mathematics appears to be so excessively rich" [Grundlagen, §§ 90-91].

a system be constructed, which appears "as a system of truths that are connected with one another by logical inference" [*Nachlaß*, p.205]). I will return to this crucial issue in the next section.

It should be noted, however, that back in 1879 Frege draws the distinction between the gradual process of "the recognition of a scientific truth" and the issue of grounding the truth, in general terms, *i.e.*, without reference to the specific case of mathematics. The formulation of the distinction in 1879 also indicates a parallel with the classical way of drawing the distinction between induction and deduction:

> The recognition of a scientific truth generally passes several stages of certainty. Perhaps first guessed from a limited number of particular cases, a universal proposition becomes more and more firmly established by being connected with other truths through chains of inference... It can thus be asked, on the one hand, by what path a proposition was gradually reached, and on the other hand, in what way it is now finally to be most firmly established.   [*Begriffsschrift*, p.iii]

At this stage Frege's distinction between "recognition of a truth" and "grounding" does not seem to coincide with the distinction he draws in the same preface between "psychological origination" and "grounding". Here Frege says that all judgment can be considered from the point of view of psychological origination, at least "for beings like us". But this is the subject matter of psychology, an empirical science. Concerning the ground of truth, Frege is interested in a distinction between truths for which a purely logical grounding is possible and truths that require extralogical grounding. The latter distinction, Frege claims, is based upon the "most perfect method of proof". But some years later, Frege's view of the former distinction seems to change. In *Grundlagen der Arithmetik* (1884), Frege tries to sharpen the original distinction between the path by which "a proposition was gradually reached" and grounding, by borrowing an expression from Leibniz. "Our concern here", writes Frege quoting Leibniz, is "not with the history of our discoveries which is different for different people, but with the connection and natural order of truths, which is always the same" [*Grundlagen*, § 17]. The reference to Leibniz allows Frege to introduce the notion of "the connection and natural order of truths", which he will use as of this moment in opposition to the "history of our discoveries".[21] Given that history and origination are issues partly intertwined, the conflation between psychological

---

[21] Earlier in the century, mathematicians like Bolzano and Cournot discussed the issue of genuine grounding proofs by relying on what I call here the "Leibnizian" assumption of an objective order of truths. For an illuminating comparison between Bolzano and A. Cournot, a French mathematician working in the mid 19th century, *cf.* [Man99]. Both Bolzano and Cournot trace their stringent requirements for scientific demonstration back to Aristotle. *Cf.* Footnote 28.

origination and history of discoveries would seem hardly avoidable, but this "puts superstition on the same footing as a scientific discovery" [*Nachlaß*, p.147], and is the confusion of which the psychologist is guilty.

Indeed, at this stage of his investigation into the logical grounding of arithmetic, for Frege, presenting pure mathematics in a rigorous system of proof amounts to advancing the view that the question of proof belongs neither to psychology nor history, but to mathematics (or to the science of logic, as he would like to show). But this may seem far from clear. In order to elucidate we need to turn to the question of proof. What does the notion of "logical proof" amount to for Frege? In the next section I shall discuss Frege's notion of proof against the background of two ideas:

(1) the Leibnizian conception of "the connection and natural order of truths" and

(2) the idea of system.

It is with respect to these that we may justify the claim that Frege's notion of "logical proof" is truly explanatory. As it will emerge, for Frege both ideas are inter-connected.

We may briefly summarize our discussion so far as follows. At about the time he composes the *Grundlagen* Frege's reasons for sharpening the original distinction between recognition of the truth and grounding seem to be twofold. Firstly, the distinction responds now to Frege's explicit efforts to sharply separate two disciplines, logic and psychology. But what is the real issue here? To begin with, Frege is engaged in the effort to clear the ground for a view of the science of logic with specific demands. The demands are to construct a system of proof which will be rigorous, and will provide a deeper grounding of the truths of arithmetic. The demands are related, as Frege relates the ideal of system to "the essence of explanation" [*Nachlaß*, p.36]. At the same time he is under growing pressure to distance himself from current philosophical treatments of logic which are too psychologistic in taking logic to be one of the empirical sciences. Thus, in order to secure the strict separation of two different perspectives concerning reasoning Frege sets up a methodological distinction between "the objective" (the logical) and "the subjective" (the psychological) [*Grundlagen*, p.x]. Indeed, the latter distinction appears as one of three guiding principles underlying *Grundlagen*. Secondly, in sharpening the distinction between recognition of the truth and grounding Frege is now also interested in distancing from the "historical mode" of investigation. Thus, while Frege's concern is, as he writes, with "the connection and natural order of truths which is always the same" [*Grundlagen*, §17], the question "by what path a proposition was gradually reached" [*Begriffsschrift*, p.iii] becomes finally relegated to the history of science.

4 Proof, the ultimate ground of truth, and "the essence of explanation"

We need to understand the significance of Frege's programme to prove that "arithmetic is a branch of logic" [*Grundgesetze*, I, 1]. In the context of the ideas motivating the *Begriffsschrift*, Frege's idea to firmly connect truth with the ground of truth –the order of reasons– goes back to Leibniz, as well as to the idea that order and perspicuity in expression can only be created by a notational system. I suggest that both ideas are part of the epistemic background against which Frege conceives his stringent requirements for proof. I shall turn to some of the specific assumptions underlying this background in what follows. But let me advance here Frege's guiding idea which includes an implicit reference to idealized "inferential capacities". In order to "fully understand" a truth we need to understand its logical relations to other truths.[22] Put in Leibnizian terms, this means that one does not fully understand a truth unless one knows the (*a priori*) proof.[23]

---

[22] Dummett [Dum81, p.83] made this idea into the turning point of his interpretation of Frege's philosophy of logic as a philosophy of language. While Dummett was interested in pointing out what he saw as the novelty of Frege's contribution, Burge in [Bur98, p.343-345] argues that the originality of Frege's position consists in his integration of traditional rationalist views "with his deep conception of what goes into adequate understanding". Understanding involves more than "mere mastering of words and concepts", it involves "mastering a deep rational and explanatory order". A requirement for full understanding of a thought is coming to understand logical structure and the latter derives from "seeing what structures are more fruitful in accounting for the patterns of inference that we reflectively engage in". According to Burge much that is original about Frege's methodology bears little relation to the Euclidean tradition, nonetheless the rationalist strand can be made to accord with pragmatic elements in his position. On the other hand, Burge does seem to agree with the standard reading according to which Frege's logic is a variant of quantificational logic.

[23] Concerning analytic truths, Leibniz [*Discourse*] pointed to the link with the notion of (*a priori*) proof. Putting Frege's view into the perspective of this Leibnizian idea is useful here for the following reason. According to the classical scheme, a proof is called "*a priori*" when it follows the "*a priori*" order" of reasons from the principles to the consequences. Thus to speak of "*a priori*" proofs would make sense only relative to the classical order of scientific demonstration. With Kant's critical philosophy the notion of the "*a priori*" suffers a radical transformation. As of 1762 Kant identifies understanding with judgment. A judgment is said to be a priori when it is "independent of experience". This characterization of the notion "*a priori*" remains negative in that it only says what the *a priori* is not. Moreover, basically constitutive judgments depend on the "*a priori*" categories of the understanding and the "*a priori*" intuition of space and time. According to the critics, it is at this constitutive level that the difficulties with the Kantian notions become clear; as Bolzano insists "[...] all this seems to rest on a distinction which is not thought out clearly enough, between that which is empirical and which is *a priori* in our

Now, let us start by focusing on Frege's notion of proof. For Frege, "the most reliable" proof, the "most solid ground" of truth is the one that follows pure logic. He writes in the preface to *Begriffsschrift*:

> The most reliable way to carry out a proof, obviously, is to follow pure logic, a way that discarding the particular characteristics of objects depends solely on those laws upon which all knowledge rests.   [*Begriffsschrift*, p.iii]

We see that already in 1879 Frege indicates his epistemological concern when describing the laws of logic as "those laws upon which all knowledge rests". Putting the emphasis on "the most reliable way to carry out a proof" will suggest here that certainty was a major concern for Frege. If we add to this the claim that mathematical truths are constructed or demonstrated, not explained, Frege's aim here would seem to be, to argue for the "foundational" role a logical proof plays with respect to arithmetic [GroBre00, p.81]. This comports well with the influential reading underlying "foundational studies" that can be traced back to Russell and to his principal followers who have long underscored the continuity of concerns in the line Frege-Russell-Hilbert-Gödel. This standard reading includes an interpretation of the most recent history of logic and the logicist conception of mathematics according to which the search for certainty was a major concern.[24] As far as the standard reading of Frege's logic is concerned, there clearly seem to be many problems.[25] Firstly, Frege insists upon exhibiting "the logical relation which binds the whole (*i.e.*, conclusion and premises) together" [*Nachlaß*, p.236] in two dimensions instead of adopting the linear notation of standard quantificational logic. Secondly, Frege demands that

---

cognitions". Indeed, Kant's first *Critique* "begins with this distinction, but it gives no proper definition of these things" [*Beiträge*, p.138-139]/[Rus80, p.A246]. In *Grundlagen*, Frege claims that one motivation underlying his project is to show that the truths of arithmetic are analytic (*a priori*) refuting the Kantian view that arithmetic relies on intuition. But the fact remains that the view of arithmetic judgments as analytic *a priori* Frege ends up with seems to be closer to the classical view defended by Leibniz. Moreover, for Frege the relevant distinction is between the logical and the extralogical. Thus whereas Kant has to grapple with the distinction between the empirical and the *a priori* ("independent of experience"), for Frege the logical order follows the objective order of truths and the domain of the extralogical remains completely unspecified.

[24] According to Russell, the logicist project began only with Frege's work and was reshaped by Russell himself so as to escape from the class paradoxes. However, also Dedekind, an important mathematician by 1890, defended a logicist conception of pure mathematics, as recent research has revealed. Ferreiros in [Fer99] advances the view that Dedekind, in spite of the fact that he did not develop any logical theory, was the main source of logicist ideas in the 1890s, a decade considered crucial for the diffusion of logicism. *Cf.* also [Fer04].

[25] Frege's logic is (almost) universally regarded as a notational variant of quantificational logic. Macbeth [Mac05] calls this "standard reading" into question. She sets out to show that "Frege's logic is something hitherto unknown to us". *Cf.* Footnote 14.

the premises of an inference be acknowledged to be true. Frege's insistence upon the greater perspicuity of his two dimensional notation which he claims takes care of "everything necessary for a correct inference" [*Begriffsschrift*, p.iii], and his conception of a primitive symbol for his basic logical relation are aspects of his logic that seem to be deeply connected. In what follows I shall focus on the second idea which is most relevant to the topic of this paper.

A mathematical explanation provides insight into the reasons (grounds) for acknowledging a mathematical truth. We can think of some clear historical reasons why Frege does not speak about mathematical proofs in "explanatory" terms. Yet, as already noted, he relates the issue of explanation with the goal of systematization when he writes that "the essence of explanation lies precisely in the fact that a wide, possibly unsurveyable, manifold is governed by one or a few sentences" [*Nachlaß*, p.36]. Also early on, Frege still employs the notion of "causal connection" in the broader (classical) sense which is related to scientific explanation.[26] However, at least as of 1884 he is aware of the Kantian distinction between causal-empirical explanation and normative justification, which had been in use in neo-Kantian circles. Following this current philosophical usage, Frege seems to reserve the term "explanation" for issues dealing with experience and causality, where causality is taken in the modern sense of "efficient causality". At least this seems to come out clearly in the context of his discussion of psychologism, where Frege makes a sharp distinction between "reasons" and "causes". For instance, he thinks of the subject matter of psychology as dealing with causal-empirical accounts of the laws of thought. But to confuse the order of reasons proper and the order of causes is to be guilty of the confusion that leads to psychologism.[27]

But this is not to say that the issue of explanation was not raised in the context of mathematics after Kant.[28] In particular, this is not to say

---

[26] *Cf.* [*Begriffsschrift*, §§ 5-12]. For an illuminating discussion of Frege's use of the notion of a causal connection "in the broader sense of a causal law", *cf.* [Mac05, p.23-24].

[27] Concerning the distinction between "reasons" proper and productive "causes", Frege writes "With the psychological conception of logic we lose the distinction between the grounds that justify a conviction and the causes that actually produce it. This means that justification in the proper sense is not possible, what we have in its place is an account of how the conviction was arrived at, from which it is to be inferred that everything has been caused by psychological factors" [*Nachlaß*, p.147].

[28] We recall the way Frege relies on Leibniz's idea of a natural order of truth [*Grundlagen*, § 17]. Compare this idea and the issue of explanatory proofs with the discussion of Bolzano's views as reported in [Man99]. According to Mancosu, Bolzano was not the only mathematician in the 19th century to raise the issue of mathematical explanation. The French mathematician-philosopher Cournot approached the topic independently but along similar lines [Man99, p. 444-51]. Both Bolzano and Cournot started from the distinction between the objective order of truths and the subjective order of

that the issue of "mathematical explanation" does not bear directly upon Frege's concerns when discussing his stringent requirements for proof. Logical proof, as Frege understands it, plays an explanatory role in the sense of offering a "deeper grounding" of the truths of arithmetic. However, in order to make this idea plausible we need to re-establish contact with the classical tradition's understanding of scientific demonstration, in particular with the issue of mathematical explanation as raised by 19th century mathematicians such as Bolzano who made explicit reference to Aristotle's conception of scientific demonstration. When distinguishing between grounding proofs and proofs that aim at establishing certainty, Bolzano clearly emphasizes the value of the former: "it must be seen as a virtue of a proof if the truth to be established derives from its own objective ground".[29] For Bolzano, as for Frege, genuine proofs must express real grounding relations. In contrast, derivations need not. Only genuine proofs may be regarded as having explanatory power. Kitcher in [Kit75] points to the link with the issue of mathematical explanation in the case of Bolzano's concerns.[30]

Earlier, I claimed that Frege's notion of proof is indebted to Leibniz. In doing so, I have assumed some continuity of the classical conception of science from the Euclidean-Aristotelian scheme to a modern rationalist view, for such ideas clearly belong to the classical tradition.[31]

---

knowledge. Cournot traces the distinction between the objective order (which he calls "the order of reason") and the subjective (described as "the logical order" in the sense of the actual scientific practice) back to Aristotle. In order to characterise the objective order Cournot opposes the relationship connecting a theorem and its ground (reason) to the relationships of force and causality. The former type of relationship is abstract and independent of time, as is the case with the truths of mathematics, an abstract realm "where the relationship of cause and effect is completely absent" [Man99, p. 445].

[29] Bolzano [*Wissenschaftslehre*, p.525] recognizes that the distinction goes back to Aristotle's distinction between *a priori* proofs of the reasoned fact (*dioti*) and *a posteriori* proofs of the fact (*oti*).

[30] Kitcher [Kit75, p.237] emphasizes that Bolzano "did offer a link with explanation. To give the grounds of a proposition is to answer a question as to why the proposition is true". For a discussion of Kitcher's view, *cf.* [Man99].

[31] Burge [Bur98, IV] scrutinizes what he calls "Frege's Euclideanism". Perhaps the main interest of the paper lies in the way Burge tackles the question he proposes to discuss. On one hand we have Frege's rationalist appeal to self-evidence concerning the laws of logic at the basis of the system. On the other hand, we have pragmatic considerations guiding Frege's theory construction. Burge asks: 'How does the appeal to self-evidence accord with the fallibilist, pragmatic elements in his position?' Concerning Frege's reliance on the rationalist tradition, however, Burge makes a much broader claim than I intend to do here. I limit myself to the connection with Leibniz. The connection between Frege and Leibniz is important for this paper. For, facing the fierce attacks against deductive logic and the synthetic mode of presentation launched by Descartes and other 17th century mathematicians, Leibniz takes it upon himself to defend the epistemic value of deduction. In this regard, Frege was following the footsteps of Leibniz. For, in the same spirit, Frege was taking upon himself the defense of the epistemic value of the "deductive mode of

The reference to Leibniz also provides another interesting point of contact with Bolzano who most strongly disagrees with the basic tenets of Kant's philosophy of mathematics. Against Kant, Bolzano writes: "...I believe that also mathematical proofs can be, and must be, carried from concepts alone" [*Wissenschaftslehre*, p.266]. From such remarks, and others, we can see that Frege's intellectual motivations echo those of nineteenth century mathematicians like Bolzano who sought to free analysis and number theory from any dependence on geometry and kinematics insisting that "the concepts of time and motion are just as foreign to general mathematics as the concept of space".[32]

5 The aim of proof and "striving after a system"

Now when it comes to considering the nature of proof, Frege –borrowing Leibniz's words– assumes "a connection and natural order of truths, which is always the same". Such "connection and natural order of truths" is conceived as an objective structure of derivation which is independent of any particular theoretical framework we may develop.[33] Belonging to the abstract "space of reasons", it is there, as it were, "waiting to be discovered". Let us see how this idea is supposed to work for Frege. To begin with, Frege tells us that it lies deep "in the nature of mathematics" always "to prefer proof" to inductive confirmation, whenever "proof is possible" [*Grundlagen*, § 2]. He points to the fact that proofs may serve different purposes:

> A proof does not only serve to convince us of the truth of what is proved: it also serves to reveal the logical relations between truths.
> [*Nachlaß*, p.204]

For Frege the aim of proof is bound up with his assumption that "the most perfect method" of proof is the deductive mode of presentation as developed in *Begriffsschrift*. This contains "fixed guidelines, along which the deductions are to run" [*Kleine Schriften*, p.235]. The "fundamental idea" of Frege's new notation is to "limit to the bare minimun the number of modes of inference, and set these up as rules" [*Kleine Schriften*, p.236]. Moreover, by "proving everything that can be proved" we must try to reduce

---

inference" in the face of psychologistic tendencies inherited from Kant's attacks against the sterility of deductive logic. (*Cf.* Footnote 12.)

[32] Demopoulos [Dem95] made this interpretative claim elaborating on Coffa's early paper; *cf.* [Cof82]. I am quoting Bolzano according to [Dem95, p.76]. Also Dummett [Dum91, p. 70] pointed to the striking resemblance between Bolzano's work and Frege's central concern "not to arrive at certainty concerning the truths of arithmetic, but to establish the ground of our acceptance of them, and, in particular, to refute the belief that intuition was among those grounds".

[33] For the view that there are more primitive laws in "logical space" than required for the construction of a system, *cf.* [*Begriffsschrift*, § 13]. *Cf.* also [Bur98, p.325].

"the number of those primitive laws as far as possible" and "require that all propositions used without proof be declared as such, so that we can see distinctly what the whole structure rests upon" [*Grundgesetze*, p.2]. In other words, there is no genuine proof unless made possible by "the most perfect method of proof". As Frege argues in his lecture notes of 1914, the search for proofs is deeply connected with the search for primitive laws or fundamental principles. The aim of proof, we recall, is to afford insight into the dependence of truths upon one another. The further these mathematical investigations are pursued, the fewer become the primitive truths to which everything can be reduced. Primitive truths can be seen as "laws that govern modes of inference" to be expressed in *Begriffsschrift* "as formulae, so that their source, whether in logic or intuition, can be investigated" [Mac05, 1.2]. That is to say, we discover by such investigations whether or not the laws at the basis of the structure are logical or otherwise.

Given Frege's logicist goals the significance of the search for basic laws should be clear. Frege's guiding idea is that the nature of mathematical truth is to be revealed by making explicit the connection to the primitive truths, the laws upon which "the whole of mathematics rests". So, we need to learn about the primitive truths. Now, considering the question of a purely logical grounding of the truths of arithmetic, we must first see, Frege tells us in the 1879 preface, how far we can get in arithmetic by inference alone, supported only by the logical laws and so that "nothing intuitive can intrude here unnoticed, everything has to depend on the chain of inference being free of gap" [*Begriffsschrift*, p.iii]. As already mentioned, for Frege, the requirement "to reveal the logical relations between truths" [*Nachlaß*, p.204] goes hand in hand with the demand to construct a system. On one hand, the aim of systematization leads to the requirement of perspicuity in expression. On the other hand, the requirement of perspicuity relates closely to the ideal of full understanding.[34] It is with this assumption that the epistemological component comes in. In order to fully understand a truth, we recall, we need to understand the inferential connections with other truths, and this requires sharp definitions of the concepts and gapless proofs that "reveal the logical relations between truths" and stop only at the ultimate ground. Primitive truths do not require proof because they are self-evident, and being self-evident we understand them by fully grasping the primitive concepts, which Frege calls the "ultimate building blocks" [*Kleine Schriften*, p.114] of a science.

---

[34] In order to argue more fully for the connection between perspicuity of expression and full understanding, it would be required to consider the evolution of Frege's ideas and "Frege's work brought to maturity" [Mac05, Chapter 4]. From a different perspective, this is also an important idea underlying Burge's paper [Bur98].

To sum up, concerning genuine proofs, Frege has stringent requirements in mind. To be more specific about them, let us simply ask, what would a "purely logical proof" guarantee for Frege "if we find everything in order" [*Grundgesetze*, p.3] with the axiomatic system he proposed?

Firstly, a logical proof must guarantee an "ultimate" grounding (*Begründung*) of a theorem (*cf.* [Fri02]). The totality of the inference chains that are formed connecting truths from premise(s) to conclusion constitutes the proof of the theorem. "We may (also) say that a proof starts from propositions that are accepted as true and leads via chains of inference to the theorem" [*Nachlaß*, p.204]. For Frege, however, the inference chains connecting truths which deserve to be called "proofs" are chains which eventually come up against the "axioms", more precisely, proofs that go all the way back to something unprovable, the "primitive truths" (taken as the axioms in the system). As Dummett points out, Frege's requirement that all legitimate inference must be from true premises is at the basis of his axiomatic treatment of logic [Dum91, p.25].

However, Frege's requirement that genuine proofs be "ultimate grounding" proofs, means more than this. For, it seems possible to accept that inferences must be from true premises, while rejecting the assumption that there is one natural ordering of truths to be captured. That is, we could posit a competing set of axioms for our basis. Secondly, Frege requires that genuine proofs be gap free, so as to reveal every presupposition that in ordinary mathematical practice may sneak in unnoticed. I shall say more on this in the concluding section of the paper.

6 The epistemic background to Frege's notion of proof and the science of logic

Frege insists that it is science itself that "demands that we prove whatever is susceptible of proof and that we do not rest until we come up with something unprovable" [*Nachlaß*, p.204-5]. Is this really the demand of science? It seems clear that Frege does not mean science as it is practiced, but "science as it should be" from the point of view of the ideal of science he is embracing. In particular, it is clear that Frege has in mind specific demands concerning the discipline of mathematics. He speaks of the "striving after a system" as justified in mathematics. But he also complains about the fragmentation he finds prevalent in 19th century mathematical practice. Even Euclid, he says, who had "an inkling of this idea of a system" [*Nachlaß*, p.205], failed fully to realize it. Next to the idea that order and perspicuity can only be created by a system we find the assumption of a "kernel of truths", which would explain our ("ultimate") understanding of mathematics:

> [Science] must endeavour to make the circle of unprovable primitive truths

as small as possible [...] for the whole of mathematics is contained in these primitive truths as in a kernel. Our only concern is to generate the whole of mathematics from this kernel. The essence of mathematics has to be defined by this kernel of truths, and until we have learnt what these primitive truths are we cannot be clear about the nature of mathematics. [*Nachlaß*, p.205]

Finally Frege relates this concern to an old question that was apparently "already being asked by Euclid". He agrees that the question belongs largely to philosophy, but he insists, it "must still be regarded as mathematical" [*Nachlaß*, p.205]. It is the job of the mathematician-philosopher to provide insight into the nature of mathematical truth, but the problem of the logical grounding of pure mathematics which includes the design of the appropriate notation, the search for proofs, the search for basic laws, and the discovery of "this kernel of truths" (the axioms of the system), is a problem to be dealt with in mathematics in "fruitful collaboration" with the science of logic.[35]

Now to return to the issue we started out with, how does Frege's demand to construct a system relate to what appears to be a more fundamental demand to explain our understanding of mathematical knowledge? In her exchange with Jeshion [Jes04], Weiner argues that Frege's demand to construct a system

> arises from a desire to identify sources of knowledge. Systematization, however, need not be understood in this way. It is a recognizable mathematical project that can be undertaken without any peculiarly philosophical motivation. [Wei04, p.122]

But is this how Frege's demand to construct a system should be understood? Weiner is right when she says that an account that assumes this would not fit with Frege's writings.[36] As I have tried to show, Frege's

---

[35] For a discussion of the notion of "fruitfulness" in Frege's work, *cf.* [Tap95].

[36] According to Weiner this is precisely Jeshion's point in emphasizing the mathematical notion of "self-evidence": she is guided by "her desire to avoid attributing to Frege any views that carry with them the taint of the purely philosophical" [Wei04, p.126]. But this seems to be based on a misunderstanding. Jeshion [Jes01, Jes04] sets out to articulate the role that the notion of "self-evidence" plays within Frege's philosophy. But I see some problems with Jeshion's starting point. Within Frege's writings there are two distinct German expressions which in English are translated as "self-evidence": "*selbstverständlich*" and "*einleuchtend*". Jeshion claims that both of these notions are needed to understand Frege's notion of grounding proof. The first one would correspond to what might be called an objective conception of self-evidence, while the second one corresponds to epistemic aspects of the reasoner including subjective elements. She claims that the "principal mark of Frege's Euclideanism is contained in his remarks on the *selbstverständlich* status (of primitive truths and that) the term '*selbständlich*' translates, literally, as 'self-standing'. This is, of course, metaphoric" [Jes01, p.948]. But this isn't metaphoric, as she claims. It is plainly a mistaken translation of the German "*selbstverständlich*".

philosophical and epistemic requirements are based upon two background assumptions, which turn out to be inter-connected:

(1) that there is an objective order of truths to be captured — this is an explanatory order, and

(2) that order and perspicuity can only be created by a system.

The link between both assumptions is Frege's notion of a "primitive truth".[37]

A quick look at the ideas that motivated the *Begriffsschrift* will show us that the stringent requirements posed by Frege upon scientific work in mathematics proved to be heuristically very fruitful. In 1897, almost two decades after the publication of *Begriffsschrift*, Frege looks back to the ideas that motivated his early work and recalling how he came upon the need to invent a new notation he writes:

> I became aware of the need for a conceptual notation when I was looking for the fundamental principles or axioms upon which the whole of mathematics rests. Only after this question is answered can it be hoped to trace successfully the springs of knowledge upon which this science thrives.
> [*Kleine Schriften*, p.235]

Frege's interest in identifying "the springs of knowledge upon which this science thrives" brings us finally to the epistemic concern underlying his project to offer a deeper grounding of the truths of arithmetic. As I pointed out, Frege's strategy here rehearses Leibniz's idea that one does not fully understand a truth unless one knows the (*a priori*) proof. Now, the aim of ultimate grounding proofs, we recall, is to offer insight into the dependence of truths upon one another until we come up with something unprovable, the primitive truths. According to Frege, such truths are primitive laws that stand on their own, *i.e.*, ones neither needing nor admitting of proof.[38] Firstly, primitive truths are not capable of proof because they cannot be grounded on other truths. Secondly, a truth is primitive, whenever it does not need proof because "it can be recognized as true independently of other truths". This is how Frege characterizes the concept "axiom". Axioms are said to be self-evident in the sense that they can be "recognized as true independently of other truths" [*Nachlaß*, p.168]. Note that a primitive truth may be recognized only by anyone who fully understands the concepts involved, perhaps only under ideal epistemic conditions. But this is the sort of idealization Frege's science of logic requires. Ultimate grounding proofs may stop only at the ultimate ground, the axioms. Such a requirement

---

[37] Concerning the notion of "primitive truths", *cf.* Leibniz's [*Monadologie*, §35].

[38] The idea that primitive laws are truths that are "neither capable nor in need of proof" [*Grundlagen*, § 3], goes back to Leibniz [*Nouveaux Essais*, IV, ix, 2].

would ensure that grounding proofs possess certain "epistemic" virtues. Yet this can only be guaranteed if the concepts involved are sharply defined and proofs are gap-free, so as to reveal every presupposition leading us all the way back to the ultimate ground. We thus return to Frege's second requirement for rigorous proofs. So, what was Frege trying to prove by means of rigorous proofs?

In *Grundgesetze*, Frege states the task of logicism, as he understands it:

> Because there are no gaps in the chains of inference, every "axiom" [...] upon which a proof is based is brought to light; and in this way we gain a basis upon which to judge the epistemological nature of the law that is proved. I have drawn together everything that can facilitate a judgment as to whether the chains of inference are cohesive and the buttresses solid [...] If we find everything in order, then we have accurate knowledge of the grounds upon which each individual theorem is based. [*Grundgesetze*, p.3]

On Frege's conception of the close bond between logic and mathematics, what the science of logic aims at proving by means of rigorous proof is the epistemic value of "the deductive mode of inference" and this insight holds whether or not "we find everything in order" with the system Frege proposed.

Primary Sources.

[*Wissenschaftslehre*]   Bernard **Bolzano**, Wissenschaftslehre, Versuch einer ausführlichen und grösstentheils neuen Darstellung der Logik mit steter Rücksicht auf deren bisherige Bearbeiter, J.E. von Seidel 1837; *english translation in:* [Geo72]

[*Beiträge*]   Bernard **Bolzano**, Beiträge zu einer begründeten Darstellung der Mathematik, Prag 1810; *english translation in:* [Rus80], [Ewa96, p.176-224]

[*Regulae*]   René **Descartes**, Regulae ad directionem ingenii, *incomplete; english translation in:* [CotStoMur85]

[*Begriffsschrift*]   Gottlob **Frege**, Begriffsschrift, eine der arithmetischen nachgebildete Formelsprache des reinen Denkens, Louis Nebert 1879; *reprinted in:* [Ang64]

[*Grundlagen*]   Gottlob **Frege**, Die Grundlagen der Arithmetik: eine logisch-mathematische Untersuchung über den Begriff der Zahl, Koebner 1884; *english translation in:* [Aus68]

[*Grundgesetze*]   Gottlob **Frege**, Grundgesetze der Arithmetik, Band I, Verlag Hermann Pohle 1893; *partial english translation in:* [Fur67]

[*Kleine Schriften*]   Gottlob **Frege**, Kleine Schriften, *in:* [Ang67]; *english translation in:* [McG84]

[*Nachlaß*]   Gottlob **Frege**, Nachgelassene Schriften, *in:* [Her$_0$KamKau69]; *english translation in:* [Her$_0$KamKau79]

[Discourse] Gottfried Wilhelm **Leibniz**, Discourse on Metaphysics, *translated in:* [Mon02]

[Monadologie] Gottfried Wilhelm **Leibniz**, Monadologie, *translated in:* [Mon02]

[Nouveaux Essais] Gottfried Wilhelm **Leibniz**, Nouveaux essais sur l'entendement humain, *english translation in:* [RemBen96]

[Russell] Bertrand **Russell**, The Autobiography of Bertrand Russell, 3 volumes, George Allen and Unwin, 1967, 1968, 1969 (page citations follow the 2000 Routledge edition)

References.

[Ang64] Ignacio **Angelelli** (*ed.*), Gottlob Frege, Begriffsschrift und andere Aufsätze, mit E. Husserls und H. Scholz' Anmerkungen, Georg Olms 1964

[Ang67] Ignacio **Angelelli** (*ed.*), Gottlob Frege, Kleine Schriften, Georg Olms 1967; *english translation in:* [McG84]

[Aus68] John L. **Austin** (*ed., trans.*), Gottlob Frege, The Foundations of Arithmetic: A Logico-Mathematical Enquiry into the Concept of Number, Northwestern University Press 1968

[Bur98] Tyler **Burge**, Frege on Knowing the Foundations, **Mind** 107 (1998), p.305-347

[Cla82] Desmond M. **Clarke**, Descartes' Philosophy of Science, Pennsylvania State University Press 1982

[Cof82] Alberto **Coffa**, Kant, Bolzano and the Emergence of Logicism, **The Journal of Philosophy** 74 (1982), p.679-689; *reprinted in:* [Dem95]

[CotStoMur85] John **Cottingham**, Robert **Stoothoff**, and Dugald **Murdoch** (*eds.*), Rules for the Direction of the Mind, The Philosophical Writings of René Descartes, Cambridge University Press 1985

[Dem95] William **Demopoulos**, (*ed.*), Frege's Philosophy of Mathematics, Harvard University Press 1995

[Dem94] William **Demopoulos**, Frege and the Rigorization of Analysis, **Journal of Symbolic Logic**, 23 (1994), p.225-246; *reprinted in:* [Dem95]

[Dum81] Michael A.E. **Dummett**, Frege: Philosophy of Language, Duckworth 1981

[Dum91] Michael A.E. **Dummett**, Frege: Philosophy of Mathematics, Duckworth 1991

[EchIbaMor92] Javier **Echeverria**, Adoni **Ibarra**, and Thomas **Mormann** (*eds.*), The Space of Mathematics: Philosophical, Epistemological, and Historical Explorations, Walter de Gruyter 1992 [Foundations of Communication]

[Ewa96] William B. **Ewald** (*ed.*), From Kant to Hilbert: A source book in the foundations of mathematics, Volume 1, Oxford Clarendon Press 1996

[Fer01]	José **Ferreirós**, Labyrinth of Thought, A history of set theory and its role in modern mathematics, 2nd ed., Birkhäuser 2001
[Fer04]	José **Ferreirós**, Uncertain Foundations, **Metascience** 13 (2004), p.79-82
[Fri02]	Michele **Friend**, What a proof guarantees for Frege, Presented at Hopos 2002, Montreal, Canada, June 2002, *unpublished*
[Fur67]	Montgomery **Furth** (*ed., trans.*), Gottlob Frege, The Basic Laws of Arithmetic, University of California Press 1967
[Gau89]	Stephen **Gaukroger**, Cartesian Logic, An Essay on Descartes's Conception of Inference, Clarendon Press 1989
[Geo72]	Rolf **George** (*ed., trans.*), Bernard Bolzano, Theory of science; attempt at a detailed and in the main novel exposition of logic with constant attention to earlier authors, University of California Press 1972
[Gra92]	Ivor **Grattan-Guinness**, Structure-similarity as a Cornerstone of the Philosophy of Mathematics, *in:* [EchIbaMor92, p.91-111]
[GroBre00]	Emily **Grosholz** and Herbert **Breger** (*eds.*), The Growth of Mathematical Knowledge, Kluwer Academic Publishers 2000
[Her$_0$KamKau69]	Hans **Hermes**, Friedrich **Kambartel**, and Friedrich **Kaulbach** (*eds.*), Gottlob Frege, Nachgelassene Schriften, Felix Meiner 1969
[Her$_0$KamKau79]	Hans **Hermes**, and Friedrich **Kambartel**, Friedrich **Kaulbach** (*eds.*), Gottlob Frege, Posthumous Writings, University of Chicago Press 1979; *translated by* Peter **Long** and Roger **White**
[Her$_1$97]	Reuben **Hersh**, What is Mathematics, Really?, Oxford University Press 1997
[Jes01]	Robin **Jeshion**, Frege's notion of self-evidence, **Mind** 110 (2001), p.937-976
[Jes04]	Robin **Jeshion**, Frege: Evidence for Self-Evidence, **Mind** 113 (2004), p.131-137
[Kem18]	Norman **Kemp Smith**, A Commentary to Kant's Critique of Pure Reason, McMillan 1918; *second edition 1923*
[Kit75]	Philip **Kitcher**, Bolzano's ideal of algebraic analysis, **Studies in History and Philosophy of Science** 6 (1975), p.229-269
[Maa06]	Jaap **Maat**, The Status of Logic in the Seventeenth Century, *in:* Benedikt Löwe, Volker Peckhaus, Thoralf Räsch (*eds.*), Foundations of the Formal Sciences IV, The History of the Concept of the Formal Sciences, College Publications 2006 [Studies in Logic 3], p.157-167 (*this volume*)
[Mac05]	Danielle **Macbeth**, Frege's Logic, Harvard University Press 2005
[Man96]	Paolo **Mancosu**, Philosophy of Mathematics and Mathematical Practice in the Seventeenth Century, Oxford University Press 1996
[Man99]	Paolo **Mancosu**, Bolzano and Cournot on Mathematical Explanation, **Revue d'Histoire de Sciences** 52 (199), p.429-455
[Man00]	Paolo **Mancosu**, On Mathematical Explanation, *in:* [GroBre00, p.103-119]
[McG84]	Brian **McGuinness** (*ed.*), Gottlob Frege, Collected Papers on Mathematics, Logic, and Philosophy, Basil Blackwell 1984

[Mon02] George R. **Montgomery** (*ed., trans.*), Gottfried Wilhelm Leibniz, Discourse on Metaphysics, Correspondence with Arnauld, Monadology, Open Court 1902

[RecAwo04] Erich H. **Reck**, Steve **Awodey** (*eds.*), Frege's Lectures on Logic, Carnap's Jena Notes 1910-1914, Translated and edited, with introductory essay by the editors, Based on the German text, edited, with an introduction and annotations by G. Gabriel, Open Court 2004

[RemBen96] Peter **Remnant** and Jonathan **Bennett** (*eds.*), Gottfried Wilhelm Leibniz, Nouveaux Essais, New Essays Concerning Human Understanding, Cambridge University Press 1996

[Rus80] Stephen B. **Russ**, The mathematical works of Bernard Bolzano published between 1804 and 1817, Open University 1980; *Ph.D thesis*

[SieSomTal02] Wilfried **Sieg**, Richard **Sommer**, and Carolyn **Talcott** (*eds.*), Reflections on the Foundations of Mathematics, Essays in Honor of Solomon Feferman, A K Peters 2002 [Lecture Notes in Logic 15]

[Slu87] Hans **Sluga**, Frege against the Booleans, **Notre Dame Journal of Formal Logic** 28 (1987), p.80-98

[Tap95] Jamie **Tappenden**, Extending Knowledge and 'Fruitful Concepts': Fregean Themes in the Foundations of Mathematics, **Noûs** 29 (1995), p.427-67

[Wei04] Joan **Weiner**, What was Frege Trying to Prove?, a response to Jeshion, **Mind** 113 (2004), p.115-27

Received: June 11st, 2003;
In revised version: June 26th, 2003; May 2nd, 2005;
Accepted by the editors: July 6th, 2005.

Benedikt **Löwe**, Volker **Peckhaus**, Thoralf **Räsch** (*eds.*)
Foundations of the Formal Sciences IV
The History of the Concept of the Formal Sciences

Classical mechanics as a formal(ised) science

Ivor Grattan-Guinness

Middlesex University at Enfield
Middlesex EN3 4SF
England
E-mail: ivor2@mdx.ac.uk

ABSTRACT. This paper reviews an ensemble of historical and especially philosophical issues that attended the development of classical mechanics from Isaac Newton's innovations in the late 17th century to a massive rise during the 18th century. Three traditions developed then, of which Newtonian mechanics was only one. All three traditions then profoundly influenced classical mathematical physics, as it rose in its various branches in the 19th century; mechanics itself expanded still further in some notable ways. Of the philosophical questions discussed, some are epistemological, including claims of reduction; others are methodological. The place of mathematical theories in physical theories is also addressed in this context.

1 Introduction

One of the senses in which the word "formal" is used in mathematics relates to the extent to which a theory is developed in a systematic way, maybe even to the point of explicit axiomatization. The advantages of this presentation include the exposure of the assumptions made, the detection of others not previously noticed, and maybe suggested strategies for further enlargement of the theory. But against them comes the possible loss of heuristic background and motivation, since the axioms or assumptions are in place precisely for that purpose rather than for self-evidence, which indeed they may lack.

This balance of advantages and disadvantages involves another one, namely, that between the needs of creating and developing a theory and those required to found and especially to justify it. Epistemologies usually focus much more on the latter category than to the former; for example in mathematics, in connection with set theory and logics, and sometimes with abstract algebras and the differential and integral calculus. But the former category deserves much more attention, and especially in applied mathematics, where philosophical issues are more complicated than in pure mathematics since they involve the physical interpretation of the theories as well as the mathematical content.

I treat here the case of classical mechanics, especially as it developed in the 18th century, covering both statics and dynamics; some notice is taken of its extensions during the following century, and of its interactions with the new field of mathematical physics. Different formalisations are noted, which indeed constituted competing traditions. I also note the place of the calculus and of physics as important neighbouring subjects, and finally consider an ensemble of philosophical questions. A substantial bibliography of historical and critical literature is appended, and some items are cited in the text as sources for particular points made: the others cover a substantial part of the period discussed here, and/or raise important (historico-)philosophical issues. I do not treat the history or philosophy of quantum mechanics or relativity theory, or the modern penchant for full-scale axiomatization of physical theories even where mechanics is included.

2 The range of mechanics

First, I review the range of phenomena that had fallen under study. By and certainly during the second half of the 18th century mechanics had become so imposing an empire that it can helpfully be divided into five branches. This division, and the attached adjectives, are my impositions, but they are faithful to the period. For more elaborate summary as of around 1800, *cf.* [Gra90a, Chapter 5-6, 8]; and for later developments see [Stä05, Vos01].

2.1 Corporeal mechanics

Corporeal mechanics concerns "ordinary-sized" objects as found and handled on the Earth, including not only sticks and stones but also continuous media such as fluids and elastic surfaces and bodies, and also sound (which was then thought to belong as a phenomenon to mechanics). I include here the various basic principles assumed under each tradition, both for statics and for dynamics, since they were largely conceived in terrestrial contexts. Underlying the other four further branches about to be summarized, these principles will be reviewed in § 3.2-3.4.

2.2 Celestial mechanics

Celestial mechanics belongs to that part of mathematical astronomy in which the heavenly bodies are taken to be point masses. Major questions included the fine details of their orbits and rotations [Wil$_0$85], and from the 1770s the desire to prove that the planetary system was stable. Another principal issue was the three-body problem, inspired by the study of the relationships between the motions of the Sun, the Earth and the Moon [Gau17, Waf76]. Comets caused especial interest but also much perplexity.

2.3 Planetary mechanics

Planetary mechanics takes as its major concern the shape of the heavenly bodies; the term "planetary" is not quite right, although in the 18th century satellites were sometimes called "planet of the second order". The Earth was, of course, the most important such body, especially with and after the demonstration of its oblateness in the 1740s [Gre95]. The next most popular body was the Moon, and not only because of three-body problem and tidal theory. Other important topics included precession and nutation of the heavenly bodies; the motions of the terrestrial tides; and topography and cartography, the latter trying to cope with the requirements both of land-dwellers with maps and of sea-farers with charts.

2.4 Engineering mechanics

Engineering mechanics marks the major link to engineering and technology. Military interests prevailed in many of these topics. Several of them can be grouped together under "friction studies"; important case studies included the stability of embankments [Köt$_0$92] and the construction of buildings or structures of various kinds, such as arches and bridges [Pon52].

The study of large bodies of fluid, usually water, formed a major sub-branch [For05]. It included both hydraulics and fluid mechanics, especially involving canals and locks, and dams and rivers; and the flow of water out of orifices, especially the properties of cavitation and contraction. The most important artefacts were water-wheels, turbines, pumps, valves and pistons, both their design and (in)efficiency of their operation. Related topics include the building and steerage of boats and ships, and the use of sails.

2.5 Molecular mechanics

Finally comes Molecular mechanics, which treated the interaction between the supposed "molecules" comprising the intimate structure of matter; indeed, the way in which extended continuous bodies were composed of its molecular components was a major question [Stä05]. This was the smallest of the five branches during the century, and usually arose in connection with the other four. For example, elasticity and friction studies would entail certain assumptions about the constitution of the bodies under examination, while fluid mechanics sometimes drew on the presumed behavior in terms of the motions of its molecules. Another context, often linking to physics, was aether theory, when this ubiquitous medium was presumed either to be purely continuous, or else to possess a "molecular" structure with molecules regarded as far smaller even than those of ordinary matter.

3 Three traditions in mechanics

3.1 Personnel

A major feature of the development of mechanics during the 18th century was the emergence of three distinct traditions, in competition not only over the question of heuristics versus formalization but also concerning the territory of legitimate application. Anyone expecting to find that the principles laid down by Isaac Newton (1642-1727) controlled *all* mechanics will receive a great surprise, especially for mathematics on the Continent, which came to dominate from mid century when the subject declined substantially in Britain. While Newtonian mechanics was very influential, two other traditions were emerging in which different principles were adopted, and relative to which Newton's laws were theorems. The differences between them emerged mainly in connection with dynamics, including its own relationship to statics.

Among the principal mathematicians, the Swiss were notable already in the late 17th century through contributions made by the Bernoulli brothers James (1657-1705) and John (1667-1748), and later from John's son Daniel (1700-1782) and sort-of student Leonhard Euler (1707-1783). Among the French, Pierre Varignon (1654-1722) was also early, to be followed from the 1730s by Jean Le Rond d'Alembert (1717-1783) and Alexis Clairaut (1713-1765).

The next generation there included Charles Coulomb (1736-1806), Gaspard Monge (1746-1818), Charles Bossut (1730-1814), Lazare Carnot (1753-1823), Pierre Simon Laplace (1749-1827), Adrien Marie Legendre (1752-1833) and Gaspard Riche de Prony (1775-1843); and also Joseph Louis Lagrange (1736-1813) when he moved to Paris in 1787 after stints in his native Turin and then in Berlin. At least a dozen other figures in these and other countries (especially Italy) made distinguished contributions in various parts of mechanics during the century.

I now review the three traditions that were developing in order to study the above range of phenomena, or at least large parts of it — and indeed competing with each other [Gra90b].

3.2 Newton's approach

Newton's approach was of course prominent, especially in celestial and planetary mechanics. His laws were at once both mathematical and mechanical. The second one was often used in the form

$$\text{force} = \text{mass} \times \text{acceleration} \qquad (1)$$

including by Newton himself; but he actually formulated in terms of a relationship between increments of impulse and increments of momentum [Coh71], [Bra95]. (The modification of this law for bodies of variable masses was made later.) The first law was well understood to apply both to static and to dynamic equilibrium. However, within dynamics the derivation of some results was sometimes problematic, until it was realized (by Euler among others) that the principle of angular momentum had to be adopted as a fourth law [Tru68, Chapter 5]. The notion of central forces and actions between bodies (balanced by reaction according to Newton's third law) was widely adopted; however, the inverse-square law was taken up with more enthusiasm in Britain than on the Continent, where other laws were also mooted [Gui99].

3.3 Kinetic and potential energies

An alternative tradition, with quite a long history, drew upon the relationship between kinetic and potential energies. (I use the modern terms: "*vis viva*" and "*forces vives*" then were popular names for the former notion, while the latter, involving "force × distance" in some form or other, received various names.) Gottfried Wilhelm Leibniz (1646-1716) used it to try to mathematicise Descartes's vortex theory of celestial motion, which Newton came to loathe. This tradition gained its best credentials in engineering and technology; by the 1780s it was elevated into a general approach, with special utility in cases of impact and percussion where disequilibrium occurred [Sco70]. A pioneer of this tradition was Coulomb [$Gil_1$71], and the main advocate was Carnot [$Gil_0$71]; both men had engineering backgrounds.

3.4

Carnot was thereby confronting not only the Newtonian tradition but also, and indeed especially, the third and newest one, which had grown up in reaction against Newton's. One stimulus was d'Alembert's puzzlement over the notion of force; he suggested that (1) should be taken as its definition. But then some new law is needed to replace it, and he offered a rather incoherent statement, now known as "d'Alembert's principle", about the relationship between the motions of masses when left in their current state of equilibrium and when affected by imposed actions such as forces or impacts [Fra85].

This tradition also adopted "the principle of least action", an optimizing law formulated during the 1740s with the help of the calculus of variations; "action" was a technical term, denoting "force × velocity × distance" in a variety of contexts: for example, for infinitesimal displacements ds it required an integral in distance s. As in other contexts in mechanics, some tricky metaphysical issues arose, concerning the relationships between force and substance; here Euler also invoked religious grounds in order to guarantee its generality. The principle was to be utilized comprehensively first by Lagrange, and in a secular spirit [Pul89].

The final main principle was that of "virtual velocities" (not "work", a word that hinted at the disliked notion of force): a refinement of d'Alembert's principle, it stated how masses move after disturbance from equilibrium. But it assumed that mechanical situations could always be reduced to equilibrate ones, and that dynamics could be reduced to statics. Various efforts were made to prove it from other statical and dynamical principles, such as that of the lever [Lin04], [Bai75].

Another marked feature of this tradition was its formulation and development exclusively in algebraic terms. Indeed, it is often called "analytical", and Lagrange's treatise Méchanique analitique (1788) was the definitive statement of it for a long time. There are no diagrams in the book, the author tells us early on, and he means it, seriously. The use of the calculus of variations in an exclusively algebraic form inspired the alternative adjective "variational" to name this tradition.

4 Neighbouring subjects

4.1 Differential and integral calculus

One major subject is the *differential and integral calculus*, which Newton and Leibniz invented independently in the 1660s and 1670s respectively. My choice of adjectives to characterize the theory reflects the history: they are Leibniz's, for his theory was published first (in 1684-1686) and developed quickly, while Newton's version, using "fluxions" and "fluents", appeared in print only in 1704 though a few manuscripts on it had circulated for some time [Gui89]. Soon afterwards the aged Newton launched a charge of plagiarism against Leibniz, which was justly rejected; but the affair polarized the international mathematical community, and Britain's isolation and eventual decline seems to have been a consequence. Ironically, mechanics would have formed a sounder context for a charge of plagiarism [Mel93], even though, as we saw above, Newton and Leibniz adopted different theories.

While Newton's mechanics was closely studied on the Continent, his calculus was largely ignored there. Instead the Leibnizian form flowered massively, especially because of its utility in all branches of mechanics; in particular, d'Alembert, Euler and others developed the calculus of *several* variables from the mid century onwards, leading to partial and not just ordinary differential equations. In addition, the study of problems in optimization led to the calculus of variations: the central role in it given to algebra led it to play a major role in analytical mechanics (§ 3.4). Indeed, from the 1770s Lagrange offered a rival foundation for the calculus, reducing it also to algebra by assuming that every mathematical function $F(x)$ could be expanded in a certain power series (the "Taylor series") for every value of its argument variable x [Gra90a, Chapter 3-4]. This is the main context where the various forms of the calculus bore upon the competition between the three traditions in mechanics; for both the calculus and mechanics were to be algebraized.

Lagrange's assumption about the series expansion turned out to be mistaken, for counterexamples were found in the early 1820s by A.L. Cauchy (1789-1857). He formulated the calculus in yet another way, forging newly intimate links with the theory of functions and with infinite series by placing all under a theory of limits, not just relying upon an it as an intuitive notion. This is the tradition that has come to dominate [Bot86, Chapter 2-3,5]; [Gra90a, Chapter 10-11]; but it lacks the intuitive feel that adorns especially the Leibnizian version, and applied mathematicians tended not to use it.

4.2 Physics

The other main neighbouring subject was physics. In strange contrast to its later status, during the 18th century it was rather minor discipline, and non-mathematical. It dealt mainly with the constitution of matter; properties of air, related to the propagation of sound and to barometry; heat theory, including gases and vapours; electrostatics (as we would then say) and magnetism; and physical optics, this being the largest single part. It overlapped with several branches of mechanics (for example, gases with elasticity), and also with physical chemistry as then understood (for example, in the detailed shape of molecules). Mechanics was the leading "classical" science of its time, very prestigious; physics was a "Baconian" science, and of much lower status [Kuh76]. Physics also overlapped with "natural philosophy" in English-speaking countries and *"Naturphilosophie"* in German ones, which were not the same as each other.

Major changes occurred at the turn into the 19th century, led by the French, who provided most of the principal physicists and mathematicians of the time. In particular, mathematization was brought to bear upon the main areas of physics, in two rather different stages [Gra90a, Chapter 7,12-14]. First, from the mid 1800s Laplace greatly extended molecular mechanics by treating *all* physical phenomena as based upon molecules construed as sources of central forces that interacted with other molecules [Fox74]; [Gra90a, Chapter 7]. The greatest success was achieved in physical optics, where Newton's ballistic theory was adopted and adapted: Jean Baptiste Biot (1774-1862) was especially active, though the main theoretical advances were effected by Etienne Malus (1775-1812). By the 1820s, however, even Laplace acknowledged that the new waval theories of Augustin Jean Fresnel (1788-1822), based upon properties of the supposed aether, possessed a greater element of deducibility and predictive power than his own molecular approach. Similarly, Laplace's molecular modelling of heat diffusion, which was adopted by Siméon Denis Poisson (1781-1840), never matched either in articulation or reception the approach developed from the mid 1800s by Joseph Fourier (1768-1830), who held that the nature of heat should not be examined but taken as known, with cold as its opposite [Wei88, Chapter 4-5]. His position can be called "positivist", for that word was coined from the late 1820s onwards by Auguste Comte (1798-1857), who took Fourier's study of heat theory as paradigmatic for his own philosophy of science [Pic93].

Two other parts of physics were electrostatics and magnetism, where molecular modelling was enriched (or contaminated?) by the assumption also of the existence of electric and magnetic fluids. Poisson was the most prominent Laplacian, with major analyzes of electrostatics and magnetism that made some use of molecular modelling; but strangely, he never attempted to molecularise electromagnetic and—dynamic phenomena after their exhibition in the early 1820s. The principal French figure here, Adrien Marie Ampère (1775-1836), did not include molecularism in his own aetherian models.

In these two stages, the first one led by Laplace and the second by others using other principles, physics acquired mathematical physics. Thereby it gained status greatly, especially with the massive attention given also to the theoretical and experimental sides. The various physical theories possessed their own principles, such as Huygens's law of double refraction in optics or Newton's law of cooling for heat; but otherwise mechanical principles were closely imitated. For example, something sort of "force" was held to be in operation, usually acting centrally and capable of decomposition into components by the cosine law; conservation of something like mechanical "*vis viva*" was advocated at times (for example, in Fresnel's interpretation of Huygens's law of double refraction); one of Ampère's main goals in electrodynamics was to establish an inverse square law of attraction between the elements of wires; he laid emphasis upon equilibrium; and so on. To convey the massive extent of these imitations or analogies, I characterize classical mathematical physics as mechanics in fancy dress.

4.3 19th century mechanics

During the 19th century mechanics continued to develop strongly, often in interaction with its fancily clad descendants but also within its own remit [Gra90a, Chapter 15-16]; [Gar99]; and above all, [KleMül96-35]. In 1803 an extraordinary lacuna in statics was filled when Louis Poinsot (1777-1859) added to the existing theory of the composition of forces a companion analysis of "couples" (his word). Similarly but much later, in 1892 A.M. Lyapunov (1857-1918) greatly expanded the repertoire of form of equilibrium that lay with in the notion of stability of dynamical systems.

Another development with great consequences for both mechanics and mathematical physics was the introduction of potential theory. A key stage was the recognition of the important book obscurely published book in 1828 by George Green (1791-1841) [Gra95]. Hitherto some theorems on equipotential surfaces had been known and used especially in planetary mechanics and magnetism; but when Green's contribution was properly recognized, much deeper and more general results were found relating interior and surface conditions over extended bodies or regions for a wide class of physical phenomena [Bac83]; Green's own context had lain in electrostatics and magnetism.

By contrast, this approach was not at all convenient for the energywork tradition, where Carnot's emphasis on disequilibrium was adopted by various French engineering scientists, such as Claude Navier (1785-1836), Jean Victor Poncelet (1788-1867) and Gaspard Gustave Coriolis (1792-1843). In this tradition potentials were not assumed to exist, and loss of kinetic energy was held to be converted into "work" (Coriolis's word from 1829, and covering "force × distance" for all contexts). The issue of equilibrium versus disequilibrium split these two traditions quite fundamentally, since it involved the generality of application: engineers recognized the importance of impact, variationists ignored it. The issue was mathematical as well as physical; given the work term $\sum_r F_r dx_r$ for forces F_r displaced by infinitesimal distances dx_r, variationists could equate it to the differential of an assumed potential P,

$$\sum_r F_r dx_r = dP \qquad (2)$$

and then integrate (2) to obtain P. But supporters of energy/work could not so proceed; instead, they worried about (non)-elastic properties of extended bodies and fluids and the effect of impacts upon them, such as the the loss (or disappearance) of energy. A reconciliation of these two diametrically opposed positions came only around the mid 19th century with the "energy physics" of Hermann Helmholtz (1821-1894), William Thomson (1824-1907) and others. For them "all" actions involved in a phenomenon, physical ones such as heat and electrolysis as well as the mechanical ones, were taken into consideration in the assumption of P; much clarity was brought to the means by which energy could be converted into work within a *very* general theory of the *"Erhaltung der Kraft"* — including, for example in the final pages of Helmholtz, the behaviour of batteries [Haa09, Can$_0$93].

5   Some philosophical questions

As is evident from this review, various philosophical issues surround the advocacy of these traditions and criteria of preference among them. I review a handful here, mostly focussed upon mechanics. They address "why" as well as "how" questions; this distinction is not always given the attention that it deserves, and when it does disagreements may again occur! In my reading of the three traditions described in § 3, Newton's answer to the "why" question for the interaction between bodies is the inverse square law, and to the "how" question the three laws; in energy/work mechanics the exchange of these two categories is held to explain the "why", though further details concerning the phenomena at hand furnish the "how"; in variational mechanics the principle of least action claims status of both "why" and "how", though d'Alembert's and virtual velocities principles may be more useful regarding the latter. The "How?" question usually predominate in the discussions below.

5.1   Certainty of Knowledge

A main hope for all philosophies has been to explain the apparent certainty of knowledge; and mechanics was a prime candidate, on account of both its high status and its heavy use of mathematics. This search for certainty helped to stimulate positivist and empiricist interpretations of mechanics. But the usual difficulties arise; namely, that some major notions are not very experiential: for example, force for d'Alembert (§ 3.4). So *reification* comes into play, as ever in theorizing. An important case is the mass-point [Kör04], made explicit by Euler though used informally before him (for example, by Newton): if it is really a point, then how can it possess mass?

5.2   Justification

All the traditions made some assumptions, both concerning "why" and "how"; in the Newtonian case, more of them than was explicitly realized (§ 3.2). These are examples of creative processes; the *justification* of the assumptions, and of the consequent theorems is a separate and later matter.

In mechanics the balance between creation and justification differs between the variational tradition and the other two. Some new results came from variational mechanics: for example, the so-called "Lagrange-Poisson brackets" canonical solutions of the equations of motion, found in late 1800s [Gra90a, p.371-386]. But normally this tradition, both in the form adopted by Lagrange and the extended version developed from the late 1820s onwards principally by William Rowan Hamilton (1805-1865), is most powerful in systematizing a large range of theories in mechanics (and mathematical physics) and results already found in the Newtonian or the energy-work traditions [Pra35]. This is somewhat akin to the method of fully axiomatizing a mathematical theory (including mechanics) as practiced from the late 19th century onwards. In such theories the answers are usually known in advance, and are re-derived in this efficient way; in the case of variational mechanics and physics one needs to know which quantity has to be optimized. The central role of algebra in variational mechanics was also crucial; general formulae were offered, *not* encumbered with special additional properties that a geometric diagram cannot help exhibiting.

In such ways this tradition was more formalized (in the sense of the title of this article) than the other two. An interesting question for all branches of mathematics, and their respective histories, is the extent to which formalization, including axiomatization, has helped or hindered the finding of new results. In classical mechanics the restrictions seem to have been quite severe, but this is surely not true in general (for parts of topology, for example).

5.3 Reductionism & Teleology

Two features of the variational tradition caused disquiet. One was reductionism, especially in the claim made for the role of the principle of virtual velocities of reducing mechanical situations to equilibrium and thereby of reducing dynamics to statics. The other is teleology, especially with the principle of least action, which posits a macro criterion for the required situation such as the actual path taken by the particle under the stated conditions. As Henri Poincaré nicely put it,

> This molecule seems to know the point to which we want to take it, to foresee the time that it will take to reach it by such a path, and then to know how to choose the most convenient path. [Poi02, p.128]

Teleology does not obtain in the other traditions, because initial conditions have to be specified in addition to the formulae describing the motion. Ironically, in overlooking this point, the Newtonian tradition has sometimes been given a deterministic interpretation: Laplace was an important advocate.

5.4 Ontology & Epistemology

The advocacy (or rejection) of reduction ties in closely with issues of ontology and epistemology; that is, which terms in a theory are held to refer, and which notions are primitive with the other ones derivative from them. As the account above shows, the history of mechanics exhibits a wide range of fundamental disagreements. The notion of force seems to have excited the most anxiety, partly for technical reasons; in particular, was the major form, "mass × acceleration" or "mass × velocity" or "mass × velocity^2", or something else? Important philosophical questions also arose, especially concerning the relationship between (whichever) force and causality, in its various forms: was force a "thing" or substance, or was it known only for its effects?

Newton seems to have subscribed to a version of the former position, while d'Alembert and supporters of energy/work adopted the second position, though for different reasons. Some of the latter even preferred work to energy as the basic notion, on empiricist grounds: while velocity was a legitimate notion, velocitysquared was hard to grasp, while work used just force and distance.

Another ontological issue arose in connection with potentials: if the work term always admits one in (2), does force become a notion subordinate to that of potential, or is (2) only mathematical talk? The same kind of question concerns the notion of action, as used in its principle of the least. The mathematician to whom such questions should most obviously be addressed is Lagrange, but he was disappointingly mute on his position and its philosophical consequences: Boudri even takes him to be a formalist, treating mechanics *only* as mathematical expressions [Bou02, Chapter 7], though surely there is more than symbols to the claim that dynamics is reduced to statics.

5.5 Desimplification

The articulation of general theories usually requires additional hypotheses specific for particular sub-branches; for example in mechanics, parts of continuum mechanics such as hydraulics and elasticity theory. These moves exemplify an important part of scientific theorizing which I call *desimplification*; the general situation was handled by omitting many factors of the phenomena in order to obtain a scenario sufficiently simplified (though still complicated) to lead to soluble solutions; then (some of) these simplifications are eliminated in the next stages. An outstanding suite of examples is provided by the so-called "simple" pendulum; for extremely precise demands from geodesy and from some contexts in astronomy led to ever more intricate analyzes; so it became a very *complicated* instrument indeed when allowance was made for possible differences between the down- and upswings of the bob, little kicks at the end of the swings, air resistance to its motion, twists in the suspending wire or bar, motion of the wire across the supporting chape, torsion in the frame holding a large pendulum, and so on and on [Wol89].

5.6 Structure Similarity

Desimplification relates to the final point treated here, which concerns the use of a mathematical theory Ma in a physical one Ph (such as mechanics). The theory Ma will have a structure of its own, and so does Ph; to what extent can they be linked? That is, is *structure similarity* asserted to hold between Ma and Ph? (*Cf.* [Gra92].) The status of the potential in (2) is an important case: does the integration of that equation make sense mechanically (or physically) as well as mathematically? Another nice example is provided by Laplace's molecular physics noted in §4.2. He construed physical phenomena as cumulative actions of a molecule by all the others of its type (such as a molecule of light receiving action from its companions) according to a central "force" f, of which the form was unknown apart from assumed to decrease rapidly as distance increased. These actions would accumulate, like a sum; so Laplace took the totality to be an integral involving f. After him Cauchy and Poisson thought that these integrals should be replaced by sums, in order to represent more accurately the supposed physical interactions. By contrast to all this, positivist Fourier not only did not make any assumptions about heat (§4.2); he also did not adopt the wave theory of heat of his day just because his solutions of the diffusion equations took the form of (an infinite series of) sine and cosine functions, nor did he treat his integral solution of the diffusion equation for infinite bodies as accumulations of anything.

Both the assertion and rejection of structure-similarity between Ma and Ph are legitimate stances: the combination of possible positions is quite extensive. The philosophy of classical mechanics and mathematical physics abounds with such issues; many of them are not well studied, and not all of them reduce to issues of formalization. Pace [Wigner 1960] mathematics is often very effective in the physical sciences — not surprisingly, since the physics often motivated the mathematics in the first place. Indeed, its uncooperativeness seems to be less severe than is envisioned in [Wil$_1$00] when structure-similarity is given a central philosophical place.

References.

[Bac83] Max **Bacharach**, Abriss der Geschichte der Potentialtheorie, Thein'sche Druckerei 1883

[Bai75] Patrice **Bailhache**, Louis Poinsot, La théorie générale de l'équilibre et du mouvement des systémes, J. Vrin 1975

[Bel73] James F. **Bell**, The experimental foundations of solid mechanics, Springer 1973 [Encyclopaedia of physics, volume VIa/1]

[Bog76] Alekseï N. **Bogolyubov** (*ed.*), Mekhanika i fizika XVIII b[eke], Izdatelstvo Nauka 1976

[Bog78] Alekseï N. **Bogolyubov** (*ed.*), Mekhanika i fizika vtoroi polovini XVIII b[eke], Izdatelstvo Nauka 1978

[Bot86] Umberto **Bottazzini**, The higher calculus: a history of real and complex analysis from Euler to Weierstrass, Springer 1986; *translated by* Warren **van Egmond**

[Bou02] J. Christiaan **Boudri**, What was mechanical about mechanics: The concept of force between metaphysics and mechanics from Newton and Lagrange, Kluwer Academic 2002; *translated by* Sen **McGlinn**

[Bra95] J. Bruce **Brackenridge**, The key to Newton's dynamics, University of California Press 1995

[Bur$_0$193] Heinrich **Burkhardt**, Entwicklungen nach oscillirenden Functionen und Integration der Differentialgleichungen der mathematischen Physik, Teubner 1908 [Jahresbericht der Deutschen Mathematiker-Vereinigung 10]

[Bur$_1$InhKöt$_1$87] Clemens **Burrichter**, Rüdiger **Inhetveen**, and Rudolf **Kötter** (*eds.*), Zum Wandel des Naturverständnisses, Schöningh 1987

[Bur$_2$93] Piers **Bursill-Hall** (*ed.*), R. J. Boscovich Vita e attivita scientifica, His life and scientific work, Enciclopedia Italiana 1993, p.281-306

[Can$_0$93] Kenneth L. **Caneva**, Robert Mayer and the conservation of energy, Princeton University Press 1993

[Can$_1$Dos81] John T. **Cannon** and Sigalia **Dostrovsky**, The evolution of dynamics: vibration theory from 1687 to 1742, Springer 1981

[Coh71] I. Bernhard **Cohen**, Newton's second law and the concept of force in the Principia, *in:* [Pal71, p.143-185]

[Dar05] Olivier **Darrigol**, Worlds of flow: a history of hydrodynamics from the Bernoullis to Prantl, Oxford University Press 2005

[Dug55] René **Dugas**, Histoire de la mécanique, Griffon 1955; *english translation in:* [Dug88]

[Dug88] René **Dugas**, A history of mechanics, Dover 1988

[Duh03] Pierre-Marie-Maurice **Duhem**, L'évolution de la mécanique, Paris (Joanin) 1903; *over-rated, english translation in:* [Duh80]

[Duh80] Pierre-Marie-Maurice **Duhem**, The evolution of mechanics, Sijthoff and Noordhoff 1980

[Düh73] Eugen C. **Dühring**, Kritische Geschichte der allgemeinen Principien der Mechanik, Grieben 1873

[EchIbaMor92] Javier **Echeverria**, Adoni **Ibarra**, and Thomas **Mormann** (*eds.*), The Space of Mathematics: Philosophical, Epistemological, and Historical Explorations, Walter de Gruyter 1992 [Foundations of Communication]

[For05]	Philipp **Forchheimer**, Hydraulik, *in:* [KleMül96-35, p.342-472] (1905)
[Fox74]	Robert **Fox**, The rise and fall of Laplacian physics, **Historical studies in the physical sciences** 4 (1974), p.81-136
[Fra83]	Craig **Fraser**, J.L. Lagrange's early contributions to the principles and methods of mechanics, **Archive for history of exact sciences** 28 (1983), p.197-241
[Fra85]	Craig **Fraser**, D'Alembert's principle: the original formulation and application in Jean d'Alembert's Traité de dynamique (1743), **Centaurus** 28 (1985), p.31-61
[Gar99]	Elizabeth **Garber**, The language of physics, The calculus and the development of theoretical physics in Europe, 1750-1914, Birkhäuser 1999
[Gau17]	Alain **Gautier**, Essai historique sur le problème des trois corps, ou dissertation sur la théorie des mouvemens da la lune et des planètes, abstraction faite de leur figure, Courcier 1817
[Gil$_0$71]	Charles C. **Gillispie** (*ed.*), Lazare Carnot, savant, Princeton University Press 1971
[Gil$_1$71]	C. Stewart **Gillmor**, Charles Augustin Coulomb: physics and engineering in eighteenth century France, Princeton University Press 1971
[Gra84]	Ivor **Grattan-Guinness**, Work for the workers: advances in engineering mechanics and instruction in France, 1800-1830, **Annals of science** 41 (1984), p.1-33
[Gra87]	Ivor **Grattan-Guinness**, From Laplacian physics to mathematical physics, 1805-1826, *in:* [Bur$_1$InhKöt$_1$87, p.11-34]
[Gra89]	Ivor **Grattan-Guinness**, Modes and manners of applied mathematics: the case of mechanics, *in:* [Row$_0$McC89, p.109-126]
[Gra90a]	Ivor **Grattan-Guinness**, Convolutions in French mathematics, 1800-1840, From the calculus and mechanics to mathematical analysis and mathematical physics, 3 volumes, Birkhäuser 1990
[Gra90b]	Ivor **Grattan-Guinness**, The varieties of mechanics by 1800, **Historia mathematica** 17 (1990), p.313-338
[Gra92]	Ivor **Grattan-Guinness**, Structure-similarity as a cornerstone of the philosophy of mathematics, *in:* [EchIbaMor92, p.91-111]
[Gra95]	Ivor **Grattan-Guinness**, Why did George Green write his essay of 1828 on electricity and magnetism?, **American Mathematical Monthly** 52 (1995), p.387-396
[Gra00]	Ivor **Grattan-Guinness**, Daniel Bernoulli (1700-1782) and the varieties of mechanics in the 18th century, **Nieuw Archief voor Wiskunde** 5 (2000), p.242-249
[Gra05]	Ivor **Grattan-Guinness**, (*ed.*), Landmark writings in Western mathematics 1640-1940, Elsevier 2005
[Gre95]	John Leonard **Greenberg**, The problem of the Earth's shape from Newton to Clairaut, The rise of geomechanics in 18th-century Paris, Cambridge University Press 1995
[Gui89]	Niccolo **Guicciardini**, The development of Newtonian calculus in Britain 1700-1800, Cambridge University Press 1989

[Gui99] Niccolo **Guicciardini**, Reading the Principia, The debate on Newton's mathematical methods of natural philosophy from 1687 to 1736, Cambridge University Press 1999

[Haa09] Arthur Erich **Haas**, Die Entwicklungsgeschichte des Satzes von der Erhaltung der Kraft, Hölder 1909

[Har82] Peter M. **Harman**, Metaphysics and natural philosophy, The problem of substance in classical physics, Barnes and Noble 1982

[Hei02] John **Heilbron** (*ed.*), Storia della scienza, volume 5, L'età dei lumi, Enciclopedia Italiana 2002; *large section on mathematics*

[Hey72] Jacques **Heyman**, Coulomb's memoir on statics, An essay on the history of civil engineering, Cambridge University Press 1972

[Isr96] Giorgio **Israel**, Le mathématisation du réel, Seuil 1996

[Jen83] Marcel **Jenni** (*ed.*), Leonhard Euler 1707-1783: Beiträge zu Leben und Werk, Gedenkband des Kantons Basel-Stadt, Birkhäuser 1983

[Jou08] Émile **Jouguet**, Lectures de mécanique, La mécanique enseignée par les auteurs originaux, volume 1, Gauthier-Villars 1908; *covers the 17th century*

[Jou09] Émile **Jouguet**, Lectures de mécanique, La mécanique enseignée par les auteurs originaux, volume 2, Gauthier-Villars 1909

[KleMül96-35] Felix **Klein** and Conrad **Müller**, Encyklopädie der mathematischen Wissenschaften, Band IV: Mechanik, Teubner 1896-1935

[Kör04] Theodor **Körner**, Der Begriff des materiellen Punktes in der Mechanik des achtzehnten Jahrhunderts, **Bibliotheca mathematica** (3) (1904), p.15-62; *also published as Universität Kiel doctorate*

[Köt$_0$92] Fritz W.F. **Kötter**, Die Entwicklung der Lehre vom Erddruck, **Jahresbericht der Deutschen Mathematiker-Vereinigung** 2 (1892), p.77-154

[Kuh76] Thomas S. **Kuhn**, Mathematical vs. Experimental traditions in the Development of Physical Science, **Journal of Interdisciplinary Science** 7 (1976), p.1-31; *also in:* [Kuh77, p.31-65]

[Kuh77] Thomas S. **Kuhn**, The essential tension, Chicago University Press 1977

[Lin04] Richard **Lindt**, Das Prinzip der virtuellen Geschwindigkeiten, seine Beweise und die Unmöglichkeit seiner Umkehrung bei Verwendung des Begriffes 'Gleichgewicht eines Massensystems', **Abhandlungen zur Geschichte der Mathematik** 18 (1904), p.145-195

[Lüt05] Jesper **Lützen**, Mechanistic images in geometric form. Heinrich Hertz's principles of mechanics, Oxford University Press 2005

[Mac83] Ernst **Mach**, Die Mechanik in ihrer Entwicklung historisch-kritisch dargestellt, Brockhaus, 1883

[Mal92] Giulio **Maltese**, La storia de «F=ma», La seconda legge del moto nel XVIII secolo, Olschki 1992 [Bibliotheca di Nuncius 7]

[Mel93] Domenico Bertoloni **Meli**, Equivalence and priority: Newton versus Leibniz, including Leibniz's unpublished manuscripts on the 'Principia', Clarendon Press 1993

[Pal71] Robert **Palter** (*ed.*), The annus mirabilis of Sir Isaac Newton 1666-1966, The MIT Press 1971

[Pic93]	Mary **Pickering**, Auguste Comte, An intellectual biography, Cambridge University Press 1993
[Poi02]	Henri **Poincaré**, La science et l'hypothèse, Flammarion 1902, *english translation in:* [Poi05]
[Poi05]	Henri **Poincaré**, Science and hypothesis, Scott 1905; *reprinted at* Dover 1952
[Pon52]	Jean Victor **Poncelet**, Examen critique et historique des principales théories ou solutions concernant l'équilibre des voûtes, **Comptes rendus de l'Académie des Sciences** 35 (1852), p.495-502, p.531-540, p.577-587
[Pop51]	Karl R. **Popper**, Indeterminism in quantum physics and in classical physics, **British Journal in the Philosophy of Science** 1 (1951), p.117-133, p.173-195
[Pra35]	Georg **Prange**, Die allgemeine Integrationsmethoden der analytischen Mechanik, *in:* [KleMül96-35, p.505-804] (1935)
[Pul89]	Helmut **Pulte**, Das Prinzip der kleinsten Wirkung und die Kraftkonzeptionen der rationalen Mechanik, Steiner 1989
[Pul05]	Helmut **Pulte**, Axiomatik und Empirie, Eine wissenschaftstheoriegeschichtliche Untersuchung zur mathematischen Naturphilosophie von Newton bis Neumann, Wissenschaftliche Buchgesellschaft 2005 [Edition Universität]
[Roc98]	John J. **Roche**, The mathematics of measurement, A critical history, Athlone Press 1998
[Row$_0$McC89]	David **Rowe** and John **McCleary** (*eds.*), History of modern mathematics, volume 2, Academic Press 1989
[Row$_1$02]	John S. **Rowlinson**, Cohesion, a scientific history of intermolecular forces, Cambridge University Press 2002
[Rüh81]	Moritz **Rühlmann**, Vorträge zur Geschichte der theoretischen Maschinenlehre und der damit in Zusammenhang stehenden mathematischen Wissenschaften, Schwetschke 1881
[Sch$_0$70]	Robert E. **Schofield** (*ed.*), Mechanism and materialism, British natural philosophy in an age of reason, Princeton University Press 1970
[Sch$_1$OppvDy04]	Karl **Schwarzschild**, Samuel **Oppenheim**, and Walter **von Dyck** (*eds.*), Encyklopädie der mathematischen Wissenschaften, Band 6, part 1 Geodäsie und Geophysik, part 2, Astronomie, Teubner 1904-1934
[Sco70]	Wilson L. **Scott**, The conflict between atomism and conservation theory 1644 to 1860, Elsevier 1970
[Spe82]	David **Speiser** (*ed.*), Die Werke von Daniel Bernoulli, Birkhäuser, Bd.2, 1982 [Die gesammelten Werke der Mathematiker und Physiker der Familie Bernoulli]; *Some important editorial introductions*
[Stä05]	Paul **Stäckel**, Elementare Dynamik der Punktsysteme und starren Körper, 1905, *in:* [KleMül96-35, p.435-684]
[Sza77]	István **Szabó**, Geschichte der mechanischen Prinzipien und ihrer wichtigsten Anwendungen, Birkhäuser 1977
[Tod73]	Isaac **Todhunter**, A history of the mathematical theories of attraction and figure of the earth from the time of Newton to that of Laplace, 2 volumes, Macmillan 1873; *reprinted by* Dover 1962

[Tru54]   Clifford A. **Truesdell**, Rational fluid mechanics 1687-1765, in: Clifford A. Truesdell (*ed.*), Leonhardi Euleri Commentationes mechanicae ad theoriam corporum fluidorum pertinentes, Orell Füssli 1954 [Leonhardi Euleri Opera Omnia II 12], p.vii-cxxv

[Tru55]   Clifford A. **Truesdell**, Editor's introduction, in: Clifford A. Truesdell (*ed.*), Leonhardi Euleri Commentationes mechanicae ad theoriam corporum fluidorum pertinentes, Orell Füssli 1955 [Leonhardi Euleri Opera Omnia II 13], p.vii-cxviii

[Tru60]   Clifford A. **Truesdell** (*ed.*), Leonhard Euler, The rational mechanics of flexible or elastic bodies 1638-1788, Orell Füssli 1960 [Leonhardi Euleri Opera Omnia II 11(2)]

[Tru68]   Clifford A. **Truesdell**, Essays in the history of mechanics, Springer 1968

[Vat93]   François **Vatin**, Le travail, Economie et physique 1780-1830, Presses Universitaires de France 1993

[Vos01]   Aurel **Voss**, Die Prinzipien der rationellen Mechanik (1901), in: [KleMül96-35, p.3-121]

[Waf76]   Craig B. **Waff**, Universal gravitation and the motion of the moon's apogee: the establishment and reception of Newton's inverse-square law 1687-1749, Johns Hopkins University 1976; *Ph.D. thesis*

[War03]   Andrew **Warwick**, Masters of theory, Cambridge and the rise of mathematical physics, University of Chicago Press 2003

[Wei88]   Burghard **Weiss**, Zwischen Physikotheologie und Positivismus, Pierre Prevost (1751-1839) und die korpuskularkinetische Physik der Genfer Schule, Lang 1988

[Why61]   Lancelot Law **Whyte** (*ed.*), Roger Joseph Boscovich, Studies of his life and work on the 250th anniversary of his birth, Allen & Unwin 1961

[Wig60]   Eugene **Wigner**, The unreasonable effectiveness of mathematics in the natural sciences, **Communications in pure and applied mathematics** 13 (1960), p.1-14

[Wil$_0$80]   Curtis A. **Wilson**, Perturbation and solar tables from Lacaille to Delambre: the rapprochement of observation and theory, **Archive for History of Exact Sciences** 22 (1980), p.53-188, p.189-304

[Wil$_0$85]   Curtis A. **Wilson**, The great inequality of Jupiter and Saturn: from Kepler to Laplace, **Archive for History of Exact Sciences** 33 (1985), p.15-290

[Wil$_1$00]   Mark **Wilson**, The unreasonable uncooperativeness of mathematics in the natural sciences, **The Monist** 83 (2000), p.296-314

[Wol89]   C.J.E. **Wolf** (*ed.*), Mémoires sur le pendule précédés d'une bibliographie, Gauthier-Villars 1889

[Zan13]   Ottavio **Zanotti-Bianco**, Le idée di Lagrange, Laplace, Gauss, Schiaperelli sull'origine delle comete, Memoria storico, **Memorie della Reale Accademia delle Scienze di Torino** 63 (1913), p.59-110

Received: May 28th, 2003;
In revised version: December 2nd, 2003;
Accepted by the editors: February 14th, 2004.

Benedikt **Löwe**, Volker **Peckhaus**, Thoralf **Räsch** (eds.)
Foundations of the Formal Sciences IV
The History of the Concept of the Formal Sciences

Husserl's argument against naturalism and his own foundation of pure philosophy

LEILA HAAPARANTA[*]

Department of Mathematics, Statistics and Philosophy
University of Tampere
Tampere, Finland
E-mail: Leila.Haaparanta@uta.fi

1 Introduction: on the distinction between pure and naturalistic philosophy

Philip Kitcher writes in his article "The Naturalists Return" [Kit92] that after Quine there was a radical change in the analytic tradition: philosophy returned to what Frege and Wittgenstein had rejected in the end of the nineteenth century and in the beginning of the twentieth century, that is, to the naturalistic view that empirical sciences are relevant in view of philosophical claims and that philosophy must take the results of empirical research into account. The concept of naturalism has various uses in contemporary philosophical discussion. According to Robert Koons [Koo00], metaphilosophical naturalism is the view that there is continuity between the methods of philosophy and those of natural science and that philosophical theories are merely a species of scientific theories [Koo00, p.62]. In my paper entitled "On the Possibility of Naturalistic and of Pure Epistemology" [Haa$_1$99], I suggested that as a metaphilosophical doctrine naturalism is the view that philosophical knowledge and philosophical chains of arguments cannot be distinguished from empirical knowledge and the chains of arguments put forward by the special sciences [Haa$_1$99, p.32]. Moreover, it often happens that naturalism and physicalism are tied together.

---
[*]I am grateful to my anonymous referees for useful comments.

Steven Wagner and Richard Warner [WagWar93] point out that the notion of naturalism is closely related to that of physicalism. They also note that "physicalism" would seem to connote a narrower view which gives priviledge to physics among natural sciences. In their view, the difference between the two notions blurs, because those who call themselves physicalists also endorse chemistry, ecology, neuroanatomy and so on [WagWar93, p.1]. Several authors suggest that physicalism is a specific version of naturalism (*cf.*, *e.g.*, [CraMor00, p.xi]). If we consider naturalism an ontological doctrine, it is the view that all that there is is nature, and that nature does not include any meanings or intentions, which could be grasped by a specific philosophical method. By the term "naturalism" philosophers often refer to an epistemological position concerning the sources of our knowledge. Roughly stated, epistemological naturalism or naturalistic epistemology is the view that all knowledge is more or less empirical. Philip Kitcher and Susan Haack, for example, have tried to define and disambiguate the concept of naturalistic epistemology in their various writings (*cf.*, *e.g.*, [Kit92, p.74-76], and [Haa$_0$93, p.118-119]).

Following my earlier study, I will here regard pure and naturalistic philosophy as metaphilosophical doctrines, that is, doctrines concerning the task of philosophy and the nature of philosophical practice. That way of understanding the distinction has its source of inspiration in Gottlob Frege's philosophy. Post-Fregean, or even post-Kantian, studies of the distinctions between analytic and synthetic judgements and knowledge *a priori* and knowledge *a posteriori* have both philosophical and metaphilosophical import, as they contribute to conceptual analysis and give us arguments for and against various positions on the nature of philosophy. Frege considers the distinctions between syntheticity and analyticity as well as aprioricity and aposterioricity in his *Grundlagen der Arithmetik* and states that those distinctions concern the justification for making a judgement, not the content of the judgement. According to Frege, when we use those concepts, we speak about different ways of justifying the taking to be true of a judgement. He characterises the concepts in the following manner: If we prove a proposition and in the proof only rely on general logical laws and definitions, then the proposition is analytic. If it is impossible to give a proof without making use of truths which are not general logical truths but which belong to a special field of knowledge, then the proposition is synthetic. A truth is *a posteriori*, if it is impossible to construct a proof without referring to facts, that is, truths which cannot be proved and which are not general, because they contain claims about particular objects. If a truth can be derived solely from general laws that do not need a proof and that cannot be proved, then the truth is *a priori* [*Grundlagen*, §3].

Frege does not tell us what he thinks of the nature of philosophical propositions. He states in the [*Grundlagen*, § 3] that if a judgement cannot be justified by means of logic and definitions, sense perception or general laws which neither need nor allow justification, then we cannot say whether the judgement we are interested in is analytic, synthetic, *a priori* or *a posteriori*. If Frege thinks that philosophical judgements can be classified by means of those concepts, as an antinaturalist he either takes them to be analytic *a priori* or he thinks that there is a special field of philosophical knowledge just as there is the field of geometric knowledge which provides us with synthetic truths *a priori* based on pure intuition. However, Frege does not consider those alternatives. No matter what his position is, his distinctions provide us with a useful starting-point when we seek to clarify the distinction between pure and naturalistic philosophy. What is especially interesting in Frege's approach is that he refers to our argumentation strategies, when he characterises distinctions between analytic and synthetic propositions, on the one hand, and between truths *a priori* and truths *a posteriori*, on the other.

Following Frege's line of thought, pure and naturalistic philosophy can thus be understood as views concerning what kind of argumentation strategies are permitted in philosophical discussion. The underlying question is what kind of judgements one is or is not allowed to make in philosophical discussion; hence, to what kind of propositions one is or is not permitted to refer when one makes a philosophical judgement. In order to find out a philosopher's metaphilosophical commitments one has to study what the propositions or beliefs are like on which a philosopher bases his or her claims. That procedure has a Fregean spirit, that is, it is close to how Frege draws his distinctions in the *Grundlagen*. The beliefs that lie under a philosopher's judgement may be pure, that is, they may concern logic, the relations between concepts or what are traditionally called essences of things, or they may be everyday beliefs or empirical, more specifically, scientific beliefs, hence, beliefs based on what the philosopher takes to be results of scientific research. On this characterisation, a naturalist is a philosopher who thinks that a philosopher is permitted or even obliged to refer to empirical beliefs, even to beliefs produced by scientific research, in philosophical discussion. On the other hand, a pure philosopher does not allow that kind of beliefs to occur in argumentation; he or she denies what a naturalist regards as permitted or as obligatory in philosophical discussion. Both these positions can be considered answers to the requirement that philosophical research must be reliable, but they disagree on what guarantees reliability. One who speaks in favour of pure philosophy may argue that a professional philosopher must not resort to empirical beliefs, because in doing so he or she relies on research that he or she has not done or is not even able to evaluate. That is to argue that, paradoxically, a philosopher who relies on scientific research is not on the sure path of science. Naturalism can also be seen as a threat to the whole field of philosophy, as it creates the impression that natural scientists can take care of the tasks that belong to philosophers. That was one of Edmund Husserl's worries.

The present paper is a further elaboration of Haaparanta [Haa$_1$99], in which the possibility of naturalistic and of pure epistemology are discussed and challenged. I will discuss one of Edmund Husserl's arguments against one version of naturalism. I argue that Husserl's programme of pure philosophy faces the same problem he sees in the version of naturalism that he himself criticises. The problem is the following: if one wishes to express in language the position to which one is committed in one's philosophical practice and, what is more important, practise philosophy according to that commitment, one needs a view from the outside of language, which Husserl takes to be problematic for the two metaphilosophical positions. I will not discuss naturalism in such fields as the philosophy of mathematics and the philosophy of logic. What I am interested in is the view on philosophical argumentation that Husserl's antinaturalistic view on the foundations of logic implies.

2 Husserl's argument against one version of naturalism

Edmund Husserl, whose phenomenology was meant to be pure philosophy, attacks various forms of logical psychologism and anthropologism in the first volume of his *Logische Untersuchungen*. Both psychologism and anthropologism are forms of naturalism. Husserl lists representatives of these doctrines such as Mill, Bain, Wundt, Sigwart, Erdmann and Lipps [*LU I*, §38]. I will focus on one argument that Husserl raises against logical anthropologism [*LU I*, §36]. According to that doctrine the concepts and laws of logic and the rules of logical inference express the structures and functions of the mind that are typical of the human race. Even if Husserl directs his criticism against anthropologism, it also hits other forms of naturalism, such as the view that can be called transcendental psychologism. Benno Erdmann, for example, whom Husserl and Frege criticised, represented that kind of naturalism. Erdmann states that apodictic judgements are judgements the contradictories of which we are not able to think and that these include the laws of logic and of pure mathematics [Erd23, p.469]. He argues that the impossibility to think of contradictory judgements shows that apodictic judgements represent the conditions of our thought but in his view we cannot conclude that those conditions are the conditions of all valid thought, hence that they would express the essence of thought. He argues that the necessity of apodictic judgements is hypothetical, as it is tied to the present state of human thought [Erd23, p.472-477]. By making the judgement that other conditions of valid thought are possible, Erdmann represents the kind of naturalistic doctrine that Husserl takes to be problematic in the argument that will be discussed below. I will show that Husserl is committed to the view that we cannot step beyond the limits of thought. This view can be called rigorous transcendental philosophy. In Husserl's arguments against naturalism there is thus an interesting point where Husserl introduces himself as a transcendental philosopher in the sense that he wants to draw the limits of thought.

In the antinaturalistic argument in which I am here interested Husserl seeks to pinpoint the problems in the following claim made by the naturalists:

(1) There can be beings that are not committed to the law of contradiction and the law of the excluded middle.

Husserl pays attention to the ambiguity of the statement. First, it can be construed as the following statement:

(2) There can be beings whose judgements contain truths that do not follow the law of contradiction and the law of the excluded middle.

According to that interpretation, it is possible that someone makes a judgement of the form "*A and non-A*" and that the judgement is true. Secondly, according to Husserl, the naturalists' thesis can be understood as the following claim:

(3) There can be beings whose course of judging is not psychologically regulated by the law of contradiction and the law of the excluded middle.

Husserl remarks that (3) is true as we are such creatures. He comments on (2) by saying that if those beings understand the words "true" and "false" as we do, hence in a way that is determined by the mentioned laws of logic, then it does not make sense to say that those laws do not hold; that is because it is precisely those laws that determine what the words "true" and "false" mean to us and to those beings. On the other hand, if the beings that naturalists are talking about use the words "true" and "false" in some other way than we do, the whole debate has to do with the meanings of words. Husserl concludes that if logical naturalists have thus changed the meanings of the words "true" and "false", they cannot speak about truth and falsity as we do, hence, in a way that is determined by the laws of contradiction and the excluded middle.

It is not easy to see Husserl's background assumptions. It may seem that by means of his various arguments against naturalism Husserl seeks to justify classical logic by showing that its rejection leads to absurdities. In the mentioned argument there are other presuppositions. I suggest the following: Husserl assumes that there is a primary or a basic language with a basic logic to which we are tied; no hierarchies of languages are allowed. Husserl seeks to show that naturalism leads to absurdities, because it is an effort to consider the basic forms of thought and the basic laws of logic from the outside, hence, without being committed to them in the considerations. According to this interpretation, Husserl assumes that the basic concepts and laws of logic draw the limits of what can be thought and that one comes up with absurdities, if one seeks to describe those limits from the outside by making the claim that those limits are determined by the structures and the functioning of the human mind and that they could be drawn differently. Husserl does not exclude the possibility that someone has a different logic. Instead, he points out that the claim that it is possible that someone has a different logic cannot be made in our logic or in that other logic.[1]

We may also interpret Husserl's argument as follows: If a naturalist says that it is possible that someone has a different logic, he or she uses the concept of possibility. We may now ask the following question: What is the logic that defines the concept of possibility in the sense of logical possibility? That must be the speaker's, hence, the naturalist's, logic. If the speaker is a member of our species, he or she must formulate his or her claim in a logic the necessity of which he or she is challenging by the very claim. If a pure philosopher like Husserl argues that in order to describe and challenge our logic, we must step beyond that logic and that we cannot even think of such a step and, moreover, that this is precisely the naturalist's problem, he himself ought to be quiet about the basic forms of our thought. In what follows, I will show by means of examples that Husserl has the very same problem in his own project of pure philosophy.

3 Husserl's argument and the programme of pure philosophy

We may now look at what Husserl, the pure philosopher, himself does and what kind of moves he accepts in his philosophical practice. The starting-point of phenomenological studies is the step from the natural to the phenomenological attitude. It means that a phenomenologist must be able to

---

[1] I have presented this argument in a closely similar form in [Haa$_1$99].

distinguish between what is pure and what is empirical. The phenomenologist's claims concerning experiences are meant to be eidetic truths, that is, truths concerning the essences of those experiences. Husserl's project of pure philosophy emphasises that very feature of the philosophical enterprise. Husserl takes pure consciousness to be the object of philosophical research. Phenomenological research seeks to find the structures of consciousness without using the method of empirical research. The idea that a philosopher must give up the natural attitude and move to a phenomenological attitude is the core of Husserl's philosophical project, for example, in the first volume of the [Ideen], and that project is an outgrowth of his earlier logical antinaturalism. When Husserl starts his new science of philosophy, he specifies its object of research by changing the attitude. Husserl's project must be understood precisely against the background that it seeks to overcome naturalistic philosophy. The natural attitude, in which we, among other things, posit the world as present to us and support our claims by means of empirical facts, is our everyday attitude towards the world and the attitude of the natural and the social sciences. Husserl demands that it must not be the philosopher's attitude; as the philosopher's attitude it would be the naturalistic attitude (*cf.*, *e.g.*, [*Phil. als str. Wiss.*, p.302 & p.315]).

In the first volume of the *Ideen*, Husserl tells us what the purely philosophical attitude is by telling us what it is not, hence, by carrying out phenomenological reductions. To put it in different words from those Husserl himself uses, the reductions mean giving prohibitions and obligations to philosophers. Husserl tells us on what kind of propositions or beliefs a philosopher is not allowed to base his or her claims. In his view, a philosopher must rely on truths concerning the essences. Using the terminology I suggested above on the basis of Frege's distinctions, Husserl gives restrictions to philosophers' argumentation strategies and excludes certain kinds of arguments from the field of philosophy.

Another example of Husserl's philosophical practice can be found in the second volume of his *Logische Untersuchungen* [*LU II*, §40-45], in [*Form. & transz. Logik*] and in [*Erfahrung & Urteil*]. His logical studies have to do with the epistemological foundations of logic. In the *Logische Untersuchungen*, Husserl proposes a doctrine of categorial perception, which, among other things, seeks to be an answer to the question concerning the epistemological origin of logical concepts. Husserl introduces the concept of categorial perception along with the concept of sensuous perception. He thinks that categories, hence logical forms, are in sensuous objects, but they are as it were hidden in those objects. By categorial perception Husserl means perception in which a subject sees sensuous objects in such a way that categories are revealed. Husserl thinks that categories have their origin in sense perception. It means that, in his view, we set logical forms into objects in naïve, unreflected acts of consciousness but in order to know anything about those forms, that is, in order to know what they are and how they function in judgements and inferences, we have to see sensuous objects in terms of those forms, that is, we have to reflect on our objects of sense perception. Categorial perception is the recognition of logical concepts and logical laws in what is given in sense perception, but that recognition is not an outcome of gathering empirical data. On the contrary, Husserl seeks to justify the claims concerning what the basic logical concepts and laws are by referring to reports on what we perceive categorially. He argues that logical categories and laws need justification and the pure epistemology of logic will tell us how our logic is based on the structure and functioning of consciousness (*cf.* [Haa$_1$88]). According to Husserl, that is neither the consciousness of one individual nor the consciousness shared by members of a specific culture or by all human beings. Husserl thinks that it is consciousness in general, transcendental consciousness, and a pure philosopher does not make the judgement that it could be different, if, for example, the human race were different from what it is.

A third example of the conditions of phenomenological research is that a phenomenologist must be able to distinguish between form and matter, that is, morphe and hyle, in the analysis of experience. A naïve observer, who is committed to everyday thought, cannot make that distinction, but a pure philosopher assumes that the distinction can be made from a reflective philosophical point of view.

Those practices that I have described are not innocent, but they have important background assumptions. The three cases, the step from the natural to the phenomenological attitude, studying the epistemological foundations of logic and distinguishing between form and matter, presuppose that a philosopher is able to distinguish between what is pure and what is empirical and could be different as far as the forms of thought are concerned, hence in the specific sense that the forms of thought do not prevent it from being otherwise. But making that very distinction presupposes a step beyond the limits of thought, and that was precisely the step that was problematic in naturalistic philosophy in Husserl's view, as we saw above. If a naturalistic philosopher comes up with a contradiction with his or her judgement "Logic could be different", Husserl's problem is that in his pure philosophy he cannot make the judgement: "Logic cannot be different from what it is", because that claim must also be made outside our logic.

If we cannot step beyond the limits of thought, we cannot distinguish between pure and empirical beliefs, or we cannot make a distinction between what is pure and what is empirical in our web of beliefs. If a transcendental philosopher wishes to act rigorously following his or her conviction that we are tied to our forms of thought, he or she cannot make and express in language that kind of distinction. A puzzling conclusion follows: In order to be a pure philosopher in the sense described above, one must deny the possibility of distinguishing between what is pure and what is empirical. If that move is not possible, it is not possible to see pure forms; instead, it is only possible to find various beliefs in the web of beliefs. But if that is the case, it is not possible to practise pure philosophy. On the other hand, if one thinks that it is possible to distinguish between what is pure and what is empirical, one must give up such pure philosophy as assumes that we are tied to our forms of thought; in order to practise pure philosophy, one has to be free from those conditions. But if practising pure philosophy requires a point of view from the outside of the forms of thought, or to use the phrase adopted in the twentieth century, from the outside of the limits of language, its proponent cannot blame a naturalist for needing the same point of view.

4   A concluding remark

The arguments presented in this paper are not meant to convince anyone that all forms of pure or of naturalistic philosophy are in trouble. What they suggest, however, is that it is illuminating to formulate the distinction between pure and naturalistic philosophy by referring to philosophers' argumentation strategies and then to have a look at those actual strategies.

Primary Sources.

[*Grundlagen*]   Gottlob **Frege**, Die Grundlagen der Arithmetik: eine logisch-mathematische Untersuchung über den Begriff der Zahl, Koebner 1884; *english translation in:* [Aus68]

[*LU I*]   Edmund **Husserl**, Logische Untersuchungen, Erster Band: Prolegomena zur reinen Logik, *in:* [Hol75]

[*LU II*]   Edmund **Husserl**, Logische Untersuchungen, Zweiter Band: Untersuchungen zur Phänomenologie und Theorie der Erkenntnis, *in:* [Pan84]

[*Phil. als str. Wiss.*]   Edmund **Husserl**, Philosophie als strenge Wissenschaft (1910 - 1911), *in:* [Ber65]

[*Ideen*]   Edmund **Husserl**, Ideen zu einer reinen Phänomenologie und phänomenologischen Philosophie, Erstes Buch: Allgemeine Einführung in die reine Phänomenologie, *in:* [Sch76]

[*Form. & transz. Logik*]   Edmund **Husserl**, Formale und transzendentale Logik: Versuch einer Kritik der logischen Vernunft, Verlag von Max Niemeyer 1929

[Erfahrung & Urteil]   Edmund **Husserl**, Erfahrung und Urteil: Untersuchungen zur Genealogie der Logik, in: [Lan64]

References.

[Aus68]   John L. **Austin** (ed., trans.), Gottlob Frege, The Foundations of Arithmetic: A Logico-Mathematical Enquiry into the Concept of Number, Northwestern University Press 1968

[Ber65]   Rudolf **Berlinger** (ed.), Edmund Husserl, Philosophie als strenge Wissenschaft, Vittorio Klostermann 1965 [Quellen der Philosophie 1]

[CraMor00]   William Lane **Craig** and James Porter **Moreland** (eds.), Naturalism: A critical analysis, Routledge 2000

[Erd23]   Benno **Erdmann**, Logik, de Gruyter 1923

[Haa$_0$93]   Susan **Haack**, Evidence and Inquiry: Towards Reconstruction in Epistemology, Blackwell 1993

[Haa$_1$85]   Leila **Haaparanta**, Frege's Doctrine of Being, Societas Philosophica Fennica 1985 [Acta Philosophica Fennica 39]

[Haa$_1$88]   Leila **Haaparanta**, Analysis as the Method of Logical Discovery: Some Remarks on Frege and Husserl, **Synthese** 77 (1988), p.73-97

[Haa$_1$99]   Leila **Haaparanta**, On the Possibility of Naturalistic and of Pure Epistemology, **Synthese** 118 (1999), p.31-47

[Hol75]   Elmar **Holenstein** (ed.), Edmund Husserl, Logische Untersuchungen, Erster Band: Prolegomena zur reinen Logik, Text der 1. und 2. Auflage, Springer-Verlag 1975 [Husserliana 18]

[Kit92]   Philip **Kitcher**, The Naturalists Return, **The Philosophical Review** 101 (1992), p.53-114

[Koo00]   Robert C. **Koons**, The incompatibility of naturalism and scientific realism, in: [CraMor00, p.49-63]

[Lan64]   Ludwig **Landgrebe** (ed.), Edmund Husserl, Erfahrung und Urteil: Untersuchungen zur Genealogie der Logik, Claassen Verlag 1964

[NeuCoh73]   Marie **Neurath** and Robert S. **Cohen**, Otto Neurath: Empiricism and Sociology, Reidel 1973

[Pan84]   Ursula **Panzer** (ed.), Edmund Husserl, Logische Untersuchungen. Zweiter Band: Untersuchungen zur Phänomenologie und Theorie der Erkenntnis, two volumes, Springer-Verlag 1984 [Husserliana 19]

[Sch76]   Karl **Schuhmann** (ed., Edmund Husserl, Ideen zu einer reinen Phänomenologie und phänomenologischen Philosophie, Erstes Buch: Allgemeine Einführung in die reine Phänomenologie, two volumes, Springer-Verlag 1976 [Husserliana 3]

[Ver29]   **Verein Ernst Mach** (ed.), Der Wiener Kreis, The Vienna Circle of the Scientific Conception of the World (Wissenschaftliche Weltauffassung, Der Wiener Kreis) (1929), in: [NeuCoh73, pp.301-318]

[WagWar93]   Steven J. **Wagner** and Richard **Warner** (*eds.*), Naturalism: A Critical Appraisal, University of Notre Dame Press 1993

Received: June 30th, 2003;
In revised version: May 30th, 2004;
Accepted by the editors: June 5th, 2004.

Benedikt **Löwe**, Volker **Peckhaus**, Thoralf **Räsch** (eds.)
Foundations of the Formal Sciences IV
The History of the Concept of the Formal Sciences

Bronze age formal science?
With additional remarks on the historiography of distant mathematics

JENS HØYRUP[*]

Roskilde University
Postboks 260
4000 Roskilde, Denmark
E-mail: jensh@ruc.dk

In memoriam Robert Merton
(1910-2003)

ABSTRACT. My paper falls in two unequally long parts, each turning around a particular permutation of the same three keywords: The first, longer and main part treats *past understandings* of *mathematics*. The second, shorter part takes up *understandings* of *past mathematics*. In both parts, the "past" spoken of focuses on the Near Eastern Bronze Age, but other pre-modern mathematical cultures also enter the argument.

1 Past understandings of mathematics

In any proper sense, a "formal science" is a science which does not positively tell us anything about the world, a glove that fits any possible hand. Understood thus, the understanding of mathematics as formal science did not impose itself before, say, the acceptance of non-Euclidean geometry or the reception of Hilbert's *Grundlagen*, and was hardly possible before Kant's definitive formulation of the distinction between analytic and synthetic propositions. Often, of course, etymology only tells us what words

---

[*]The paper was prepared during a stay at the *Max-Planck-Institut für Wissenschaftsgeschichte*, Berlin. I use the opportunity to express my sincere gratitude for the hospitality I received.

do not mean any longer. In the present case, however, the derivation of "formal" from "form" (and ultimately from εἶδος, of which this Latin term is a loan-translation) is worth taking into account. This allows us to trace explicit formulations of precursor ideas back to classical Antiquity.

1.1 Aristotle and others

According to Aristotle, "the mathematicals" (τὰ μαθηματικά, *i.e.*, geometrical shapes, numbers, and ratios) are properties of real-world objects or collections of objects. These properties are isolated through a process of "abstraction" or "removal" of other aspects of the same objects or collections.[1] In [*Metaph.*, M.I-III] the term used is ἀφαίρεσις, "taking away", and it depends on the discipline which applies its perspective whether physical or mathematical properties are the essential ones (the others being "accidents"); from the point of view of geometry, the process that brings forth the sphere from the bronze sphere is thus (in modern terms) an objective one. In [*Phys.*, II] the term is χωρίζω, "to separate", and the process is performed "according to thought" (τῇ νοήσει), the physicality of the object remaining in any case essential — perhaps because the topic of the whole work is, exactly, *physics*.

Whether objective or subjective, this bringing-forth of the mathematicals by a process of abstraction does not suggest that mathematics –posterior, not prior to actual reality– is in any way a "formal" science. But there is more to it. As pointed out in [*Metaph.*, B (998a2-5)] "the [sensible] circle touches the ruler not at a point but [along a line] as Protagoras used to say in refuting the geometricians"; if the mathematicals were mere properties as we understand that term, it seems strange that a sensible line and a sensible circle should touch each other along a (sensible) line but their corresponding mathematical properties only in a point (which then, seemingly, should be the mathematical property of the sensible line of touching). Moreover, in [*Cael.*, II.14 (297b 21-23)][2] we find that

> the earth is either spherical, or it is spherical according to nature, and one should say each thing to be such as it professes to be according to nature and subsistence, not however as it is by violence or against nature.

Sphericality –clearly a mathematical property– is thus "by nature". A nature, however, is what we might term an ideal (and essential) property, an aim or τέλος which may not be achieved completely (or not at all). As explained in [*Phys.*, II (199a32-199b4)]:[3]

---

[1] See [Høy02b] for substantiation of the following condensed observations and for references to earlier work.

[2] As everywhere in the following where no translator is indicated, the translation is mine.

[3] With a slight correction of Charlton's translation in ⟨angles⟩.

> Mistakes occur even in that which is in accordance with art. Men who possess the art of writing have written incorrectly, doctors have administered the wrong medicine. So clearly the same is possible also in that which is in accordance with nature. If it sometimes happens over things which are in accordance with art, that that which goes right is for something, and that which goes wrong is attempted for something but miscarries, it may be the same with things which are natural, and monsters may be ⟨failures⟩ at that which is for something. [...]

Finally, there is an interesting passage in [*An. post.*, I (79a6-10)]:[4]

> Of this kind [*viz.*, studied by more than one science] are all objects which, while having a separate substantial existence, yet exhibit certain specific forms. For the mathematical sciences are concerned with forms; they do not confine their demonstrations to a particular substrate. Even if geometrical problems ⟨treat of⟩ a particular substrate, at least they do ⟨not do so *qua* treating of a substrate⟩.

This passage, like [*Metaph.*, M], presupposes that it depends on the discipline investigating an object what constitutes its *form*; if geometry is that discipline, then its sphericity may be its form *even if* (according to the *De caelo* passage, and in the likeness of "substantial" forms) *that form is not achieved completely*. The same mathematical form, for its part, may apply (together with everything that holds for a geometrical sphere) to all spherical objects irrespective of their remaining properties (their form or nature according to the perspective of physics, as well as their accidents). All in all, we are not far away from what could reasonably be described as a formal view of mathematics.

This view remained familiar to all those who grew up intellectually with Aristotelianism between the twelfth and the seventeenth centuries. It has to do with the Scholastic endorsement of the Plotinian "substantial forms" just mentioned[5] — an endorsement which presupposes that objects may possess other forms that do not define them as substances, *e.g.*, the mathematical form of a bronze sphere, the eightiness of a flock of sheep.[6] The view is more unexpected for those who have grown up with the modern standard view of Platonism and Aristotelianism, to whom it might give the impression that Aristotle's view of mathematics is, after all, closer to Platonism than we would expect. However, if we take a closer view of Aristotle's various refutations of what he claims to be Plato's view of numbers (ideal as

---

[4] Corrections of the translation in ⟨angles⟩.

[5] As discussed by Anderson [And69], Thomas Aquinas takes the existence of mathematical form (and mathematical matter) quite seriously.

[6] Mathematical forms, however, were not the only non-substantial forms discussed by St. Thomas and other scholastics.

well as mathematical and sensible[7]) and accept that Aristotle was better informed than we poor readers of his dialogues will ever be, the reverse conclusion may seem more plausible: namely that the supposedly "Platonist" view of geometrical shapes, *etc.*, as ideals which the sensible world emulates imperfectly is a post-Renaissance construction resulting from the superimposition of the Aristotelian view on Plato's reminiscence theory as it can be read from the dialogues mixed up with the "Platonism" of Plotinus and Porphyrios, already formulated in Aristotelian terms. Genuine Platonism may have been further removed from regarding mathematics as a formal science — but whether it really was is difficult to know.

1.2 Scribal cultures I: Middle Kingdom Egypt

It may seem even more difficult to grasp what the anonymous scribes of the Near Eastern Bronze Age thought about the mathematical techniques they were using. Yet whereas Plato left no mathematical writings (and no Greek mathematical writings survive that can be unquestionably dated to his or earlier eras), we know some of the materials by which the scribes were taught.

Let us first examine the situation in Middle Kingdom Egypt. The most important mathematical source is without doubt the Rhind Mathematical papyrus — according to its colophon a copy of an original written under Amenemhet III (c. 1844-1797 BCE). Its introductory passage [Cla99, p.122] promises nothing like a distinction between mathematical and other kinds of knowledge:

> Accurate reckoning [or Rules for reckoning] for inquiring into things, and the knowledge of all things, mysteries [...] all secrets.

What follows, however, treats of neither "mysteries" nor "all things"[8] but only mathematical calculation.[9] At first comes a tabulation of the solutions to the problem to express 2 as a sum of aliquot parts of n, n being an odd number between 3 and 101 — a table which serves as a calculational aid for all that follows. Then comes a tabulation of the numbers 1-9 divided by 10, also (as all fractional numbers in Middle Kingdom and later

---

[7] *E.g.*, [*Metaph.*, N (1090b27-35)], "no mathematical theorem applies to [ideal numbers], unless one tries to interfere with the principles of mathematics and invent particular theories of one's own"; further, those who invented "two kinds of number, the Ideal and the mathematical as well, neither have explained nor can explain in any way how mathematical number will exist and of what it will be composed".

[8] Admittedly, the words of the introductory phrase can be translated in different ways — according to various translators it speaks of "secrets", "obscurities" or "mysteries"; but there is no doubt as to the general tenor.

[9] A few non-mathematical entries in end of the copy, counted as #85-87, did not belong in the original, *cf.* [Pee23, p.128f].

Pharaonic mathematics) expressed as sums of aliquot parts; problems applying this table to the division of n loaves among ten men, $n = 1, 2, 6, 7, 8$ and 9; and a sequence of problems dealing with pure numbers or indeterminate "quantities". #35-84 treat concrete computations involving the various metrological systems, a geometry section (#41-60) presenting also the standard rules for determining triangular, rectangular and circular areas and the volumes of rectangular and circular prisms. Sacred or magical uses of numbers, on the other hand, are excluded, though certainly not absent from Egyptian scribal wisdom in general.

The territory that is covered is thus the whole of mathematics understood as the determination of concrete quantities together with the purely arithmetical and geometrical tools for this.[10] Expressed in global terms, the cognitive territory delimited by the papyrus corresponds to the Hellenistic-Biblical phrase "in number, measure, and weight" (which, through Augustine and Isidore, was to remain the current characterization of the mathematization of the world until the sixteenth century). The organization of the text –first all-purpose computational aids, then preparatory pure-number- and pure-quantity-problems, then problems dealing with measurable entities– shows that arithmetic was really understood as a glove that could be fitted onto any determination of a quantity.

If we believe the introductory phrase to be sincere when claiming to deal with "all things", then we may conclude that the author-scribe assumed *everything* to be made according to "measure, number, and weight", and that his text was meant to present in the beginning something like a body of formal knowledge which was next to be applied to everything — or at least to everything falling within a scribe's horizon. We know, however, that a professional scribe (but not necessarily a teacher of scribal mathematics) was supposed to deal with matters that could not be reduced to quantitative measure; if not necessarily for the author of the text then at least for scribes in general we must therefore assume that the body of "formal" knowledge encompassing arithmetic and geometric rules was only supposed to be generally applicable to a restricted area. This area constitutes what we usually refer to as "Egyptian" (or better, "Pharaonic") mathematics. We do not know how it was called (though "reckoning"/ḥsb is a fair guess); but there is no doubt that it was an Egyptian concept, not a splitting of the Egyptian cognitive world imposed artificially from outside. Students of ethnomathematics may be right that mathematics is *our* concept when applied

---

[10]This conclusion also fits the Moscow Mathematical Papyrus [Str30], although this second major source for Middle Kingdom mathematics (a copy of a collection of problems with solutions, apparently some kind of corrected examination paper) does not on its own entail quite such far-reaching conclusions.

to non-literate cultures — but it covers the thinking of Middle-Kingdom scribes quite well.

1.3 Scribal cultures II: Old Babylonian era

From Old Babylonian Iraq (2000-1600 BCE) we possess no global presentation of mathematics similar to the Rhind Papyrus — Old Babylonian mathematics was much too extensive to be contained in a single or a couple of clay tablets, and so far archaeologists have not stumbled on an undisturbed library which might inform us in a similar way.

The sources we have fall into three groups:

1. Table texts: tables of reciprocals and multiplication (and squares *etc.*); tables for metrological conversion; and tables of technical constants.

2. Tablets for rough mathematical work — at least as a rule student exercises.[11]

3. Problem texts.

The problem texts can be grouped in two intersecting ways. Firstly, they may start by a statement and next tell the procedure to be used ("procedure texts"), or they may give the statement only and perhaps the solution ("catalogue texts"); on the other hand, they may contain a collection of sundry problems ("anthology texts"); a sequence of systematically organized problems belonging to a single domain ("theme texts"); or a single or a couple of problems. Only the second division concerns us here (but catalogue texts are always theme texts[12]).

The table texts contain nothing but a single or combined table — except for those excerpts which were copied as a writing exercise and perhaps as a step in the memorization process, and which may combine the table with other matters being learned; this tells us something about the organization of teaching but little about the view of mathematics. The tablets for rough mathematical work regularly carry numbers on one face and a writing exercise (a proverb) on the other; this corresponds to a modern student's use of a single notebook for all subjects, and again tells us something about school organization: firstly, that the same students were learning to write Sumerian and to make mathematical computations; secondly, that basic

---

[11]That these have to be singled out as a separate group was pointed out in recent years by Eleanor Robson — *cf.*, *e.g.*, [Rob99, p.246-251] and [Rob00, p.23-30]. The tablets from one sub-group contain pure numerical computations, the others a geometric figure (a triangle, a square with diagonals, *etc.*), with some pertinent numbers; specimens belonging to the last group are to be found in [NeuSac45, p.42-44].

[12]For a further subdivision of this category, *cf.* [Rob99, p.9], where it is also argued that the texts in question were not meant as "catalogues" *stricto sensu*.

arithmetic was learned at a moment when literary education was quite advanced.

For the moment we shall leave out of the picture the contents of the problem texts, and hence also the single-problem texts as a group. Most informative in the first instance is the organization of anthology and theme texts.

Anthology texts, as explained, contain mathematical problems from different mathematical domains, in which respect they are similar to the Rhind and Moscow papyri. Like these, moreover, they contain *nothing but* mathematics (understood as the determination of numerical or measurable quantities). Theme texts were evidently even more restricted in scope. Mathematics *in toto* was thus a self-contained cognitive territory.

The anthology texts contain nothing but problems, nothing like the initial table and pure-number introduction of the Rhind Papyrus — tables were produced and arithmetical exercises were performed separately. However, the corpus of mathematical problem texts (anthology texts, theme texts and single-problem texts together) shows itself to be linked specifically to the corpus of tables in a different way, namely through its number notation. All mathematical texts employ the same sexagesimal place value system as the tables. This was a floating-point system, and thus only of any use in contexts where the order of magnitude could be presupposed. It could not and did not serve in accounting or other economic texts — it would not allow a judge to decide whether a debt was $\frac{1}{10}$ shekel of silver or 360 shekels. Its original practical use will have been for intermediate calculations (just like the engineers' use of the slide rule, also floating-point); but it is also employed in all mathematical problem texts.[13] Together with the stock of formulas for geometric calculation, sexagesimal place-value arithmetic was thus a framework of formal knowledge that held together the cognitive territory of "Old Babylonian mathematics". All this is not very different from what we could derive from the organization of the Egyptian mathematical papyri.[14] But in the present case a look at what has been called "Old Babylonian algebra" may bring to light another level. The traditional reason for using this expression is that a certain class of problems and problem solutions can be translated more or less homomorphically into modern equation algebra (at least as well as the theorems of [*Elements*, II.1-10] can be translated into algebraic identities), and that the technique

---

[13] Many of the texts that deal with real-world entities (or pretend to do so) express data and the final result in non–place-value notation and use the place value notation only for intermediate calculations.

[14] Actually, one table of technical constants (TMS III) tells to give coefficients for "anything at all", much in the vein of the Rhind Mathematical Papyrus, and with the same obvious restriction to what can be submitted to calculation. See [Rob99, p.20 n.8].

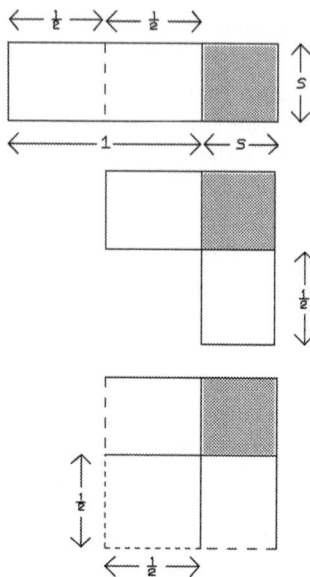

Figure 1. The procedure of BM 13901 #1, in slightly distorted proportions.

they make use of was therefore believed initially to be a numerical technique of the same kind. This turns out on closer inspection to be a misconception. Actually, the technique is geometrical — more precisely, based on what we might characterize as a geometry of measured and measurable lines in a rectangular grid.[15] As a first step in the argument we may look at a very simple problem, BM 13901 #1:[16]

---

[15] *Cf.*, for example, [Høy02a].

[16] Translation and discussion [Høy02a, p.50-52]. I use a "conformal translation", firstly, in order to avoid a reading through modern mathematical concepts, secondly, so as to distinguish operations or concepts which were kept apart by the Babylonians even though the traditional arithmetical interpretation conflates them (the two different "additions" found in the present text, two different "subtractions", four different "multiplications", and two different "halves"). Numbers are transliterated according to Thureau-Dangin's generalization of the degree-minute-second system, ´, ´´, ... indicating decreasing, `, ``, ... increasing sexagesimal order of magnitude and ° 'order zero' (1°15´ thus stands for $1\frac{1}{4}$, 1` for 60). In the interest of readability I have omitted the square brackets indicating damage to the text — no reconstructions are subject to doubt in the texts I quote.

Obv. I

1. The surface and my confrontation[17] I have accumulated:[18] 45′ is it. 1, the projection,[19]

2. you posit. The moiety[20] of 1 you break, 30′ and 30′ you make hold.

3. 15′ to 45′ you append: by 1, 1 is equal. 30′ which you have made hold

4. in the inside of 1 you tear out: 30′ the confrontation.

The problem deals with a "confrontation", a square configuration identified by its side s and possessing an area. The sum of the measuring numbers of these is stated to be 45′ ($=\frac{3}{4}$). The procedure can be followed in Figure 1: The left side s of the shaded square is provided with a "projection" (line 1), which creates a rectangle ⊏⊐ $(s,1)$ whose area equals the length of the side s; this rectangle, together with the shaded square area, must therefore also equal 45′. "Breaking" the "projection 1" (together with the adjacent rectangle) and moving the outer "moiety" so as to "hold" a small square □(30′) = 15′ does not change the area (line 2), but completing the resulting gnomon by "appending" the small square results in a large square, whose area must be 45′ + 15′ = 1 (line 3). Therefore, the side of the large square — that which is "equal by" (the square area) 1 — must also be 1 (line 3). "Tearing out" that part of the rectangle which was moved so as to "hold" leaves 1 − 30′ = 30′ for the "confrontation", [the side of] the square configuration.

---

[17] As seen by the Babylonians, a square configuration (*mithartum*, literally "[situation characterized by a] confrontation [of equals]") was numerically parametrized by and hence identified with its side — it "was" its side and "had" an area, whereas our "has" a side and "is" an area. The "confrontation" thus stands for the configuration as well as for the length of its side.
[18] "To accumulate" is an additive operation which concerns or may concern the measuring numbers alone of the quantities to be added. It thus allows the addition of (the measuring numbers of) lengths and areas, of areas and volumes, or of bricks, men and working days. Another addition ("appending") is concrete and therefore by necessity homogeneous. It serves when a quantity a is joined to another quantity A, augmenting thereby the measure of the latter without changing its identity (as when interest, in Babylonian "the appended", is joined to my bank account while leaving it as mine).
[19] The "projection" (*wāṣītum*, literally something which protrudes or sticks out) designates a line of length 1 which, when applied to another line L as width, transforms it into a rectangle ⊏⊐ $(L,1)$ without changing its measure.
[20] The "moiety" of an entity is its "necessary" or "natural" half, a half that could be no other fraction — as the circular radius is by necessity the exact half of the diameter, and the area of a triangle is found by raising exactly the half of the base to the height. It is found by "breaking", a term which is used in no other function in the mathematical texts.

At first glance, this does not look much like anything we know as "algebra". However, if we move to the level of principles, the text has much in common with equation algebra as trained in a modern school.

Firstly, the method is *analytical*: It presupposes that the solution exists, and moves stepwise from the (moderately) complex relation that is given until the unknown side of the square has been isolated. The steps are also quite similar to those through which we would solve the following problem:

$$x^2 + 1 \cdot x = \frac{3}{4} \iff x^2 + 2\frac{1}{2} \cdot x + (\frac{1}{2})^2 = \frac{3}{4} + (\frac{1}{2})^2$$
$$\iff x^2 + 2 \cdot \frac{1}{2} \cdot x + (\frac{1}{2})^2 = \frac{3}{4} + \frac{1}{4} = 1$$
$$\iff (x + \frac{1}{2})^2 = 1$$
$$\iff x + \frac{1}{2} = \sqrt{1} = 1$$
$$\iff x = 1 - \frac{1}{2} = \frac{1}{2}$$

Finally, the Old Babylonian procedure is reasoned but naïve, just like the sequence of algebraic operations — once the meaning of the terms and the nature of the operations is understood, no explanation beyond the indication of the steps themselves seems to be needed for the correctness of the procedure to be evident. In contrast, an exposition which made explicit the reasons for the validity of each step could be termed "critical", in the Kantian sense (asking for the possibility and limits of their validity).

Modern equation algebra, of course, has further characteristics. It is used to solve structurally similar problems belonging to many different ontological domains, but it is based on a neutral representation — the realm of pure numbers. In this sense, even equation algebra is a self-contained body of formal knowledge. Can anything similar be said about "Old Babylonian algebra"?

It is tempting to give a negative answer at least inasmuch as the neutral representation is concerned, accustomed as we are (Hilbert notwithstanding) to seeing geometrical entities as more ontologically loaded than numbers. As a matter of fact, however, the answer should be affirmative. This becomes clear if we investigate the terminological usage of the "algebra" texts.

The basic entities they deal with are uš, sag̃, a.šà and *mithartum*. The entities uš and sag̃ stand for the length and width (more properly "front") of a rectangle, respectively, a.šà for the area of a rectangle or square (or other geometric figure). All three are Sumerian terms, and they are consistently written in Sumerian even though they were pronounced in Akkadian

(Babylonian), as *šiddum*, *pūtum* and *eqlum*. However, if the length of a wall or a carrying distance is referred to, *šiddum* is written in syllabic Akkadian; moreover, even though the non-technical meaning of a.šà/*eqlum* is "field", mathematical problems which take real fields as a pretext use a different and less adequate word, seemingly in order not to interfere with the technical word for the area. *mithartum*, the term designating the square configuration parametrized by the side, is Akkadian. This does not imply, however, that it was the word employed in Akkadian-speaking practical surveying; here, a square field was seemingly supposed to possess 4 *fronts* (*pāt*, plural of *pūtum*) — but since real fields were almost always very oblong, the textual support for this assumption is indirect.

We may conclude that the rectangles and squares of the "algebra" texts were seen as belonging to a different category than the rectangular and quadratic pieces of real land dealt with, for instance, by surveyors or architects. If the numbers of modern algebra are abstract, namely by being distinguished from numbers of something, then the geometry of the Old Babylonian "algebra" is also abstract. There remains the question of whether this "algebra" was also "used to solve structurally similar problems belonging to many different ontological domains". The answer, once again, is positive. Let us first look at problem 12 from the tablet (BM 13901) whose first problem we have just discussed:[21]

Obv. II

27. The surfaces of my two confrontations I have accumulated: 21′40″.

28. My confrontations I have made hold: 10′.

29. The moiety of 21′40″ you break: 10′50″ and 10′50″ you make hold,

30. 1′57″21‴40⁗ is it. 10′ and 10′ you make hold, 1′40″

31. inside 1′57″21‴40⁗ you tear out: by 17″21‴40⁗, 4′10″ is equal.

32. 4′10″ to one 10′50″ you append: by 15′, 30′ is equal.

33. 30′ the first confrontation.

34. 4′10″ inside the second 10′50″ you tear out: by 6′40″, 20′ is equal.

35. 20′ the second confrontation.

---

[21] From [Høy02a, p.71]. I tacitly correct some erroneous numbers in the text.

We are thus told that the sum $\square(s1) + \square(s2)$ of two square areas is $21'40''$, while the area $\sqsubset\!\!\sqsupset (s1, s2)$ of the rectangle contained by the two sides is $10'$ (*Cf.* Figure 2, left). Formally, the problem is of the fourth degree, but it is easily solved as a biquadratic in several ways. The text chooses to represent the *surfaces* $\square(s1)$ and $\square(s2)$ by the *sides* (say, L and W) of a rectangle, whose surface $\sqsubset\!\!\sqsupset (L, W)$ is then found as $\sqsubset\!\!\sqsupset (\square(s1), \square(s2)) = \square(\sqsubset\!\!\sqsupset (s1, s2)) = \square(10') = 1'40''$, while the sum $L + W$ of the sides is known to be $21'40''$ — see Figure 2. This is a standard problem and solved according to the standard, as shown in the right part of the diagram: we may imagine the rectangle to be prolonged by the width, in such a way that the total length equals the known magnitude $L + W = 21'40''$. This segment is bisected and "made hold", which produces a square with side $10'50''$ and surface $1'57''21'''40''''$. Part of this square is identical with the original rectangle $\sqsubset\!\!\sqsupset (L, W)$, which is "torn out". The shaded remainder is a square with surface $17''21'''40''''$ and hence side $4'10''$. Adding this to the side of the square gives us the length L of the rectangle; "tearing it out" from the other side gives W. The numbers $s1$ and $s2$ are then found as the "equalsides" of L and W.

This is a case of representation inside the standard representation — areas being represented by lines. Another, more complex case is found in the problem TMS XIX #2 [Høy02a, p.195-200]: It deals with a rectangle for which is given, beyond the area, the area of another rectangle whose length is the cube erected on the original length, and whose width is the original diagonal; this is a bi-biquadratic problem (thus of the eighth degree), and it is solved as a cascade of quadratic problems.

Much simpler is the problem YBC 6967 [Høy02a, p.55-58], which deals with two numbers (*igûm*, meaning "the reciprocal", and *igibûm*, "its reciprocal") belonging together in the table of reciprocals; their product must hence be 1 or, as supposed in the actual case, 60.[22] Since this number is spoken of as a "surface" (and since the operations are the usual geometric ones), the numbers are represented by the sides of a rectangle. The text runs as follows:

Obv.

1. The *igibûm* over the *igûm*, 7 it goes beyond

2. *igûm* and *igibûm* what?

---

[22] In principle, any power of 60 is possible; but due to the floating-point character of the notation, only the values 1 and 60 can be distinguished.

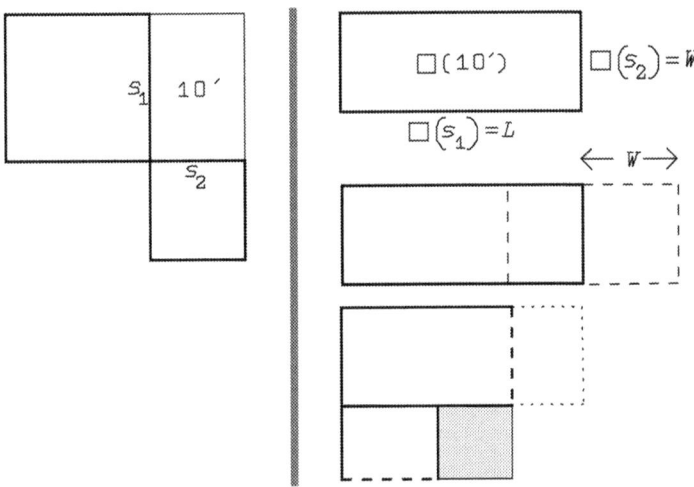

Figure 2. The procedure of BM 13901 #12. solved as a cascade of quadratic problems.

3. You, 7 which the *igibûm*

4. over the *igibûm* goes beyond

5. to two break: 3°30´;

6. 3°30´ together with 3°30´

7. make hold: 12°15´.

8. To 12°15´ which comes up for you

9. 1` the surface append: 1`12°15´.

10. The equalside of 1`12°15´ what? 8°30´.

11. 8°30´ and 8°30´, its counterpart, lay down.

Rev.

1. 3°30´, the made-hold,

2. from one tear out,

3. to one append.

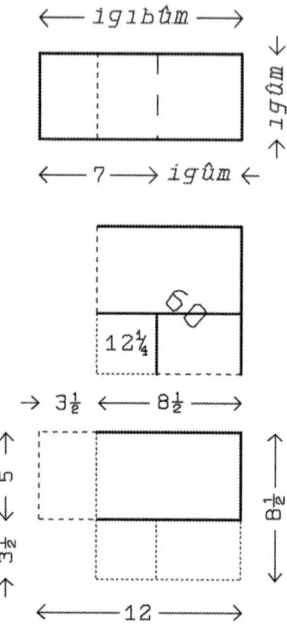

Figure 3. The procedure of YBC 6967.

4. The first is 12, the second is 5.

5. 12 is the *igibûm*, 5 is the *igûm*.

The procedure can be followed in Figure 3; as in Figure 1, the excess of the length over the width is bisected and the outer moiety moved so as to hold together with the inner moiety a quadratic complement. By joining the gnomon to this complement we get a larger square with surface $72\frac{1}{4}$, whose sides (the "equalside" and its "counterpart") are found to be $8\frac{1}{2}$; removing the moiety $3\frac{1}{2}$ of the excess that was joined to the width leaves the width (the *igibûm*); joining it where it originally belonged gives the length (the *igibûm*). What we have is thus a perfect example of algebraic representation, only inverted in respect of what we are accustomed to: geometric entities represent numbers, not vice versa.

Entities belonging more definitely to "real life" are also represented. In the problem contained in the text TMS XIII [Høy02a, p.206-209], we are told the difference between the rates (volume units per shekel) at which a given quantity of oil was bought and sold, together with the total profit resulting from the transaction. This is another problem of the second degree — the area of the rectangle contained by the two rates turns out to be their difference divided by the total profit and multiplied by the total amount of oil. In AO 8862 #8 we know the sum of a number of workers, the number of days they work, and the number of bricks they produce according to a fixed rate per man-day (the number of bricks being thus proportional to the number of man-days). The text contains no description of the procedure, but there is no doubt that the numbers of men and days are to be represented by the sides of a rectangle, whose area is proportional to the number of bricks.

All in all we thus see that Old Babylonian "algebra", just as modern equation algebra, was "used to solve structurally similar problems belonging to many different ontological domains [while being] based on a neutral representation". It was hence in itself a body of formal knowledge, nested within the larger body of "Old Babylonian mathematics".[23]

One related question then presents itself: did the authors of the Old Babylonian texts see this body of "algebraic" knowledge as a distinct body? This question can be answered from the theme texts — these, indeed, single out groups of problems which the authors saw as belonging together. On the whole, the answer turns out to be affirmative. Let us first look at the text BM 13901. It contains 24 problems, all of which deal with "algebraic" problems about one or more squares (although the biquadratic #12, as we have seen, is reduced to an "algebraic" problem about a rectangle). This we may compare, on one hand, with the text TMS V, on the other with BM 15285. TMS V is a catalogue text, also dealing with squares.[24] Unlike BM 13901, however, it includes two types of linear problems (a multiple of the side being given; and the difference between the areas of two squares being given together with either the sum of the sides or their difference). It thus does not go beyond the "algebraic" realm.

BM 15285 is another catalogue dealing with squares[25] — more precisely, with a square of given side which is subdivided in various ways into smaller figures (squares, right isosceles triangles, circles and circular segments. No single problem of algebraic character is treated. Together with BM 13901 and TMS V it thus suggests that "algebra" was clearly distinguished from

---

[23] It should be noticed that the problems about rectangles that turn up in Seleucid and other Late Babylonian texts do not serve in a similar way to represent entities belonging to a variety of categories; they are formulated within the same kind of geometry as their Old Babylonian counterparts, and they are solved by an analytic procedure — but on this essential account they differ from modern equation algebra as well as Old Babylonian "algebra".
[24] Published in [BruRut61, p.35-49].
[25] Now published in [Rob99, p.208-217].

"non-algebra". As a matter of fact, however, things are slightly more complicated.

This is demonstrated by BM 85200+VAT 6599, a procedure text whose formal theme is (rectangular prismatic) *excavations*.[26] In most problems, one or two dimensions can be eliminated, and these problems are then solved by means of the usual algebraic second- and first-degree techniques. Some, however, are genuine irreducible cubic problems; their solution is obtained by means of factorization,[27] or through use of a table referred to as "equilateral, one appended" and listing the numbers $n^2 \cdot (n+1)$. These solutions remain analytical inasmuch as they start from the presupposition that the solution *exists*; but they come from systematic trial-and-error,[28] not from a string of successive operations; we would hardly consider them "algebraic", even though we might translate the problems into third-degree equations. The author of the text, however, appears to have regarded them as members of the same group as the indubitably "algebraic" first- and second-degree problems.[29] That self-contained body of knowledge within which the "algebraic" problems of BM 13901 and TMS V belonged was thus not made up exactly as we would expect it to be — but none the less such a selfcontained body existed. Not only Old Babylonian mathematics as a whole but also the "algebraic" domain (delimited in this particular way) was thus treated in practice (and, who knows, perhaps also somehow spoken of) as a closed body of ("formal") knowledge applicable to a variety of particular concrete domains.

1.4 Riddle collections — and a hypothesis

Does this mean that mathematics (at least the mathematics of literate cultures) was always or inherently treated as a formal science? The answer is no.

This is illustrated, for instance, by two well-known mathematical riddle collections: book XIV of the *Greek Anthology* [Pat18], and the *Propositiones ad acuendos iuvenes* [Fol78], ascribed in some manuscripts to Alcuin of York. Both combine properly arithmetical problems with other matters —

---

[26] *Cf.* [Høy02a, p.137-162].

[27] For instance, finding three factors p, q and r of the number $10'4''48'''$ which fulfil the conditions $p+q=1$, $r=p+6'$. Starting from simple factors, the solution can be found fairly quickly by systematic trial and error — but only because an exact solution exists.

[28] Or pretended trial-and-error; since the author constructed the problems backward froma known solution, he would know the correct factorization in advance. The text only gives this correct factorization and does not tell how it is arrived at.

[29] The texts YBC 4657, 4662, and 4663 [NeuSac45, p.59-73], similarly dealing with "excavations", contain simple problems that can be solved by direct computation along with second- and irreducible third-degree problems. *Cf.* [Høy02a, p.346-348].

the former with non-mathematical riddles and with oracles, the second with non-mathematical riddles and riddles with mock solutions.

Such collections, of course, were not put together by any kind of professional mathematicians, nor as the basis for systematic mathematics teaching; they represent the view of the more or less informed *outsider*. The fact that they constitute the most obvious exceptions to the assumption that mathematics was always inherently a *practically formal* science suggests a more restricted hypothesis; that *mathematics when taught as an autonomous subject within an institutionalized school system* tends to be treated as practically formal knowledge.[30] For this view, direct reasons can be given: only if mathematical knowledge can be applied to several distinct domains is there any reason to teach it autonomously; but if it is taught under such conditions by teachers whose primary practice is to teach mathematics, it is almost unavoidable that such principles as can be applied to several domains are presented in a relatively abstract form, that is, as practically formal knowledge.

2 Understandings of past mathematics

Let us now change the perspective and take it for granted that mathematics *is* a formal science — that is, $2+2=4$, independently of the kind of units we are counting and adding, whether sheep or galaxies. The mathematician, therefore, tends to see professionally the same proposition $2+2=4$ as what is *essentially* at stake in both situations — similarly to Aristotle's geometer considering a bronze sphere.

What happens when this principle is applied to the historical sources? Let us look at proposition II.6 from Euclid's [*Elements*]:

> If a straight line be bisected and a straight line be added to it in a straight line, the rectangle contained by the whole with the added straight line and the added straight line together with the square on the half is equal to the square on the straight line made up of the half and the added straight line.

This may not be quite easy to grasp, but a diagram helps — see Figure 4. The straight line that is bisected is AB, the added line is BD. Euclid's proof makes use of the diagram and shows in this way that the proposition holds true; however, for a modern mind trained in symbolic algebra it follows

---

[30] Beyond the Middle Kingdom Pharaonic and the Old Babylonian examples, one may refer to Chinese mandarin education of the Han epoch, whose mathematics we find presented in the *Nine Chapters on Arithmetic* [Vog68], or of the Indian Jaina tradition, whose mathematics we find in Mahāvīra's *Gaita-sāra-sangraha* [Rañ12]. Other surviving Indian medieval sources of importance are more closely linked to astronomy, but their contents is so much broader than astronomical computation that it points to the existence of a structure similar to what we find in Mahāvīra.

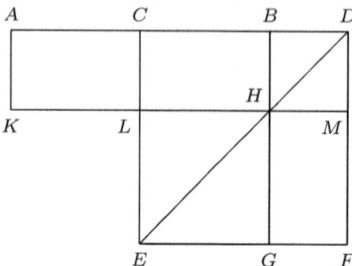

Figure 4. The diagram of *Elements* II.6.

much easier from the identity

$$(2a+e)e + a^2 = (a+e)^2$$

if $AC = CB = a$, $BD = DM = e$.

If sheep or galaxies do not matter for the truth of $2+2=4$, does it then matter whether we think of geometry or algebra in the Euclidean case? Is it really the *same thing* mathematically speaking?

The dilemma can be sharpened by a look at Old Babylonian mathematics. Let us consider the *igibûm*-problem of YBC 6967, discussed above.

We would solve this problem by means of a system of two algebraic equations, for instance like this:

$$\tilde{n} \cdot n = 60, \tilde{n} - n = 7$$

$$\frac{(\tilde{n}-n)}{2} = 3\frac{1}{2}$$

$$\left(\frac{(\tilde{n}-n)}{2}\right)^2 = 12\frac{1}{4}$$

$$\left(\frac{(\tilde{n}+n)}{2}\right)^2 = \left(\frac{(\tilde{n}-n)}{2}\right)^2 + \tilde{n} \cdot n = 12\frac{1}{4} + 60 = 72\frac{1}{4}$$

$$\frac{(\tilde{n}+n)}{2} = \sqrt{72\frac{1}{4}} = 8\frac{1}{2}$$

$$\tilde{n} = \frac{(\tilde{n}+n)}{2} + \frac{(\tilde{n}-n)}{2} = 8\frac{1}{2} + 3\frac{1}{2} = 12$$

$$n = \frac{(\tilde{n}+n)}{2} - \frac{(\tilde{n}-n)}{2} = 8\frac{1}{2} - 3\frac{1}{2} = 5$$

We might also have made a substitution,

$$\tilde{n} = n + 7$$
$$(n+7) \cdot n = n^2 + 7n = 60$$
$$n^2 + 2 \cdot 3\tfrac{1}{2} \cdot n + 3\tfrac{1}{2}^2 = 60 + 12\tfrac{1}{4}$$
$$etc.$$

but the symmetric scheme has the advantage of corresponding exactly to the Old Babylonian procedure, which however was geometric (and which has the same structure as the proof of [*Elements*, II.6]), the two numbers and their product being represented by the side and the area of the rectangle ▭ $(\tilde{n}, n)$.

The numerical steps in the tablet coincide with those of the symbolic-algebraic solutions, and the underlying principle that is made use of coincides with the identity expressed in [*Elements*, II.6]. Does this mean that it is "the same" mathematics which we find in all cases, even though Euclid proves geometrical identities which we may translate into algebraic identities, whereas the Babylonian mathematics teacher asks for and shows the solution of a problem?

Mathematicians, we might expect, trained as they are as practitioners of a formal science, would tend to give an affirmative answer. Historians, including historians of ideas and independently of whether they come from the Aristotelian-idiographic[31] or the hermeneutic tradition (searching either for the particular or trying to penetrate the conceptual world of the source material "from within") should tend to react in the opposite way. In real life, this may be an oversimplification. What we find is, however, that those who have claimed explicitly that the history of mathematics can only be written by mathematicians –first and foremost, Hieronymus Georg Zeuthen[32] and André Weil [Wei78]– are also among those who most outspokenly have argued in favour of identification. Claiming that only mathematicians are competent to write the history of mathematics thus in practice amounts to a claim that the history of mathematics should be treated as a formal science.

2.1 Second (didactical) thoughts

Before leaving this discussion we may give it a didactical twist. If we are to make students understand some general principle, we present it through a

---

[31] The "statements [of poetry] are of the nature rather of universals, whereas those of history are singulars" [*Poet.*, 1451b7].
[32] *Cf.* [LütPur94].

variety of concrete representations — no elementary school trains addition *only* on cows, and engineering students who have trained differentiation with respect exclusively to a variable called x will mostly be at a loss when the physics teacher asks them for a differentiation with respect to a variable called t and standing for time.[33] *Transfer* is only achieved when students encounter the same underlying structure in many different situations.

The question about mathematical sameness can therefore be turned into a didactical problem: Would it facilitate the practical insight into what algebra is and can be used for if students encountered algebraic reasoning and structures in several different versions — modern symbolic, Babylonian "cut-and-paste" procedures, and perhaps Euclidean proofs of identities? Would it be worth trying? If the answer is yes, then history serving didactics has to find a middle ground between the formal and the hermeneutic understanding of its role: it must show, *both* the difference between the two or three approaches and the underlying sameness. Omitting sameness means that the historical material becomes irrelevant for the purpose of mathematics teaching; omitting difference reduces the same material to a mirror in which we see nothing but our own face (a mirror from which, as Georg Lichtenberg once pointed out, "no saint will look out if a monkey looks into it").

Yet another twist is possible. In the preceding paragraphs this was formulated as if it dealt with the teaching of mathematical structure only, but we need not stop at that. Algebra when used by others than pure mathematicians is a technique for solving problems dealing with real-life entities — prices, income distributions, acceleration of charged particles in electric fields, *etc*. Essential for this is representation. As long as we only train it as a *representation* by means of pure numbers, students (and former students having become mature physicists, economists, and so on) tend to forget that the representations are representations and not the entities themselves. Working also (early in their educational career) on cases where numbers or prices are represented by geometric magnitudes they might get a better insight in what goes on when mathematical techniques are used in practice, and perhaps get slightly better at avoiding the pitfalls that inhere in the process.

Primary Sources.

[*Metaph.*] **Aristotle**, Metaphysica, *in:* [Tre33]
[*Phys.*] **Aristotle**, Physica, *in:* [Cha70]

---

[33] I speak from personal experience.

[Cael.]   **Aristotle**, De caelo, *in:* [Gut39]
[An. post.]   **Aristotle**, Analytica posteriora, *in:* [TreFor60]
[Poet.]   **Aristotle**, Poetica, *in:* [Byw24]
[Elements]   **Euclid**, Elements, *in:* [Hea26]

References.

[And69]   Thomas C. **Anderson**, Intelligible matter and the Objects of Mathematics in Aquinas, **The New Scholasticism** 43 (1969), p.555-576

[BruRut61]   Evert M. **Bruins** and Marguerite **Rutten**, Textes mathématiques de Suse, Paul Geuthner 1961 [Mémoires de la Mission Archéologique en Iran XXXIV]

[Byw24]   Ingram **Bywater** (*ed., trans.*), Aristotle, De poetica, *in:* [Ros24]

[Cha70]   William **Charlton** (*ed., trans.*), Aristotle's Physics, Books I and II, Clarendon Press 1970

[Cla99]   Marshall **Clagett**, Ancient Egyptian Science, A Source Book, Volume III, Ancient Egyptian Mathematics, American Philosophical Society 1999 [Memoirs of the American Philosophical Society 232]

[Fol78]   Menso **Folkerts**, Die älteste mathematische Aufgabensammlung in lateinischer Sprache: Die Alkuin zugeschriebenen *Propositiones ad acuendos iuvenes*, Überlieferung, Inhalt, Kritische Edition, Springer 1978 [Österreichische Akademie der Wissenschaften, Mathematisch-Naturwissenschaftliche Klasse, Denkschriften, Band 116, 6. Abhandlung]

[Gut39]   William Keith Chambers **Guthrie** (*ed., trans.*), Aristotle, On the Heavens, Heinemann 1939 [Loeb Classical Library 338]

[Hea26]   Thomas L. **Heath** (*ed., trans.*), The Thirteen Books of Euclid's Elements, 2nd revised edition, 3 volumes, Cambridge University Press 1926

[Høy02a]   Jens **Høyrup**, Lengths, Widths, Surfaces: A Portrait of Old Babylonian Algebra and Its Kin, Springer 2002 [Studies and Sources in the History of Mathematics and Physical Sciences]

[Høy02b]   Jens **Høyrup**, Existence, Substantiality, and Counterfactuality, Observations on the Status of Mathematics According to Aristotle, Euclid, and Others, **Centaurus** 44 (2002), p.1-31

[KnoRow94]   Eberhard **Knobloch** and David **Rowe** (*eds.*), The History of Modern Mathematics, volume III: Images, Ideas, and Communities, Academic Press 1994

[LütPur94]   Jesper **Lützen** and Walter **Purkert**, Conflicting Tendencies in the Historiography of Mathematics: M. Cantor and H.G. Zeuthen, *in:* [KnoRow94, p.1-42]

[NeuSac45]   Otto **Neugebauer** and A. **Sachs**, Mathematical Cuneiform Texts, American Oriental Society 1945 [American Oriental Series 29]

[Pat18] William R. **Paton** (*ed., trans.*), The Greek Anthology, 5 volumes, Heinemann 1918 [Loeb Classical Library 67, 68, 84, 85, 86]

[Pee23] T. Eric **Peet** (*ed., trans.*), The Rhind Mathematical Papyrus, British Museum 10057 and 10058, Introduction, Transcription, Translation and Commentary, University Press of Liverpool 1923

[Raṅ12] M. **Raṅgācārya**, (*ed., trans.*), The Gaṇita-sāra-sangraha of Mahāvīrācārya with English Translation and Notes, Madras Government Press 1912

[Rob99] Eleanor **Robson**, Mesopotamian Mathematics 2100-1600 BC, Technical Constants in Bureaucracy and Education, Clarendon Press 1999 [Oxford Editions of Cuneiform Texts 14]

[Rob00] Eleanor **Robson**, Mathematical Cuneiform Tablets in Philadelphia, Part 1: Problems and Calculations, **SCIAMVS** 1 (2000), p.11-48

[Ros24] William David **Ross** (*ed.*), The Works of Aristotle, volume XI, Clarendon Press 1924

[Str30] Vasilij V. **Struve** (*ed.*), Mathematischer Papyrus des Staatlichen Museums der Schönen Künste in Moskau, Herausgegeben und Kommentiert, Springer 1930 [Quellen und Studien zur Geschichte der Mathematik, Abteilung A: Quellen, Band 1]

[Tre33] Hugh **Tredennick** (*ed., trans.*), Aristotle, The Metaphysics, 2 volumes, Heinemann 1933 [Loeb Classical Library 271]

[TreFor60] Hugh **Tredennick** and Edward S. **Forster** (*eds., trans.*), Aristotle, Posterior Analytics and Topica, Heinemann 1960 [Loeb Classical Library 391]

[Vog68] Kurt **Vogel** (*ed., trans.*), Chiu chang suan shu, Neun Bücher arithmetischer Technik, Ein chinesisches Rechenbuch für den praktischen Gebrauch aus der frühen Hanzeit (202 v.Chr. bis 9 n.Chr.), Friedrich Vieweg & Sohn 1968 [Ostwalds Klassiker der Exakten Wissenschaften, Neue Folge, Band 4]

[Wei78] André **Weil**, Who Betrayed Euclid?, **Archive for History of Exact Sciences** 19 (1978), p.91-93

Received: March 10th, 2003;
In revised version: December 17th, 2003;
Accepted by the editors: February 6th, 2004.

Benedikt **Löwe**, Volker **Peckhaus**, Thoralf **Räsch** (*eds.*)
Foundations of the Formal Sciences IV
The History of the Concept of the Formal Sciences

Medieval logic as a formal science
A survey

CHRISTOPH KANN[*]

Philosophisches Institut
Heinrich-Heine-Universität Düsseldorf
Universitätsstraße 1
40225 Düsseldorf, Germany
Email: kann@phil-fak.uni-duesseldorf.de

> ABSTRACT. The paper discusses in how far medieval logic can appropriately be characterized as a formal science. In this respect, the special medieval approach to logic as a *scientia sermocinalis* is examined as well as its main doctrines, namely the theories of supposition and of consequences, and the famous characterization of logic as an *ars artium* or *scientia scientiarum*. It is pointed out that medieval logic is not devoted to the setting up of formal systems or any metalogical analysis of formal structures. Logic in the medieval sense of the discipline is necessarily connected with semantical aspects of natural language. Accordingly, we are confronted with a discipline going far beyond the formal structures of discourse. The classification of medieval logic as a formal science is appropriate only under selected perspectives.

Very few topics in the philosophical tradition are as common and wellknown as the distinction between form and matter. Form and matter are distinguished not only in contexts of metaphysics and the philosophy of nature, but also in the philosophy of science, especially in matters of scientific classification. The notion of a formal science is widely used for those disciplines whose sentences consist of formally true statements.

Sometimes we speak of formal sciences in contrast either to non-formal sciences, which can be identified with Carnap's "Realwissenschaften"[1] or to natural sciences. One of the most familiar and paradigmatic instances of a

---

[*] I thank Angelika Koelzer and Martin Schäfer for helpful comments on this paper.
[1] [Car61, p.81], *cf.* also the notion of "materiale Wissenschaften" in [Lay73, p.461].

formal science besides mathematics is logic. The notion of formal logic is based on the premise that the validity of an argument is a function of its structure or logical form. It is worth asking whether logic during all epochs, in all stages of its doctrinal development, can equally be classified as a formal science. Since medieval logic is still less known than the ancient and modern versions of the discipline –though it makes up the longest continuous epoch within the history of logic– the present paper is devoted to the question of in how far medieval logic can appropriately be characterized as a formal science.

First I will examine how logic was classified and characterized within the range of disciplines by the medieval authors themselves. Then I will go on to examine to what extent medieval logicians in their main conceptions give us explicit or implicit indications of the view that their discipline was concerned with formal topics and aspects. Separate paragraphs will be devoted to formal and material or descriptive constituents of language, to formal and semantical aspects of supposition theory, and to the distinction of formal and material consequences. In a final step I will discuss in which way the most famous characterization of logic in the Middle Ages, namely that of an *ars artium* or *scientia scientiarum*, suggests an understanding of logic as a formal science.

1   The rôle of logic in the Middle Ages

Regarding the rôle of logic within the framework of arts and sciences during the Middle Ages, we have to distinguish two related aspects, one institutional and the other scientific. As to the first aspect, we have to remember that the medieval educational system was based on the seven liberal arts, which were divided into the *trivium*, *i.e.*, three arts of language, and the *quadrivium*, *i.e.*, four mathematical arts. The so-called trivial arts were grammar, rhetoric, and logic, and during a period of several centuries virtually every educated person, at least every university graduate, received a training in these matters, especially in logic. Students in the medieval faculty of arts probably spent more time studying logic than any other discipline. This first –institutional– aspect concerning the rôle of logic is explained by the second –scientific– aspect. The trivial disciplines provided techniques of analysis and a technical vocabulary that permeate philosophical, scientific and theological writings. Logic, as mentioned before, was referred to and was generally regarded as the art of arts and the science of sciences. The increasing cultural dominance of the universities with their obligatory *disputationes* and their hierarchy of examinations on the one hand and the outstanding status of logic on the other were corresponding features of the educational world of the 13th century.

The core of the logic curriculum from the 12th century onwards was provided by the logical works of Aristotle. These represented the material for the study of types of predication, the analysis of simple propositions or statements[2] and their relations of inference and equivalence, the analysis of modal propositions, of the structure and the types of the syllogism, dialectical topics, fallacies and scientific reasoning as based on the demonstrative syllogism. Medieval logicians, however, realized that there were other, non-Aristotelian, approaches to logical subjects, questions and methods that could be investigated. The new approaches primarily included works on the signification and the supposition of terms — a distinction showing some similarity to the modern distinction between meaning and reference. The theory of signification deals with the capability of descriptive terms to function as signs, *i.e.*, their property of being meaningful. The theory of supposition was concerned with the types of reference that terms in their function as subject and predicate obtain in the context of different propositions. Another emphasis was put on consequences or valid inference forms. These innovations were by no means regarded as an alternative to tradition, but supplemented the Aristotelian *logica antiqua* under the heading of *logica moderna* or *logica modernorum*.

The medieval logicians themselves did not classify their discipline as a *scientia formalis* –to my knowledge the expression was not used in the Middle Ages– but as a *scientia sermocinalis*, *i.e.*, a science of argumentative speech, which was the overarching framework of the trivial arts. The *scientia sermocinalis* itself is one of three types into which science was divided, *e.g.*, by Peter of Spain in his well-known [*Tractatus*, p.29, 14-16]. The differences (*differentiae*) of science, as Peter states, are *naturale*, *morale*, and *sermocinale*, a division which resembles the Stoic division into natural philosophy, ethics, and logic.[3] William of Sherwood, another important logician of the 13th century, offers the same scientific differences, but –in contrast to Peter of Spain– as the result of a twofold division:[4] Since there are two sources (*principia*) of things, nature and the soul, there will accordingly also be two kinds (*genera*) of things. The things whose source or principle is nature are the concern of natural science. The others, whose source or principle is the soul, are again divided into two types. Since according to Sherwood the

---

[2] In medieval logic "*propositio*" and "*enuntiatio*" both stand for a sentence signifying something true or false and are mostly used as interchangeable terms. However, using the term "*propositio*" we have to avoid the modern understanding of proposition, or propositional content, as what is asserted or what is expressed by a sentence.

[3] The *scientiae morales* and *naturales* as the counterpart to the *scientiae sermocinales* were sometimes brought together under the integrating concept of *scientiae reales*; *cf.* [Scho92, col.1508].

[4] *Cf.* [*Introductiones*, p.2, 1-12].

soul is created without virtues or knowledge, it performs certain operations by means of which it attains to the virtues, and these are the concern of ethics or *scientia moralis*. The soul performs different operations by means of which it attains to knowledge, and these are the concern of the science of argumentative speech or *scientia sermocinalis*. At this point we meet the same threefold division of science that occurs in Peter of Spain. It is worth mentioning that the first division regarding the nature of things is metaphysical while the second division regarding the different sorts of things whose source is the soul is epistemological. The sciences whose principle is the human soul are understood as concerning basic human activities or operations, and the specific differences among them are obtained from the goals of these activities, namely virtues on the one hand and science on the other.

The term *"scientia sermocinalis"* which stands for the subtle analysis of ordinary language came into use in the late 12th or early 13th century. The designation of logic as a scientia sermocinalis was commonly accepted during the 13th century, but it was not the only one. The term *"logica"* as derived from the Greek "λόγος" can mean both *"sermo"* and *"ratio"*. Accordingly, logic was regarded either as a *scientia sermocinalis* or as a *scientia rationalis*. The medieval authors offer considerations supporting both titles. While logicians like William of Sherwood and Peter of Spain stressed the feature of logic as a linguistic science as mentioned above, other authors in the 13th century like Robert Kilwardy and St. Bonaventure called it linguistic and rational alike. In the 14th century the notion of logic as a rational science became predominant. An important reason lies in the fact that logic was about second intentions, which were higher-level concepts like "genus", "species", "predicate", *etc*. We make use of second intentions to classify our concepts or first intentions of things in the world. Second intentions reveal both universals and logical structures and were regarded as mental constructs or rational objects reached through abstraction, which means reflection on general features and relations of things and on actual pieces of discourse.

2   The analysis of the proposition

Since logic in the 13th century is focussed on the syllogism as the predominant mode of argumentation, most manuals like Peter of Spain's *Tractatus* and William of Sherwood's *Introductiones in logicam* provide us with a large and detailed treatment of the proposition as the immediate and constitutive basis of the syllogism. What is Sherwood's way of treating the proposition? There are two different approaches. The first can be identified with the well-known scholastic methodology of *definitio* and *divisio* according to which Sherwood's initial explanation of the proposition is followed by a detailed division including different types of non-assertive statements. The second approach is based on an equally well-known epistemological principle presented by Sherwood at the very beginning of his treatise on syncategorematic words: In order to obtain an understanding of something, we are dependent on analysis, *i.e.*, a subdivision into parts or constituents.[5]

In the initial paragraph of his *Syncategoremata*, Sherwood analyzes the proposition by distinguishing between two kinds of parts, namely principal and secondary ones. The principal parts, as Sherwood states, are the substantive (*nomen substantivum*) and the verb, for it is these parts which are necessary for an understanding of the proposition. Secondary parts of the proposition are the adjective, the adverb, conjunctions, and prepositions, because they are not necessary for the existence of a statement.

2.1 Formal or syncategorematic constituents of the proposition

Some secondary parts of the proposition, as Sherwood continues, are determinations of principal parts in respect of the things belonging to them. For example, when I say *"homo albus"* the word *"albus"* signifies that some thing which is a man is white. Other secondary parts are determinations of the principal parts (*i.e.*, noun and verb), insofar as these are subjects or predicates. For example, when I say *"omnis homo currit"* the word *"omnis"*, which is a universal sign, does not signify that some thing which is man is universal, but rather that *"homo"* is a universal subject. Secondary parts of this kind are called syncategorematic words. Sherwood inserts an etymological reduction — a mode of explication of a word that is often used by the medieval authors: The name *"syncategorema"*, as Sherwood explains, depends on *"sin"* which means *"con"* and "significative" or "predicative" — as if to say "conpredicative", for a syncategorematic word is always joined with something else in discourse.[6]

What we learn from Sherwood is the fact that *syncategoremata* are not any determinations of nouns and verbs or their significates, but determinations concerning nouns and verbs in their function as basic parts of the proposition, *i.e.*, in their logical-syntactical function as subject or predicate. In other words: Quantifying prefixes like *"omnis"* are not regarded as a kind of adjectival determinants of only the term following them, but conceived as operating on the proposition as a whole and thereby exercising some logical function. Subject and predicate of the proposition are usually named integral parts, which comes close to essential parts, and this notion

---

[5] [*Syncategoremata*, p.48]; revised critical edition with German translation and commentary by Christoph Kann and Raina Kirchhoff is in preparation.

[6] *Cf.* [*Syncategoremata*]; the term *"syncategorema"* dates back to antique grammar, namely to Priscianus' [*Institutiones*, II 15 (p.54, 5-7)].

is intended to mean that a proposition is made up of them and not of others. In the example *"omnis homo currit"* the syncategorematic term *"omnis"*, which itself is a secondary part of the proposition, determines the principal parts of the proposition in their function as subject or predicate. And this is an essential feature of syncategorematic words or of words in their syncategorematic use.

The most suitable way to develop an appropriate understanding of *syncategoremata* is to start with the complementary notion of *categoremata*. Categorematic terms are those which can function in their usual signification as subject or predicate term in a proposition. One of the usual examples given by medieval authors is the proposition *"homo currit"* which is composed of two categorematic terms. The *syncategoremata*, which do not meet the criterion of independent signification, must be connected with (at least) one suitable pair of *categoremata* in order to become an element within a proposition. This can be demonstrated by means of examples like *"omnis homo currit"*, *"homo non currit"*, or *"homo currit contingenter"*. *Syncategoremata* affect the function of signifying of categorematic terms appearing after them in the same proposition. Apart from simple combinations of a single subject with a single predicate, complex or hypothetical propositions (these are synonyms in medieval logic) can be determined by *syncategoremata*, e.g., *"si homo currit, animal currit"* or *"Plato currit, et Socrates currit"*. We can roughly distinguish two perspectives which were relevant for medieval approaches to the *syncategoremata*: The first, earlier perspective –predominant during the 12th and 13th centuries– was dedicated to the consignificative function of the *syncategoremata* themselves, while later inquiries, chiefly in the 14th century, were rather focussed on questions concerning the influence of *syncategoremata* on *categoremata* and their contextual reference, especially within *sophismata*.

Medieval logicians tended to extract from the vast number of *syncategoremata* those which are most relevant for logical purposes, disregarding most prepositions, conjunctions and other non-signifying words. Their main subject of interest was the quantitative or distributive signs (*"omnis"*, *"uterque"*, *"nullus"*, *"aliquis"*), exceptive or exclusive signs (*"praeter"*, *"solum"*, *"tantum"*, *"nisi"*), which include negations, affirmative or negative signs like *"est"* and *"non"*, modal signs like *"necessario"* and *"contingenter"*, the junctors *"si"*, *"et"*, and *"vel"*, and the auxiliary verbs *"incipit"* and *"desinit"*. Though medieval logicians mainly deal with syncategorematic words of logical relevance, we should refrain from assuming that *syncategoremata* can simply be identified with logical form or with logical operators, as Bocheński [Boc61, p.156f], Moody [Moo53, p.16-18], and Pinborg [Pin72, p.60f] do. On the other hand, the fact that not all *syncate-*

goremata are logical operators should not be criticized as an insurmountable deficiency of the theory, as Patzig [Pat81, p.13] does. Such criticism ignores the original intentions of the doctrine which has its roots in grammar and its genuine application in semantics. Or, to put it another way, the distinction of *categoremata* and *syncategoremata* was not intended to isolate formal elements of discourse in order to establish a formal science, but to provide us in a first stage with grammatical and in a second stage with semantical distinctions in an overarching science of discourse or of normal language.

2.2 Material or descriptive constituents of the proposition

The distinction of significant and non-significant parts of the proposition is closely connected to the medieval doctrine of the matter of statements (*materia enuntiationis*). This doctrine dates back at least to the 11th century[7] and it can frequently be found in logic textbooks of the 12th, 13th and 14th centuries, *e.g.*, in Abaelard, Peter of Spain, William of Sherwood and Albert of Saxony. What does it actually mean when we speak of the matter of a statement or a proposition, and how can this matter be relevant for logic under the aspect of a formal science? The matter of propositions is constituted by the subject and the predicate, that is by the semantic relationship of the terms that function as subject and predicate in a proposition. Three kinds of matter of the proposition are distinguished, natural, contingent, and separate.[8] We speak of natural matter when the subject receives the predicate by its very nature, as in *"homo est animal"*, since it belongs to man's nature to be an animal. The matter is contingent if the subject receives the predicate contingently, as in *"homo currit"*, since it is contingent whether a man is running or not. Finally, the matter is separate if the predicate is naturally separated from the subject, as in *"homo est asinus"*, since man as an *animal rationale mortale* is essentially excepted from being a donkey.

The main divisions of the proposition in medieval logic –namely the qualitative and quantitative divisions, the divisions into assertoric and modal statements, into *enuntiatio categorica* and *hypothetica*, and into *enuntiatio una* and *plures*– reflect the proposition's syntactical or formal structure. The reason is that medieval logic manuals –at least in the 13th and 14th centuries– follow the Aristotelian framework, according to which the main subject of logic is the syllogism. In order to analyze the syllogism, we have to go back to its immediate constituent, the proposition, and then to the proposition's immediate constituents, which at the same time are the ultimate meaningful constituents, namely the single words. Since therefore

---

[7] *Cf.* [*Dialectica*, p.54, 20-55, 32].
[8] *Cf.* [*Introductiones*, p.20, 256-22, 270; 234, n. 31]; *cf.* also [Jac80, p.61-64].

the treatment of the proposition can be subsumed under the treatment of the syllogism, the proposition will mainly be distinguished and treated under formal aspects. The main reason for studying the syllogism is to learn to set up valid demonstrations, which are necessarily in syllogistic form. But, of course, the defining properties of a demonstration go beyond formal considerations. The further specifications of a syllogistic demonstration concern its matter, *i.e.*, the nature of the premises or the propositions respectively. The distinction of the matter of the proposition, however, is obviously concerned with semantic features. Nevertheless, the treatment of the propositional matter within the framework of syllogistics and its constituents makes good sense insofar as it is of a certain relevance to the proposition's quantity. Since propositions in natural and in separate matter are always valid universally, particular propositions in a proper sense can only occur in contingent matter. Thus the matter of statements –a semantic category– has influence on their mutual quantitative relations –a formal aspect–, as becomes evident in Sherwood's explanations. Sherwood stresses the fact that whenever a particular statement is true in natural matter –*e.g.*, "*aliquis homo est animal*"– its subcontrary cannot be true (against the rule that subcontraries can be true at the same time), because whatever is affirmed of one particular in a proposition in natural matter has to be affirmed of all particulars. Similarly, as Sherwood adds, whatever is separated from or negated of one particular in separate matter is separated from or negated of all. When, for instance, the proposition "*aliquis homo non est asinus*" is true, its subcontrary "*aliquis homo est asinus*" cannot be true, violating the rule of subcontraries by virtue of the essential separation of "*homo*" and "*asinus*". To sum up, in natural matter and in separate matter, according to Sherwood, a particular proposition interchanges with (*convertitur cum*) a universal one. Therefore, subcontraries in these two matters cannot be true at the same time, and, furthermore, the truth of the particular subalternate entails (*infert*) the truth of the universal subalternant. These truths are not –as Sherwood concludes– dependent on the nature of the particular proposition, that is on its formal feature, but on the nature of its matter.

3   The theory of supposition

In contrast to the above-mentioned syncategorematic words, such as the copula, quantifiers and so on, the descriptive or categorematic signs, which function as subject or predicate in a proposition, were called terms (*termini*). The medieval treatises on the properties of terms (*proprietates terminorum*) rest upon an initial distinction: A term's property of being meaningful on its own or of having a prepropositional reference is called its signification (*significatio*). This property belongs to categorematic words by virtue of their capability to serve as language signs. On the other hand, the property of supposition (*suppositio*) is acquired by an already meaningful term when it functions as subject or predicate of a proposition. Supposition theory was used to describe what a categorematic term in its function as subject or predicate of a proposition means in a particular context, and it could serve to test inferences or diagnose fallacies. When medieval logicians claimed a statement *de virtute sermonis* or literally to be false, they maintained that the theory of supposition enables us to analyze the statement's true meaning, which may be covered up by misleading grammatical features. The supposition of a term was defined by William of Ockham (and others) as a term's standing for something else in a proposition in such a way that the term is truly predicated of that thing (or of a pronoun pointing to the thing).[9] On the basis of definitions like this, some authors constructed a theory of truth-conditions for categorical sentences. For instance, a universal affirmative proposition was considered to be true if and only if its predicate supposits for everything for which the subject supposits. Other cases were handled analogously. In general, the theory of supposition was used for two remarkably different purposes. On the one hand it served as a tool for semantic distinctions, on the other hand it constituted a kind of theory of quantification.

3.1 Supposition and semantic analysis

Roughly speaking, the semantical distinctions run as follows: If we take the proposition "*homo est animal*", the term "*homo*" stands for (*supponit pro*) its normal referents, as when "*homo*" is taken for individual human beings like Socrates, Plato and so on. In this case, "*homo*" has personal supposition (*suppositio personalis*). In contrast, in the proposition "*homo est disyllabum*", the term "*homo*" does not stand for what it usually signifies, namely men, but for the word "*homo*" itself and has material supposition (*suppositio materialis*). A third case is represented by the proposition "*homo est species*", where the word "*homo*" stands neither for its significates nor for the word itself or for the design of the word, but for the universal or for the concept expressed by it and has simple supposition (*suppositio simplex*). Simple supposition was a highly controversial issue, as we can infer from the disputed status of the universals or concepts themselves.

Medieval authors often started with a division of proper and improper supposition in order to distinguish the genuine uses of a term from, *e.g.*, its metaphorical use. The subsequent division of proper supposition into personal, material, and simple supposition represents the three basic types

---

[9] *Cf.* [*Summa Logicae*, p.193, 11-14].

of contextual reference. Nevertheless, authors like Walter Burleigh do not start with the threefold distinction of personal, material, and simple supposition, but with the twofold distinction of a *suppositio materialis* and a *suppositio formalis* and divide the latter into *suppositio personalis* and *suppositio simplex* in a second step. The idea underlying this version is to distinguish the material supposition as a nonsignificative use of a term from two significative uses under the heading of formal supposition, namely one for concrete significates in the sense of, *e.g.*, single human beings and the other for a general form or universal nature. Obviously, this position corresponds to a realistic assumption according to which a universal nature can be regarded as a significate of the general term "*homo*". The position mentioned first, however, which assumes personal supposition –the standing for individual objects in the physical world– as the only significative use of a term represents a nominalistic position.

Though material supposition shows affinities to 20th-century quotation devices, it cannot be identified entirely with the modern notion of the mention of a word in contrast to its use. The idea of mentioning a word usually indicated by quotation marks is closely connected to the assumption that by quotation marks a new term ("*homo*") with quotation marks is generated in order to refer to the original term (*homo*) without quotation marks. While the modern approach is based on the distinction of two different language signs, one of which is introduced to refer to the other, the medieval theory of supposition is based on the quite different idea of assuming different modes of use (*acceptio sive usus*) of one and the same term, and one of these modes is the material use (*cf.* [Kan95]).

To sum up, the distinction of personal, material, and simple supposition represents a semantic approach aiming to point out the different types of contextual reference of subject terms which depends on the predicate terms they are conjoined with. The theory of supposition, however, provides us with a different approach to questions of contextual reference, which is regarded as syntactical rather than semantical and which at first glance fits in much better with the features of logic as a formal science.

3.2 Supposition and syntactic analysis

Within the analysis of propositions whose general terms stand in personal supposition or are used significatively for a term's singular referents, the theory of supposition provided a tool for distinctions nowadays treated by quantification theory (*cf.* [Wei79]). Basically three modes of personal supposition were distinguished, (1) determinate, (2) confused and distributive, and (3) purely confused supposition. These modes were analyzed and explained by means of the descent (*descensus*) to singulars.

(1) A term stands in determinate supposition (*suppositio determinata*) when it is conjoined with the existential quantifier "some" or "*aliquis*" as in the proposition "*aliquis homo currit*". Here the term "*homo*" stands or supposits for all its single referents, so that one can infer a disjunctive set of singular propositions. The subject terms name all of the individuals for which the general term stands, and the respective predicate terms are identical with that of the particular proposition. Therefore, assuming that the only men are Socrates, Plato and Cicero, it follows that if some man is white, Socrates is white or Plato is white or Cicero is white.

(2) When a term in a general proposition is combined with a universal quantifier, *e.g.*, in the proposition "*omnis homo currit*", "*homo*" has confused and distributive supposition (*suppositio confusa et distributiva*). This kind of supposition given by the universal quantifier to the term immediately following it means that the term stands for all its individual instances in such a way that the descent to singular propositions yields a conjunction of propositions. Thus from our example "*omnis homo currit*" we may infer the conjunction "*Sortes currit, et Plato currit, et Cicero currit*".

(3) A third type of personal supposition is the merely confused supposition (*suppositio confusa tantum*) which occurs when the predicate term of a universal affirmative proposition stands for all its individual referents. Here the reduction to singulars is effected not by a disjunction or conjunction of singular propositions but rather by a proposition with a disjoint predicate. So, if we take the supposition of "*animal*" in our example "*omnis homo est animal*", we may infer that every man is this animal or that animal or that other animal. In contrast, it does not follow that every man is this animal, or every man is that animal, and so on.

The second mode, *i.e.*, confused and distributive supposition, is called mobile if one is entitled to carry out the descent to singulars as in the example "*omnis homo currit*". Otherwise, confused and distributive supposition is immobile, as for instance in the proposition "*omnis homo praeter Socratem currit*". Due to the phrase "*praeter Socratem*" the descent in this case is possible only in a deficient or restricted manner.

The main point about the merely confused supposition lies in the fact that it implies recognition of the problem of multiple quantification and of the extension of the scope of one quantifier to include another. Medieval logicians obviously refrained from using quantifying prefixes like "*omnis*", "*nullus*", and "*aliquis*" as a kind of adjectival determinants of exclusively the term following them, but –as already pointed out in the context of syncategorematic words– regarded them as also operating on the supposition of both terms combined in a proposition. For example, the case was considered in which every man is looking at himself, but at no other man. Here from the true proposition "every man is looking at a man" (*omnis homo videt hominem*) we cannot infer the proposition "there is a man that every man is looking at", although the converse implication would be valid: From the proposition that "there is a man that every man is looking at" we can infer the proposition "every man is looking at a man". This gives rise to the theory of ascent (*ascensus*) as the procedure corresponding to descent, which leads medieval authors to subtle questions concerning the equivalence of the propositions underlying the descent and those resulting from it (*cf.* [Spa88]). Another subject of interest is the question whether there are just these three modes of descent presented here, or whether other modes should be assumed, especially that of a proposition with a conjunct predicate — a mode which was introduced and discussed by several logicians in the 14th century as a *descensus conditionatim* (*cf.* [Rea91]). I have to leave these particular difficulties aside here, since my present intention is of a more general kind.

The fact that supposition theory on the first level of division is a tool to analyze significative and nonsignificative uses of terms and on the second level, namely that of descent to singulars, is a theory to analyze quantification resulted in the view that supposition theory is more adequately viewed as two separate theories (*cf.* [Sco66, p.30]). Under the overarching perspective of contextual reference, however, there are good reasons to regard supposition theory as a unified theory, integrating semantic and syntactic aspects and at the same time formal and non-formal aspects of language. With regard to logic as a formal science we have to emphasize that supposition of terms in general was investigated as occurring in natural discourse, and no artificial language adapted to the uses of logic was constructed. Altogether, there was no fixed system, but rather an open-ended set of rules governing the different types of supposition in order to handle all casual and special instances of contextual reference.

4 The theory of consequences

In medieval logic, complex propositions composed of two or more categorical propositions joined by any sentential connective were called hypothetical. According to a customary etymological explanation a hypothetical statement is a complex statement in which one proposition in the literal sense of "*hypo*" and "*thesis*" is "put under" another. Hypothetic statements in this broad sense were classified, depending on the connective involved, as copulative, *i.e.*, conjunctive, disjunctive, conditional, causal, temporal, and local. In each case, the function of the connective was usually defined by stating truth rules. The truth-value of these hypothetical statements is a function of the truth-values of the categorical propositions of which the hypothetical statement is composed. Accordingly, the conditional form, which is essential for the present context, could clearly be distinguished from other types of connections by only formally syntactic criteria, *i.e.*, the content or matter of the propositions being joined were not taken into account. Thus, the conjunctive was said to be true if and only if both component propositions were true, and the disjunctive was said to be true, if one of its components was true. A valid conditional or inference, which is the medieval *enuntiatio hypothetica* in the strict sense, was called a *consequentia* and its components were distinguished as *antecedens* and *consequens*.[10] As far as truth conditions are concerned, Sherwood said that for the truth of the conditional statement it was not required that its parts be true, but only that whenever (*cum*) the antecedent was true the consequent was true.[11] Peter of Spain claimed that the antecedent could not be true without the consequent, adding that every true conditional was necessary and every false conditional impossible.[12]

The notion of *consequentia* was already discussed in the context of conditional statements in the 11th century. However, it was not until the 14th century that consequences became the subject of separate treatises called *De consequentiis*, which provide a concise formulation of the rules governing the validity of a conditional argument. It has been presumed that the inquiry of consequences grew out of the study of dialectical or topical arguments, but the point is still controversial. Unlike modern logicians medieval authors seem to have paid only little attention to the distinction between consequences as rules of inference, as conditional statements, or as arguments which may be valid or invalid.

Throughout the 14th century consequences were divided into two main classes, formal and material. A consequence was usually called formal if it was valid on account of the logical form of the component sentences, or under all transformations of the categorematic terms, *i.e.*, the matter or content of the propositions. In formal consequences, as authors like Ralph Strode or Robert Fland explain, the consequent is understood in the antecedent formally.[13] In contrast, a materially valid consequence was defined as one which does not hold for all terms arranged in the same way, *i.e.*, not on formal grounds alone, and here the validity is dependent on the subject and predicate terms involved, as in the proposition "*si homo currit, animal currit*". We have to point out, however, that there was another approach to the notion of a material consequence, since there were authors like

---

[10] For the theory of consequences *cf.* [Jac93, p.101-259]; *cf.* also [Sch$_1$88].
[11] *Cf.* [*Introductiones*, p.22, 285-287].
[12] *Cf.* [*Tractatus*, p.9, 15-18].
[13] For a criticism of this psychological or epistemic account of consequences, *cf.* [Boh01], especially p.154-158.

William of Ockham who maintained that a material consequence involved the independence of antecedent and consequent. According to this position there finally remained just two instances of the material consequence, namely the two paradoxes of strict implication, *i.e.*, that anything follows from an impossible proposition (*ex impossibile quodlibet*) and that a necessary proposition follows from anything (*ex quolibet sequitur necessarium*). At the same time, these authors concentrated on the area of inferences in which *antecedens* and *consequens* are semantically related to each other, so that the antecedent actually indicates a sufficient condition for the truth of the consequent, while *vice versa* the consequent is dependent upon the antecedent. In Ockham's view and in contrast to the modern approach to the subject, the main interest is in formal inferences which –in this challenging use of the term– rest upon semantic reasons for validity.

In the 14th century the theory of consequences tended to replace syllogistics as the central and paradigmatic form of argumentation. Authors like John Buridan, Albert of Saxony, and Paul of Venice regarded the syllogism as just one among different types of consequences and incorporated syllogistics in their comprehensive treatises on *consequentiae*. These treatises contain a mixture of quite different types of rules. Some rules are of general kind, *e.g.*, "if A is the antecedent of B, and B is the antecedent of C, then A is the antecedent of C". Other rules are propositional and truth-functional, *e.g.*, "since a conjunction is true if and only if its conjuncts are true, from 'A and B' we may infer A". Moreover, *modus ponens*, *modus tollens* and De Morgan's laws are given. Medieval philosophers did not present any kind of systematization of consequential rules, but mere collections with respect to paradigmatic difficulties in disputational practice. Just like supposition theory, the theory of consequences also reveals a certain interest in formal and material aspects of argumentation alike.

5 *Dialectica est ars artium*: Logic and its special status

During the Middle Ages until early modern times logic was characterized as *ars artium* or, as was sometimes added, *scientia scientiarum*. This formula which ascribes a special status to logic with regard to other disciplines, seems to go back to Augustine [*Ord.*, X.III, 38], who had characterized logic as *disciplina disciplinarum*. The formula was not only often repeated in the medieval logical tradition, especially in the initial paragraphs of logic treatises and compendia, but also caused numerous and valuable reflections or comments. A prominent instance in the 14th century is the Buridan-commentator John Dorp who –after reflecting on logic as *ars* and *scientia*, as *scientia speculativa* and *practica*, and as *logica utens* and *docens*– starts his detailed interpretation of the *ars artium* formula by a reflection on the genitive case of the word *"artium"*.[14] According to Dorp, the construction *"ars artium"* ascribes to logic a certain exceeding (*excessus*) in comparison with all other disciplines, which should not be understood as an exceeding in the sense of highest perfection but in the sense of highest generality. Generality again can be understood in a twofold manner, namely as generality of perspectives on the one hand and as generality of application or use on the other. John Dorp sees the special status of logic in this generality of application, and he regards logic as a universal tool for other disciplines. This picture of logic as an instrument of scientific inquiry started to gain widespread popularity after the recovery of Aristotle's *Analytica posteriora* in the early 13th century. Dorp's further reflections concern two commonly used additions to the *ars artium*-formula, namely *"et scientia scientiarum"* and *"ad omnium methodorum principia viam habens"*, which both also occur in Peter of Spain.

Dorp's question, whether the *ars artium*-formula was incomplete without the addition *"et scientia scientiarum"* or not, can be answered by means of a distinction in the notion of science itself: If *"scientia"* is understood in a broad and unspecific sense, it is almost synonymous with *"ars"*, and consequently the addition in question is redundant. If, however, *"scientia"* is used in the narrow sense of a *habitus speculativus*, the addition would be mistaken since the notions of art and science in their strict meanings cannot tally with the notion of logic at the same time. Therefore, in the present context "science" is taken in its broad sense. Though the addition of *"et scientia scientiarum"* assuming *"scientia"* in the broad sense is scarcely needed, it does not seem redundant either, since according to Dorp it is suitable to stress the prominent position of logic.

The second additional clause to the formula, namely the phrase *"ad omnium methodorum principia viam habens"*, emphasizes the Aristotelian characterization of logic as an *organon* or tool concerning all disciplines. I will disregard Dorp's manifold explanations of the assumption that every science bases the construction of argumentation on the genuine principles of that science and at the same time owes these principles to logic. Rather I will focus on one remark made by Dorp in discussing the question whether or not the further addition *"aliarum a se"* should be added to our formula. By means of this addition some authors want to exclude logic itself from those sciences that logic paves the way for. If the addition *"aliarum a se"* was apt, it would be necessary to name some discipline apart from logic for which could be claimed *vice versa* that logic owes its principles to it or that it paves the way for logic. Since, however, such a discipline does

---

[14][*Compendium*, Tract. I, Diffinitio logicae]. *Cf.* [Kan94, p.338-340].

not exist, Dorp concludes that the phrase *"aliarum a se"* is not redundant but rather mistaken. As a result, logic must be subsumed under those disciplines to which it lends the way and the principles. Finally the special status ascribed to logic rests on its feature of universal application in the field of arts and sciences. The fact that logic obtains this function and is able to provide us with an interdisciplinary instrument of argumentation rests upon its abstractive character, *i.e.*, the fact that it is not restricted to any individual subject matter. It is worth asking whether logic could be regarded as a formal science just in this respect of extreme abstraction from content or in the respect that logic represents an indispensable tool in quite different doctrinal areas.

6 Concluding remarks

If we understand logic as a formal science, we focus on restricted aspects of language or discourse. In this case, as Strawson puts it, logic is the study of the "general forms of the proposition" and of "certain relations of dependence or independence [between propositions] as regards truth value" [Str92, p.36]. Moreover, for Read, logic is formal

> when it uses schematic letters to identify the formal structure of arguments, leaving only the logical expressions ('logical constants' as they are often called) in place. [Rea95, p.61]

In the Middle Ages, however, we are confronted with a remarkably different understanding and use of logic, since medieval logic not only analyzes the complex syntax and semantics of the natural language usually without the help of symbolic techniques, but also includes general questions which are nowadays subsumed under the heading of philosophy of language.

Moreover, logic –like the *scientiae sermocinales* in general– in the Middle Ages is understood as a methodological discipline. A generally accepted view was that logic is about discriminating the true from the false by means of arguing. Its aim is not to analyze the nature of things, but to reflect upon operations by means of which the human soul attains to knowledge. The operation leading to science is language in its specific form of inferential or concluding discourse. Logic, grammar, and rhetorics are intended and understood as reflexive disciplines concerned with the analysis of discourse. This analysis is not merely descriptive. Since the aim of the *trivium* is teaching to speak correctly, elegantly and truly, the *scientiae sermocinales* are critical and normative as well. Furthermore, by reflecting on operations leading to knowledge, logic had a clear-cut cognitive orientation in terms of finding the truth and of proceeding from the already known to the yet unknown.

Though according to Boehner the theories of supposition and of consequences reveal a "perfect sense for the formality of logic" [Boe63, p.315], medieval logic is in no way devoted to the setting up of formal systems or any metalogical analysis of formal structures. And its concern is by no means restricted to the syntactical reconstruction of the formal elements of discourse. We cannot speak of any kind of priority either. Logic in the medieval sense of the discipline is necessarily connected with semantical aspects of basic relevance. It is assumed that one can work within natural language, or some purified version of it. The semantics of natural language is always the basis for establishing any logical principles, and a deductive system of a purely syntactic nature simply is never intended, perhaps chiefly because an artifically constructed logical language, in which the syntax can be specified independently of semantics, is never envisioned. The fact that medieval philosophers used to accept the concept of logic as an overarching discipline of discourse was a predominant reason for integrating syntactical with semantical approaches and for regarding this discipline as a key to all other sciences.

The initial question of in how far medieval logic can be regarded as a formal science after all cannot be answered in an unambiguous and definitive way, but requires weighing the pros and cons.[15] Since logic during the Middle Ages is centred around Aristotelian syllogistics and the syllogism itself was traditionally characterized by means of its formal validity, medieval logic appears as a formal science. But when we take a look at the 14th century, syllogistics are integrated into an overarching doctrine of inferences, among which the syllogistic or formally valid inference was just one type of argumentation among others, since there were also topical inferences or material consequences holding in virtue of some extrinsic feature, such as the meanings of their terms. Furthermore, the medieval theory of fallacies concerns different types of deceptive arguments or illegitimate inferences, and thus its subject matter is not purely formal. Actually, we are confronted with a discipline going far beyond the formal structures of discourse. Therefore I do not agree with Boehner who maintains that to "speak of 'formal logic' is, in scholastic terminology, a *nugatio* or tautology" and that "medieval logic is interested only in the formality or structure of discourse" [Boe52, p.xvi]. And I do not agree with Moody who regards medieval logic as a science that is restricted to the aim of formalizing the usage of language and that intends to formulate the logical syntax for scientific discourse.[16] As a discipline of disputational practice medieval logic was rather concerned with features that in our times Ryle treated under

---

[15] *Cf.* [Nis52], especially p.108.
[16] *Cf.* [Moo53, p.10-16 *et passim*].

the heading of "informal logic" — an area in which the standards of formal logic, "the ideals of systematization and rigorous proof", are not at work [Ryl56, p.111]. Accordingly the idea that logic in the Middle Ages was exclusively understood as a discipline concerned with formal validity was rejected by King [Kin01][17]. Formal validity, as King emphasizes, was taken just as one specific kind of validity, while medieval logicians considered validity in general. Though medieval logic partially is formal (admittedly without being formalized), it nevertheless reveals informal or, as King [Kin01, p.135] prefers to say, "nonformal" features.

When we ask what the formal sciences are and how they have been perceived through history, we have to note that we make use of a scientific classification that did not prevail in the Middle Ages. During the Middle Ages logic neither employs its own esoteric language nor was it exclusively concerned with formal features of discourse. For the reasons given here, medieval logicians themselves would have refrained from regarding their discipline as a formal science as understood nowadays. In modern logic questions of syntax and semantics as well as constructions of formal systems and the inquiry of their interpretability are strictly distinguished, while in medieval logic syntactic and semantic questions are closely related to each other. When the study of logic is stimulated by the question of the foundation of arithmetic –a classic topic that gave rise to formal logic in its modern version– different intentions are in operation as compared with those that lead to a *scientia sermocinalis*. Medieval logicians are concerned with questions of the logical form of argumentation within natural language as it is used in philosophical and theological matters. Nevertheless, it makes good sense to treat medieval logic under the heading of formal sciences and their historical development, since it is part of the history of a discipline which in its predominant features and intentions is nowadays generally accepted as a formal science.[18] To sum up, the classification of medieval logic as a formal science is appropriate only under selected perspectives. The distinction of formal and non-formal does not fit in too well with the medieval view of the sciences. The notion of a *scientia sermocinalis* (or *scientia rationalis*) which obtained the status of a most general tool transcending the individual subjects and questions of the particular disciplines fits in much better with the central features of medieval logic outlined here.

---

[17] Especially p.135*sq*.

[18] For the difference between formal logic and theory of logical form, *cf.* [Per84a], especially p.17: "A formal logic codifies the possible forms of rational discourse as such. A theory of logical form examines the structures inherent in a given language used by ordinary persons in a particular time and place. A formal logic is normally developed as a formal system with axioms, theorems and rules for the manipulation of uninterpreted signs. A theory of logical form need not be presented as a formal system."

Primary Sources.

[An. post.]	**Aristotle**, Analytica posteriora, *in:* [TreFor60]
[Ord.]	**Augustine**, De ordine, *in:* [Ste80]
[Compendium]	**John Dorp**, Perutile compendium totius logicae Joannis Buridani cum praeclarissima solertissimi viri Joannis Dorp expositione, Venice 1499; *facsimile edition:* Minerva, 1965
[Dialectica]	**Garlandus Compotista**, Dialectica, *in:* [dRi59]
[Summa Logicae]	**Guilielmus de Ockham**, Summa logicae, *in:* [BoeGálBro74]
[Tractatus]	**Petrus Hispanus**, Tractatus, *in:* [dRi72]
[Institutiones]	**Priscianus**, Institutiones grammaticae, *in:* [Kei55]
[Introductiones]	**William of Sherwood**, Introductiones in logicam, *in:* [BraKan95]
[Syncategoremata]	**William of Sherwood**, Syncategoremata, *in:* [ODo41]

References.

[Bia91]	Joël **Biard** (*ed.*), Itinéraires d'Albert de Saxe: Paris-Vienne au XIVe siècle, actes du colloque organisé le 19-22 juin 1990 dans le cadre des activités de l'URA 1085 du CNRS à l'occasion du 600e anniversaire de la mort d'Albert de Saxe, Vrin 1991 [Études de philosophie médiévale 69]
[Boc61]	Joseph M. **Bocheński**, A History of Formal Logic, University of Notre Dame Press 1961
[Boe52]	Philotheus **Boehner**, Medieval Logic, An Outline of Its Development from 1250 to c. 1400, Manchester University Press 1952
[Boe63]	Philotheus **Boehner**, History of Scholastic Logic, **Encyclopedia Britannica** 14 (1963), p.312-317
[BoeGálBro74]	Philotheus **Boehner**, Gedeon **Gál**, and Stephen F. **Brown** (*eds.*), William of Ockham: Summa logicae (= Opera Philosophica I), Franciscan Institute 1974
[Boh01]	Ivan **Boh**, Consequence and Rules of Consequence in the Post-Ockham Period, *in:* [Yrj01, p.147-181]
[BraKan95]	Hartmut **Brands** and Christoph **Kann** (*eds.*) William of Sherwood: *Introductiones in logicam*, textkritisch hrsg., übersetzt, eingeleitet und mit Anmerkungen versehen, Meiner 1995
[Car61]	Rudolf **Carnap**, Der logische Aufbau der Welt, 2nd edition, Meiner 1961
[CraSpe94]	Ingrid **Craemer-Ruegenberg** and Andreas **Speer** (*eds.*), Scientia und ars im Hoch- und Spätmittelalter, Springer 1994 [Miscellanea Mediaevalia 22]
[Jac80]	Klaus **Jacobi**, Die Modalbegriffe in den logischen Schriften des Wilhelm von Shyreswood und in anderen Kompendien des 12. und 13. Jahrhunderts, Brill 1980

[Jac93]   Klaus **Jacobi** (*ed.*), Argumentationstheorie, Scholastische Forschungen zu den logischen und semantischen Regeln korrekten Folgerns, Brill 1993

[Kan94]   Christoph **Kann**, Wissenschaftstheoretische Differenzierungen zur Logik bei Johannes Buridan, *in:* [CraSpe94, p.329-340]

[Kan95]   Christoph **Kann**, Materiale Supposition und die Erwähnung von Sprachzeichen, *in:* [LenPos95, p.231-238]

[Kei55]   Heinrich **Keil**, Grammatici Latini II, Teubner 1855

[Kin01]   Peter **King**, Consequence as Inference: Mediaeval Proof Theory 1300-1350, *in:* [Yrj01, p.117-145]

[Kre88]   Norman **Kretzmann** (*ed.*), Meaning and Inference in Medieval Philosophy, Kluwer 1988

[Lay73]   Rupert **Lay**, Grundzüge einer komplexen Wissenschaftstheorie, volume 2, Knecht 1973

[LenPos95]   Hans **Lenk** and Hans **Poser** (*eds.*), Neue Realitäten, Herausforderung der Philosophie (XVI. Deutscher Kongreß für Philosophie, 20.-24. September 1993, TU Berlin, Sektionsbeiträge I), Akademie-Verlag 1995

[Moo53]   Ernest A. **Moody**, Truth and Consequence in Medieval Logic, North-Holland 1953

[Nis52]   Louise **Nisbet**, Formalism of Terminist Logic in the Fourteenth Century, **Tulane Studies in Philosophy** 1 (1952), p.107-112

[ODo41]   J. Reginald **O'Donnell**, The *Syncategoremata* of William of Sherwood, **Medieval Studies** 3 (1941), p.46-93

[Pat81]   Günther **Patzig**, Sprache und Logik, 2nd edition, Vandenhoeck & Ruprecht 1981

[Per84a]   Alan R. **Perreiah**, Introduction, *in:* [Per84b, p.17-33]

[Per84b]   Alan R. **Perreiah** (*ed.*), Paulus Venetus: *Logica parva*, Translation of the 1472 Edition with Introduction and Notes, Philosophia-Verlag 1984

[Pin72]   Jan **Pinborg**, Logik und Semantik im Mittelalter, Ein Überblick, Frommann-Holzboog 1972

[Rea91]   Stephen **Read**, *Descensus copulatim*: Albert of Saxony vs. Thomas Maulfelt, *in:* [Bia91, p.71-85]

[Rea95]   Stephen **Read**, Thinking about Logic, An Introduction to the Philosophy of Logic, Oxford University Press 1995

[dRi59]   Lambertus Marie **de Rijk** (*ed.*), Garlandus Compotista: *Dialectica*, first edition of the manuscripts with an introduction on the life and works of the author and on the contents of the present work, Van Gorcum 1959

[dRi72]   Lambertus Marie **de Rijk** (*ed.*), Petrus Hispanus: *Tractatus* called afterwards *Summule logicales*, Van Gorcum 1972

[RitGrü92]   Joachim **Ritter** and Karlfried **Gründer** (*eds.*), Historisches Wörterbuch der Philosophie 8, Wissenschaftliche Buchgesellschaft Darmstadt 1992

[Ryl56]   Gilbert **Ryle**, Dilemmas, the Tarner lectures 1953, Cambridge University Press 1956

[Sch$_0$92] Jakob Hans Josef **Schneider**, *Scientia sermocinalis/realis*, *in:* [RitGrü92, col. 1508-1516, (1508)]

[Sch₁88]	Franz **Schupp**, Logical Problems of the Medieval Theory of Consequences, with the Edition of the '*Liber consequentiarum*', Bibliopolis - Edizioni di filosofia e scienze 1988
[Sco66]	Theodore Kermit **Scott**, Sophisms on meaning and truth by John Buridan, Appleton-Century-Crofts 1966
[Spa88]	Paul Vincent **Spade**, The Logic of the Categorical: The Medieval Theory of Descent and Ascent, *in:* [Kre88, p.187-224]
[Ste80]	Michael Payne **Steppat**, Die Schola von Cassiciacum: Augustins *De ordine*, Bock + Herchen 1980
[Str92]	Peter F. **Strawson**, Analysis and Metaphysics, An Introduction to Philosophy, Oxford University Press 1992
[TreFor60]	Hugh **Tredennick** and Edward S. **Forster** (*eds., trans.*), Aristotle, Posterior Analytics and Topica, Heinemann 1960 [Loeb Classical Library 391]
[Wei79]	Hermann **Weidemann**, Wilhelm von Ockhams Suppositionstheorie und die moderne Quantorenlogik, **Vivarium** 17 (1979), p.43-60
[Yrj01]	Mikko **Yrjönsuuri** (*ed.*), Medieval Formal Logic, Obligations, Insolubles and Consequences, Kluwer 2001

Received: September 9th, 2003;
In revised version: April 24th, 2004;
Accepted by the editors: April 26th, 2004.

Benedikt **Löwe**, Volker **Peckhaus**, Thoralf **Räsch** (*eds.*)
Foundations of the Formal Sciences IV
The History of the Concept of the Formal Sciences

Logic, Physics, and Prediction in Hellenistic Philosophy: x happens, but y?

DARYN LEHOUX

Classics and Ancient History
University of Manchester
Humanities Lime Grove
Oxford Road
Manchester, M13 9PL, United Kingdom
E-mail: daryn.lehoux@manchester.ac.uk

> ABSTRACT. What is the relationship between logic and the natural sciences in antiquity? Are the two clearly distinct, or do they overlap to some extent? If so, how? To answer these questions, this paper examines the ways in which formal solutions to the problems of predictive inference developed in the ancient world, and the ways in which these formal solutions were eventually influenced by ancient physics. The interaction between the formal and natural sciences placed constraints on explanations and justifications offered in the predictive sciences, but it also determined certain constraints –derived ultimately from physical considerations– on the framing of the predictive syllogism. This doesn't quite blur the distinction between the formal and natural sciences in antiquity, but it does show how the two were mutually interdependent when faced with the problems of prediction in particular.

Introduction

When we predict a future event, assuming we are not just guessing madly, what we are doing is performing a rational operation that begins with some observed or reported *sign*, whether that sign be a newspaper report of a company merger, or an observation of a red sky in the morning. From there we reason out likely or possible events that will follow the sign: stock prices will go up, the weather will be foul, or whatnot.[1] The rational operation that

---

[1] Modern philosophy of science does not normally treat prediction as a type of sign inference. I do so here (a) in order to be clear on the practice of prediction (how is it that

gets us from the sign to the prediction can be as complex as painstakingly working out a host of possible scenarios and contingencies, or as simple as following a rule of thumb.

Ancient logicians likewise used the term *sign*, but in a broader (and technical) sense, to refer to any apparent object that was used to infer something about non-apparent (or unperceivable) states of affairs.[2] The predictive sign is thus one species of sign, which allows us to infer something about states of affairs in the future from states of affairs perceived now.[3] Predictive signs indicate what we might call their *sequentia* (the predictable events that normally follow them) by formalized and normative connections, often via actual or implied conditional statements.[4] And indeed *sign*, σημεῖον, is the technical term in Stoic logic for the antecedent of a conditional that *reveals* its consequent. As Sextus Empiricus puts it:

> They (the Stoics) say that the *sign* is the proposition, in the antecedent of a valid conditional, that reveals the consequent.[5]

The predictive sign is, as Sextus tells us, one kind of sign in this class.[6] But what role do predictive signs play in predictive conditionals, and how are these signs related to the consequents of those conditionals in ancient

---

a prediction gets made? What kinds of data do we start from, and how do we reason them through to a conclusion about the future?), and (b) in order to facilitate the treatment of historical sources which *do* treat prediction as sign-inference (this is the approach taken in [All01] and [Mac02], for example). In contrast to my approach, modern philosophers of science have tended to be more concerned with questions about how or to what extent prediction, explanation, and/or theory confirmation work (or should work) together, how to understand probabilistic predictions, or how to understand the asymmetry of our knowledge of the past vs. that of the future.

[2] This is speaking very generally. Different schools narrowed this definition down in different ways. *Cf.* [Bur$_1$82]; [All01]; [Bar$_1$02].

[3] On predictive signs, *cf.* [Leh00]; on the distinction with linguistic signs, *cf.* [Man87]. On the problems with confusion of Stoic σημεῖα with linguistic signs, *cf.* [Kah69].

[4] The conditional sentence was, in Stoic logic, seen as the logical statement *par excellence*. Netz [Net99, p.259] has also shown that by Hellenistic times, even Greek mathematical propositions were frequently framed as conditionals or quasi-conditionals. On Stoic use of conditionals, *cf.* [Bob98]; [Kah69]; [Hay69].

[5] φασὶ σημεῖον εἶναι ἀξίωμα ἐν ὑγιεῖ συνημμένῳ καθηγούμενον, ἐκκαλυπτικὸν τοῦ λήγοντος [*Adv. math.*, viii.245]. (The translations here and throughout are my own unless otherwise indicated). Sextus thinks that the definition of sign used by the Stoics can only apply to the case where, in addition to the conditional being valid, both its antecedent and consequent terms must also be true. I note that this is Sextus' reading and not a Stoic constraint. On the Stoic definition of a sign as Sextus reports it, the antecedent in "if the Earth flies, the Earth has wings" may still be a sign, even though both terms are false.

[6] Other kinds in this class include, for example, the way a bronchial discharge might reveal an ailment to a physician [*Adv. math.*, viii.253]. In this paper I will be looking at this definition as it applies to predictive signs, but a parallel case can be made for symptoms, *etc.*, and for the Stoic σημεῖον in general.

philosophy? It is in the Hellenistic schools that we see the fullest treatment of these questions. But the answers that the various schools give operate just at the borderline of logic and physics, such that each of these branches of philosophy seems to be impinging on the other. What emerges is a nuanced picture that shows us that these two sciences, physics and logic, were intricately interdependent in Hellenistic philosophy (and this interconnection had had implications for Hellenistic medicine, as well).[7] Problems in logic could have reverberations in physics, and physical considerations could place constraints on logic. This interdependence runs so deep, in fact, that it may not be possible to disentangle which (if either) of the two disciplines, logic or physics, has primacy over the other in the Hellenistic schools. Looking at this in the most general way, take the following question: Is logic built up from our experiences about the world, or do the deep structures of our thinking constrain the ways we experience the world? By looking at the Hellenistic case, we can see that the question may not be quite so black and white.

Exegetical Interlude: Reading Cicero on Prediction

Cicero is one of the most important sources for Hellenistic philosophy generally, and for Stoicism and Academic Scepticism in particular. Although himself an Academic Sceptic, he had a great deal of respect for Stoicism, and although he did not accept their system *in toto*, he did signal his agreement with much of Stoic philosophy, particularly in ethics, and his philosophical works sometimes give us the fullest account we have of some aspects of Stoicism. Much of my analysis in the present paper depends on my reading of three of Cicero's latest works: *De natura deorum, De divinatione,* and *De fato*. Parts of my analysis are in agreement with new approaches in the history of philosophy (by Bobzien, for example) and other parts, particularly my reading of *De divinatione*, are unique. In these latter passages, I give my reasons for my reading, although always aiming to keep my justification economical to avoid getting too far off topic. That being said, let us look at the sources and the debates. The meatiest ancient treatment of the issues around prediction and the predictability of future events is found in the two latter books of Cicero's great religious-philosophical trilogy that is comprised of *De natura deorum, De divinatione,* and *De fato*.[8] The first of the three works, *De natura deorum,* covers a broad range of topics on the general subject of the gods, their relationship to people, and the cosmos as a whole. In *De natura deorum,* Cicero's interlocutors move rather quickly over a broad range of issues, and some topics are only touched on lightly

---

[7]On the case of medicine, *cf.* [All01, p.87-146].
[8]On ancient treatments of prediction and fate, *cf.* [Bob98]; [Mig96]; [Cra88].

in that work. The latter two works in the trilogy pick up on some of these threads and attempt to tease them out in more detail. So we see *De divinatione* beginning with a deliberate look back at *De natura deorum,* and then taking up a more careful re-examination of one of the arguments that Balbus (the Stoic) had laid out in the first work. There Balbus posed the following argument as the "first" argument for the existence of the gods:[9]

> *If there is divination then there are gods. But there is divination. Therefore the gods exist.*

And a little later in the dialogue, he tells us that divination as an art has been built up through a long series of observations: "we are warned by many different omens, many entrails, and many other things, which long-standing use has noted in order to establish the art of divination."[10] Balbus' claim for an observational derivation gets but a short reply from Cotta (the Sceptic) in [*De nat. deor.*, III.14]. But the idea is picked up again and examined much more closely in *De divinatione.*

De divinatione seeks to establish whether in fact the gods *do* give men presentiments of the future. What arguments can proponents of divination give to support their belief, and do these arguments stand up to close examination? As the debate unfolds, a set of claims gets made by Quintus (the Stoic proponent of divination) that eventually ties divination to the Stoic belief in *fate*.

But Cicero has other fish than fate to fry in book II of *De divinatione* and so he leaves this question of fate for the final book of the trilogy, *De fato.*

The whole trilogy has the feel of a naturally developing dialectical debate: issues get raised, arguments get made and contested, and gradually the shifting cast of interlocutors homes in on the central pivot-points of the arguments. Dropped threads get picked up, and over the course of the three books, we see much of the Stoic theological/philosophical nexus get a careful exposition, and just as careful a scrutiny. As we shall see, not all of the complex edifice of Stoicism gets categorically rejected by Cicero, but certainly he presents some powerful objections to particular arguments, even if, as he tells us at the end of *De natura deorum,* he isn't necessarily convinced by those objections to reject the "most probable" arguments of the Stoics. In the course of this examination, Cicero looks at a tangled set of issues that are situated just at the point where Hellenistic physics overlaps with Hellenistic logic.

---

[9] He credits this argument to Cleanthes [*De nat. deor.*, II.12-13].

[10] *multa praeterea ostentis multa extis admonemur, multisque rebus aliis quas diuturnus usus ita notavit ut artem divinationis efficeret* [*De nat. deor.*, II.166].

I. Logic moving physics

Logic has long tentacles in Hellenistic philosophy. Specific approaches to logical problems had the potential to bring about specific concommitments in the physics, and ultimately the ethics of the various schools. To take an example, as the debates unfolded between the Stoics and Epicureans a tangle of problems emerged as an interdependent set: the question of whether logic was two-valued or many-valued had immediate implications for physical questions such as causation, determinism, and free will.[11]

The issues cash out specifically around predictive conditionals, arguments of the form:

> *If x happens, then y will happen. But x happened. Therefore y will happen.*

On the Hellenistic understanding of what such conditionals entail, it turns out that if your logic is bivalent, then you may well end up with a causal physics, and what is worse, you may actually get roped into a completely deterministic universe *via* such conditionals.[12] The Stoic position is maybe the best to begin with for showing how this is so. Starting from the bivalency of their logic, the Stoics argue that since all propositions are either true or false,[13] then the future-tense proposition in the consequent, *y will happen*, must be either true or false, even before it has happened. Suppose it is true. Then *y* will happen, unavoidably. But if *y* will unavoidably happen, then it is predetermined to happen.

This may seem to be moving a little too quickly for modern tastes between the very different realms of logical necessity and physical necessity.[14] But there is an important adjunct to the strictly logical argument that moves us (quite elegantly as it turns out) into the realm of the physical, by leveraging Stoic ideas on causation as a mediating term between bivalence and the peculiar-and completely physical-Stoic idea of deterministic fate:

---

[11] Philodemus [*Phil. Rhet.*, P.Herc. 1427, col VI.10-18] says that some Stoics thought that logic by itself can "not effect anything on its own unless it is linked to propositions from ethics or physics" [Bar$_0$99, p.66]. This paper explores how some of those links were made.

[12] The evidence for this and the debates that follow come largely from Cicero's [*De fato*, 14-16, 18-28]. On the relationship between bivalence and determinism, *cf.* [Bob98]. On Stoic logic generally, *cf.* [Bar$_0$BobMig99]; [Bob98]; [Bob86]; [Fre$_1$74]; [Bal85]; [Kne$_1$Kne$_0$62]; [Mat53]; [Bec57]; [Mue69].

[13] It is unclear whether they thought "all propositions are either true or false" to have itself been a proposition. On the vicious circularity of this, and the ways around it, *cf.* for example [WhiRus35, p.38f]. Interestingly, Whitehead and Russell's set-theoretical reading sees the circularity here as *identical in kind* to that in the Sceptical knowledge claim that one knows nothing.

[14] But then, the very distinction between logical and physical necessities may have come about *as a result of* these debates between Sceptics and Stoics; *cf.* [Fre$_0$96, p.7].

> If [says Chrysippus] there is an uncaused motion, then not every proposition (what the dialecticians call an ἀξίωμα) will be either true or false, for what does not have an efficient cause will be neither true nor false. But every proposition is either true or false, therefore there is no motion without a cause. But if this is so, then all things that happen, happen through antecedent causes.[15] If this is so, all things happen through fate, and thus it follows that whatever happens, happens through fate.[16]

All this from the seemingly innocent little belief in bivalence! To be sure there are a couple of moves in this argument that may be nonobvious to the modern philosopher: *what does not have an efficient cause will be neither true nor false,* or *what happens through antecedent causes happens through fate.* The truth of the latter proposition comes about through the Stoic definition of *fate* as just being the unbreakable chain of antecedent causes:

> I mean by *fate* that which the Greeks call εἱμαρμένην, that is, the succession and sequence of causes, such that cause connected to cause brings about something by means of itself. This is an everlasting truth, running through all eternity. And since this is so, nothing has happened which was not going to be, and in the same way nothing will be if nature does not comprehend the efficient causes of it.[17]

And this equation of fate with antecedent causes seems, if Cicero's account is reliable, to have been acceptable to Carneades (an Academic and opponent of Stoicism), and to Cicero himself.[18]

As for the former proposition, that *what does not have an efficient cause will be neither true nor false,* this seems not to have been disputed by

---

[15] On antecedent causes, *cf.* [*De caus. pr.*]; [Han98b, p.43f]; [Fre₁80]. *Cf.* also [*De caus. cont.*, 1.1-2.4]; [*De fato*, 41]. Cicero switches his terminology between *causae antegressae* [*De fato*, 20], *causae antcecedentes* [*De fato*, 9, 41], *causae antepositae* [*De fato*, 41], *causae praepositae* [*De fato*, 41], and *causae efficientes* [*De div.*, I.125]. Such terminological flexibility is characteristic of Cicero. As he himself tells us at [*De fin.*, III.52]: *re enim intellecta in verborum usu faciles esse debemus*, "when the meaning is understood, we should be flexible in the use of words." On Cicero's terminological flexibility, *cf.* [Glu95, p.117-119].

[16] *si est motus sine causa, non omnis enuntiatio, quod* ἀξίωμα *dialectici appellant, aut vera aut falsa erit; causas enim efficientis quod non habebit, id nec verum nec falsum erit; omnis autem enuntiatio aut vera aut falsa est; motus ergo sine causa nullus est. quod si ita est, omnia quae fiunt causis fiunt antegressis; id si ita est, fato omnia fiunt. efficitur igitur fato fieri quaecumque fiant* [*De fato*, 20-21].

[17] *fatum autem id appello quod Graeci* εἱμαρμένην *id est ordinem seriemque causarum, cum causae causa nexa rem ex se gignat. ea est ex omni aeternitate fluens veritas sempiterna. quod cum ita sit, nihil est factum quod non futurum fuerit, eodemque modo nihil est futurum cuius non causas id ipsum efficentes natura contineat* [*De div.*, I.125-6]; *cf.* also [*De fato*, 41]. Long and Sedley [Lon₀Sed87, vol.2, p.337, 339] read this passage as using fate as the starting point of an argument for divination. But the argument is actually running the other way around here, and seems also to have done so for Chrysippus, if the evidence in Cicero's *De fato* is to be believed.

[18] *Cf.* [*De fato*, 31, 41].

the ancient logicians of any of the schools. Even the Epicureans accepted it and so chose a multivalent logic on the one hand, and a physics with uncaused motions (the swerve) on the other. The Epicureans essentially try to do an end run around the problem by attacking the necessity of the implication *if x happens, then y will happen*. The implication only holds as necessary, say the Epicureans, if everything happens according to a strict chain of causes.[19] But what if there are some random (and therefore uncaused, as Cicero points out)[20] events in the universe? If there are some uncaused events in the universe, then the consequent of the implication *then y will happen* can be neither true nor false in the sense that y is not absolutely necessary, even given a true antecedent, *x happened*. The closest we get to an explicit explanation of *why* this relationship between efficient causation and bivalency should hold absolutely is in Alexander of Aphrodisias [*Aphr. De fato*, 192], where he says that it is an impossibility that identical circumstances (identical with respect to causation and what-is-caused in particular) should not lead to identical results unless there is an uncaused motion.

An alternate tack is taken by Cicero in *De fato*. It consists of introducing the class of *fortuitous antecedent causes (causae fortuito antegressae)*, which bring about y, but not by *necessity* in the strict sense. This way we can keep bivalent logic, and say that *y will happen* has always been either true or false, but that when it actually turns out one way or the other, it so happened fortuitously, if still by a causal chain. y was not necessary, even if it was always true that it was going to happen, nor was it uncaused. We keep both causation and bivalence.

> We see that Epicurus should not fear, in admitting that every proposition is either true or false, that it will be necessary for everything to happen through fate. For it is not because of an eternal flowing of necessary causes in nature that it is true when it is said that "Carneades will go down to the Academy," but neither is it without a cause. For there is a difference between *fortuitous* antecedent causes and causes having in themselves a *natural* efficient power.[21]

The Stoics also insist that even if *y will happen* is true, it is not *necessarily*

---

[19] On Epicurean sign theory, *cf.* [All01, Study IV]; [Bar$_0$89]; [Lon$_0$88].

[20] *Cf.* [*De fato*, 22]. Lucretius [*De. rer. nat.*, 2.254-5] can be read as corroboration of this uncaused-motion reading, where Lucretius says that the swerve "breaks the law of fate, lest cause follow cause to infinity" *(fati foedera rumpat, ex infinito ne causam causa sequatur)*.

[21] *licet enim Epicuro concedenti omne enuntiatum aut verum aut falsum esse non vereri ne omnia fato fieri sit necesse; non enim aeternis causis naturae necessitate manantibus verum est id quod ita enuntiatur 'descendit in Academiam Carneades', nec tamen sine causis, sed interest inter causas fortuito antegressas et inter causas cohibentis in se efficientiam naturalem* [*De fato*, 19].

true.[22] And Cicero seems to think that their argument can be reconciled with his own account of fortuitous antecedent causes, and the two sides here "differ in words, not in fact".[23]

So we see that specific approaches to logic have a way of forcing the physics of the various schools to move in their particular and characteristic directions. The different schools took the implications of bivalency very seriously in their physics. But when looking at their arguments, one gets a nagging chicken-and-egg feeling, where it is difficult to determine whether the logic was adopted before the physics was fully worked out, or whether the basic physical commitments had already been made by each school when the problems in logic started to emerge. And probably the answer is somewhere in the middle.

II. Physics moving logic

The converse of the bivalency/fate argument is situated in questions around the epistemological status of the predictive conditional. In his *De divinatione*, Cicero[24] argues that, in effect, predictive conditionals, in order to have some force in the real world, must be structured so that a *causal* relation obtains between the antecedent and the consequent terms. He lobs this argument against the position taken by Quintus in book I of *De divinatione*, where Quintus justifies the efficacy of divination not by calling on a causal account, but instead by offering an enumerative argument that lists successful instances of divination, almost *ad nauseam*. Cicero's rejection of the enumerative argument for divination, and the insistence on a causal substrate, effectively puts constraints on the predictive conditional

---

[22] Perhaps to avoid the Diodoran definition of what is possible as only that which is or will be the case. *Cf.* [Bob98, p.106f].

[23] *Cf.* [*De fato*, 41f].

[24] I am not convinced of the usefulness of Beard's [Bea86] attempt to separate Marcus the interlocutor's position from Cicero's own, insofar as Marcus *is* Cicero himself (as indicated by (a) the name of the character, and (b) the way that both Quintus and Marcus use Cicero's own work). The Marcus/Cicero distinction is meant to circumvent the question of how to reconcile Marcus's "position" in *De divinatione* with Cicero's position in such works as *De haruspicum responso*, and *De domo sua*, but (as Beard herself would agree) there is no reason to see Marcus as offering an actual position on *divination itself* in the work: he may be seen to be addressing *arguments* more than positions. Such a reading gets us around the problems of trying to reconcile Marcus's "position" in *De divinatione* with Cicero's position elsewhere without having to posit a shadow Cicero in *De divinatione* That being said, I will refer to the author of *De divinatione* as Cicero, and his interlocuting self as Marcus, if only for the sake of being clear on the distinction between what Cicero is trying to achieve with the dialogue as a whole, and what he as an interlocutor is arguing, sentence by sentence, in the book. For quite different attempts to reconcile *De divinatione* with Cicero's other works, *cf.* [Goa68]; [Ras98]; [Pea79]; [Gui84].

such that some kind of causal relation must hold between the antecedent and the consequent terms. The predictive conditional *if x, then y* is only acceptable now if (a) x causes y, (b) y causes x, (c) both x and y are together caused by something or someone else, z or (d) x is caused by a prescient third party (a god, as it turns out) who sends the sign to warn us of y, however y is caused. For the sake of clarity, we can separate these justifications into (a)-(c) causal justifications narrowly construed, and (d) a divine beneficence justification.[25] Quintus tries, unsuccessfully, to piggyback the divine beneficence justification onto his enumerative account, but Cicero rightly treats these as two different arguments.

Cicero begins *De divinatione* with Marcus saying that all nations and all peoples have, for all time, believed in divination. Then he flags the central problem that will occupy him in the dialogue, whether divination is justified by appeal to experience, or by appeal to a rational, causal account of its efficacy: "Now, this is what I conclude: that the ancients were more convinced by the success of divination, than they were by learned argument."[26] Marcus wishes, in this inquiry, to determine "how much trust we should have in auspices, sacred rites, and religion".[27]

Notice that Marcus is limiting his inquiry here to the official branches of Roman religion: auspices, sacred rites, and *religio*,[28] and that he is going to explore this issue by looking at the central question of the justification of divinatory prediction in particular. Here we should also note his emphasis on the two types of justification possible for divination: (a) justification based on empirical observation (where we are "convinced by its success"), and (b) justification based on reason (where we are "convinced by learned argument"). This point is picked up by Quintus at the very outset of his speech in book I. Here Quintus, whose wording closely mirrors Marcus's, quite deliberately takes the position of the *veteres* mentioned by Marcus. Compare:

> Marcus: Now, this is what I conclude: that the ancients were more convinced in these matters by (their) experiences with divination, than they were by learned argument.[29]
>
> Quintus: But of these things I conclude it more proper to look at experiences than at causes.[30]

---

[25] I note that (c) and (d) are not mutually exclusive.

[26] *atque haec, ut ego arbitror, veteres rerum magis eventis moniti quam ratione docti probaverunt* [*De div.*, I.5].

[27] *quantum auspiciis rebusque divinis religionique tribuamus* [*De div.*, I.7].

[28] On the relationship of *De divinatione* to contemporary Roman religion, *cf.* [Kro00]; [Bea86]; [Sch₀86]; [Gui84].

[29] *atque haec, ut ego arbitror, veteres rerum magis eventis moniti quam ratione docti probaverunt* [*De div.*, I.5].

[30] *quarum quidem rerum eventa magis arbitror quam causas quaeri oportere* [*De div.*,

Quintus is making a strong claim that he can defend divination by enumerating (true) facts, and that he does not require a recourse to causes. He is making this claim in spite of the fact that (as a Stoic) he *does* think there is a force that enables divination to work: *est enim vis et natura*...[31] But he is defending an argument that he feels will succeed *even in the absence* of a proof for that *vis et natura*. Accordingly, Quintus constructs an argument that is built on a voluminous and detailed enumeration of historical and contemporary incidences of true predictions. His is a deliberately inductive argument,[32] and it is the validity of this induction Marcus attacks. What Quintus' account of divinatory prediction is lacking, Marcus will claim in book II, is exactly the thing Quintus thought he could leave out of the account: a causal mechanism.

The belief in divination, says Quintus, does not entail an agreement on causes. Simple empiricism, like that employed by the physician in the observation and use of roots and herbs [*De div.*, I.13, 16], will confirm the efficacy of divination. As he lines up example after example of successful divinatory prediction, again and again Quintus separates his account from one that would require a knowledge of causes:

> And your [Cicero's] *Prognostics* is full of presentiments of these things, but

---

I.12]. Compare these passages also to where Marcus clarifies for Quintus the intent of Cotta's argument in *De natura deorum: etenim ipse Cotta sic disputat ut Stoicorum magis argumenta confutet quam hominum deleat religionem.* "for Cotta argues more in order to refute the arguments of the Stoics than to abolish *religio* from men" [*De div.*, I.8]. In all three places, Cicero is using a *magis ... quam* construction to highlight what he sees as his real intention, lest the subtlety of his argument go unnoticed, which I think he rightly had reason to fear.

[31] [*De div.*, I.12].

[32] Which is odd, since Philodemus reports that inductive arguments were unacceptable to the Stoics [*Phil. De sign.*, 3]. Philodemus even uses the same example –people die when pierced through the heart– as Sextus does in *supporting* such inference [*Adv. math.*, viii.153]. And Cicero himself elsewhere makes a distinction between arguments that proceed from facts (which he says are always no more than probable arguments), and arguments that proceed from *propriae notae*, particular signs, which are certain (as smoke is the certain sign of fire) [*De part. or.*, 34]; *cf.* also Aristotle, [*An. pr.*, II.27]; [*Rh.*, I.2]. It is this distinction that, I think, fuels Cicero's attack in *De divinatione*. Bobzien thinks that Quintus' inductive argument can be attributed to the Stoics (*cf.* [Bob98, p.88]; compare also [Han98a, p.259]), but Quintus *nowhere* tells us that this part of his argument is actually Stoic. Indeed, as he says at *De divinatione*, the enumerative argument is employed by Cratippus the Peripatetic (and compare also [*De div.*, I.82, I.110f, I.118f]). In support of my reading, I note that Frede shows that, for the Stoics, there are insurmountable obstacles to trying to turn a series of observations into a conditional rule [Fre$_1$80, p.247-8]. I note that I am being admittedly anachronistic in calling enumerative arguments "inductive." I also note that the equation of enumerative arguments and induction is neither straightforward nor unproblematic. *Cf.* [vFr83]; [Lan00, p.115f]. Lastly, by *enumerative* here, I refer to what Cicero called *dinumeratio*, rather than *enumeratio*, "recapitulation" (*cf.* [*De part. or.*, 59]).

Hellenistic Philosophy   135

> who can ascertain the *causes* of these presentiments?[33]
>
> I do not ask *why,* but I know *what* happens.[34]
>
> Likewise, what the cleft in the liver, or what the entrails mean, I accept. What the cause is, I do not know.[35]
>
> And indeed, portents, and also auspices, omens, and signs are not *causes* of why something happens after them...[36]
>
> I do not perceive the cause ... for the god does not wish me to know it.[37]
>
> You ask *why* everything happens, and quite rightly so, but that is not now under discussion. Does it happen or does it not happen? *That* is at issue.[38]
>
> If I can not examine on what account everything happens, and only can show that these things I have mentioned happened, is this not enough to answer Epicurus and Carneades? ... For in everything, great age brings about an extraordinary knowledge by means of continuous observation... since what (event) happens after what (event) is seen with repeated observation, and also what (event) is a sign of what thing.[39]

Indeed, in almost every instance where he asserts the efficacy of divination, Quintus feels the need to add that causes are irrelevant to his argument. Sheer enumeration of instances is enough, he feels, to bring us to conviction. Although Quintus does mention some causal accounts, as of prophesy through dreams [*De div.*, I.110f] and of extispicy [*De div.*, 118f], these do not form the core of his argument, nor, more importantly, the focus of Marcus's attack. Coming so late in his argument, these instances, together with his grand conclusion at [*De div.*, I.125f] serve rather as elaborations on a point that Quintus sees as sufficiently proven by enumeration. Quintus thinks he has proven the existence of divination without recourse to causation. It is only once he is satisfied that he has sufficiently established divination in this way, that he now goes on to talk about the Stoic view of the causal mechanism underpinning divination, *i.e.*, that the universe was so constituted from its beginning that certain signs should precede certain events. But I cannot overemphasize that the causal mechanism comes in

---

[33] *atque his rerum praesensionibus prognostica tua referta sunt. quis igitur elicere causas praesensionum potest?* [*De div.*, I.13].

[34] *non quaero cur, quoniam quid eveniat intellego* [*De div.*, I.15].

[35] *similiter quid fissum in extis quid fibra valeat accipio; quae causa sit nescio* [*De div.*, I.16].

[36] *etenim dirae, sicut cetera auspicia, ut omnia, ut signa, non causas adferunt cur quid eveniat [...]* [*De div.*, I.29].

[37] *non reperio causam [...] non enim me deus ista scire [...] voluit* [*De div.*, I.35].

[38] *cur fiat quidque, quaeris; recte omnino, sed non nunc id agitur: fiat necne fiat, id quaeritur* [*De div.*, I.86].

[39] *si nihil queam disputare quam ob rem quicque fiat, et tantum modo fieri ea quae commemoravi doceam, parumne Epicuro Carneadive respondeam? [...] adfert autem vetustas omnibus in rebus longinqua observatione incredibilem scientiam; [...] cum quid ex quoque eveniat et quid quamque rem significet crebra animadversione perspectum est* [*De div.*, I.109].

after the fact of Quintus's inductive argument for divination.[40] Indeed, he no sooner mentions the Stoic causal account than he slips back into enumeration, and even the *argumenta ... ducuntur a fato*, "arguments drawn from fate," as Quintus calls them, are heavily dependent on a long series of observations of signs and consequents rather than a direct perception of causes.[41] And it is the lack of a really viable causal account that Marcus sees as contentious. Marcus quite simply feels that Quintus has gone at the whole thing backwards, trying to bootstrap mechanism onto anecdote, rather than the other way around. And Marcus's argument is, at bottom, an epistemological objection to induction.

After some "skirmishing with light arms" [*De div.*, II.26], Marcus begins his real attack on Quintus' argument. And he focuses his attack on the fact that Quintus has only amassed a list of examples, and failed to give us a causal account, a *why*, of divination.

> Since you cannot *explain* anything, you attack using an amazing number of fantastic examples [...] One should use *arguments* and *reasons* for the existence of these things, not just examples.[42]
>
> It is not possible for these things to be observed with certainty, as I showed above. [...] and what connection do they have with things by nature?[43]
>
> It seems to me that you betray the city of philosophy while defending its fortifications. Insofar as you want haruspicy to be true, you overthrow all of natural philosophy.[44]

And Cicero betrays his hand again in [*De fato*, 33], when he says that Carneades refused to believe that divination was possible, *even for Apollo*, unless the events predicted were the result of a necessary causal chain. Marcus simply will not accept that divination can be justified except by a causal explanation, and indeed, the causes seem to need to be the type of cause that the Stoics are so worried about: necessary causes.

---

[40] Contrast this with Chrysippus' justification of fate reported by Eusebius, [*Praep. evan.*, IV.3.1-2]. There Eusebius credits Chrysippus with an argument for fate that rests on the *prior* acceptance of divination. But whether divination was just accepted as given in Chrysippus's argument, or whether Chrysippus had some prior argument for its existence, as Quintus does here, Eusebius does not say. In any case, Quintus is bolstering his argument for divination with fate, whereas Chrysippus is proving fate from divination.

[41] *Cf.* [*De div.*, I.128].

[42] *sed tamen cum explicare nihil posses, pugnasti commenticiorum exemplorum mirifica copia [...] argumentis et rationibus oportet quare quidque ita sit docere, non eventis [...]* [*De div.*, II.27].

[43] *haec observari certe non potuerunt, ut supra docui. [...] cum rerum autem natura quam cognationem habent?* [*De div.*, II.33].

[44] *urbem philosophiae mihi crede proditis dum castella defenditis; nam dum haruspicinam veram esse vultis, physiologiam totam pervertitis* [*De div.*, II.37].

Another aspect of the set of problems that emerge in Cicero's treatment of prediction can be seen by comparing it to what Sextus Empiricus does with predictive signs. Like Cicero, Sextus sees a distinction between the *fact that* and the *why* of how signs indicate:

> Every sensible, as the name shows, is apprehended by sense. But the sign, *qua* sign, is not apprehended by sense, but by the mind.[45]

What is remarkable about Sextus' claim here is that it is expressly *not* leveled against predictive signs, which Sextus classes as *reminiscent*, but rather against *indicative* signs. The distinction is just this: reminiscent signs call to mind things that are temporarily non-evident, such as a scar being the sign of an old (and so no longer visible) wound, whereas indicative signs call to mind things that are by their very nature non-evident, such as the existence of reason being a sign of the existence of the (inherently invisible) soul. Sextus explicitly includes predictive signs as reminiscent: "on seeing a wound to the heart we predict impending death."[46] Moreover, Sextus tells us that we come to know the connections between reminiscent signs and their sequentia through repeated observation: "[...] having observed them often paired together [...]"[47] This may seem at first to be a complete reversal of Cicero's argument: where Cicero is arguing that the repeated observation of connections is insufficient as a grounding for divination, Sextus is saying that prediction (or more properly, reminiscence) based on observed prior co-occurrence is perfectly legitimate, and that we only run into trouble when we try and move from what is observed to what is merely inferred, that is, when we use indicative signs. But on a more careful reading, we can see how Sextus and Cicero overlap to a certain extent.

What they share is an emphasis on the distinction between two types of sign, one type based on past observations, and one type based on rational or theoretical considerations. But this distinction does *not* play out in the formalism of sign inference, strictly speaking. We still stand the inferences up as conditionals in their original form,

If x happens, then y will happen. But x happened. Therefore y will happen.

---

[45]καὶ μὴν πᾶν αἰσθητόν, ὡς ἡ κλῆσις παρίστησιν, αἰσθήσει ληπτόν ἐστι, τὸ δὲ σημεῖον ὡς σημεῖον οὐκ αἰσθήσει λαμβάνεται, ἀλλὰ διανοίᾳ [*Adv. math.*, viii.207].

[46]καρδίας τε τρῶσιν θεασάμενοι μέλλοντα θάνατον προγιγνώσκομεν [*Adv. math.*, viii.153].

[47][...] πολλάκις ἀλλήλοις συνεζευγμένα παρατηρήσαντες [...] [*Adv. math.*, viii.152]. It may be relevant to mention here that Sextus was initially trained as an Empiricist physician. But note that this wound-to-the-heart example is the exact same one Philodemus uses to show why the Stoics rejected induction, and other medical examples are fairly common in texts on sign inference, since both prognosis and diagnosis are fairly clear examples of probably common sign inferences.

The difference now is that we must, say Sextus and Cicero, pay very close attention to the *relationship* between the antecedent and consequent terms of the conditional. How is the prediction of y related to the current instantiation of the sign, x? If x is a sign of y because they have often been observed together in the past, then we have the kind of sign Cicero rejects but Sextus accepts. And if x is a sign of y because they are somehow related in the causal nexus, then we have the kind of sign that Cicero says Quintus needs and Sextus says is unacceptable. The validity of the inference has now moved beyond the formal structure of the argument here, and takes into account extra-formal considerations about the content of the argument. The acceptability of the predictive inference has been constrained by physics for Cicero, and empirical evidence for Sextus.

Conclusion

Now we have come full circle. Look back at the definition Sextus gave us of the Stoic *sign:*

> They (the Stoics) say that the *sign* is the proposition, in the antecedent of a valid conditional, that reveals the consequent. [*Adv. math.*, viii.245]

We can see now that Sextus' wording is quite careful here, and some of the nuances of the definition begin to emerge. The sign is a sign if and only if it *reveals* its consequent. But this revealing is not just a formal relation. It is deeply dependent on a physical understanding of the relationship between the terms. So just as in the case where logical bivalency had direct consequences for physical causation and fate, we see that logic and physics have a good deal of interdependence in Hellenistic philosophy where issues around prediction are concerned.

Primary Sources.

[Aphr. De fato]	**Alexander of Aphrodisias**, De fato, *in:* [Sha83]
[An. pr.]	**Aristotle**, Analytica Priora, *in:* [Bek31]
[Rh.]	**Aristotle**, Rhetorica, Ars Rhetorica, *in:* [Bek31]
[De div.]	**Cicero**, De divinatione, *in:* [Ax38], *also in:* [Pea79]
[De fin.]	**Cicero**, De finibus bonorum et malorum libri quinque, *in:* [Rey98]
[De fato]	**Cicero**, De fato, *in:* [Ax38]
[De nat. deor.]	**Cicero**, De natura deorum, *in:* [PlaAx33]
[De part. or.]	**Cicero**, De partitione oratoria, *in:* [Rac42]
[Praep. evan.]	**Eusebius**, Praeparatio evangelica, *in:* [dPl79]
[De caus. cont.]	**Galen**, De causis continentibus, *in:* [Sch$_1$+69]

[De caus. pr.]	**Galen**, De causis procatarticis, *in:* [Han98b]
[De. rer. nat.]	**Lucretius**, De rerum natura libri sex, *in:* [Bai63]
[Phil. De sign.]	**Philodemus**, De signis, *in:* [DLa$_1$DLa$_0$78]
[Phil. Rhet.]	**Philodemus**, Φιλοδήμου περὶ ῥητορικῆς, libri primi et secundi, *in:* [Lon$_1$77]
[Adv. math.]	**Sextus Empiricus**, Adversus mathematicos, *in:* [Bur33-49]

References.

[Alg+99]	Keimpe **Algra**, Jonathan **Barnes**, J. **Mansfeld**, and Malcolm **Schofield**, (*eds.*), The Cambridge History of Hellenistic Philosophy, Cambridge University Press 1999
[All01]	James **Allen**, Inference from Signs: Ancient Debates about the Nature of Evidence, Clarendon Press 2001
[AnnGri89]	Julia **Annas** and Robert H. **Grimm** (*eds.*), Oxford Studies in Ancient Philosophy, Supplementary Volume: 1988, Oxford University Press 1989 [Oxford Studies in Ancient Philosophy]
[Ax38]	Wilhelm **Ax** (*ed.*), De divinatione, Teubner 1938 [M. Tvlli Ciceronis Scripta qvae manservnt omnia 46]
[Bai63]	Cyril **Bailey** (*ed.*), Lucretius, De rerum natura libri sex, Clarendon Press 1963
[Bal85]	Mariano **Baldassarri**, Introduzione alla logica stoica, Noseda 1985
[Bar$_0$89]	Jonathan **Barnes**, Epicurean Signs, *in:* [AnnGri89, p.91-134]
[Bar$_0$99]	Jonathan **Barnes**, Logic and Language, Introduction, *in:* [Alg+99, p.65-76]
[Bar$_0$BobMig99]	Jonathan **Barnes**, Susanne **Bobzien**, and Mario **Mignucci**, Logic, *in:* [Alg+99, p.77-176]
[Bar$_0$+82]	Jonathan **Barnes**, J. **Brunschwig**, Myles F. **Burnyeat**, and Malcolm **Schofield**, (*eds.*), Science and Speculation, Cambridge University Press 1982
[Bar$_1$02]	Jeffrey **Barnouw**, Propositional Perception: Phantasia, Predication, and Sign in Plato, Aristotle, and the Stoics, University Press of America 2002
[Bea86]	Mary **Beard**, Cicero and Divination: The Formation of a Latin Discourse, **Journal of Roman Studies** 76 (1986), p.33-46
[Bec57]	Oskar **Becker**, Zwei Studien zur antiken Logik, Harrassowitz 1957
[Bek31]	Immanuel **Bekker** (*ed., trans.*), Aristotelis opera, Graece, ex recensione Immanuelis Bekkeri, Reimer 1831
[Bob86]	Susanne **Bobzien**, Die stoische Modallogik, Königshausen und Neumann 1986
[Bob98]	Susanne **Bobzien**, Determinism and Freedom in Stoic Philosophy, Clarendon Press 1998

[Bur₀33-49] Robert Gregg **Bury** (*ed., trans.*), Sextus Empiricus, 4 volumes, Heinemann 1933-49 [The Loeb Classical Library 273, 291, 311, 382]

[Bur₁82] Myles F. **Burnyeat**, The Origins of Non-Deductive Inference, *in:* [Bar₀+82]

[Cra88] William Lane **Craig**, The Problem of Divine Foreknowledge and Future Contingents from Aristotle to Suarez, E.J. Brill 1988

[Fre₀96] Dorothea **Frede**, How Sceptical Were the Academic Sceptics? *in:* [Pop96, p.1-26]

[Fre₁74] Michael **Frede**, Die stoische Logik, Vandenhoeck & Ruprecht, 1974

[Fre₁80] Michael **Frede**, The Original Notion of Cause, *in:* [Sch₀Bur₁Bar₀80, p.217-249]

[Fre₁Str96] Michael **Frede** and Gisela **Striker** (*eds.*), Rationality in Greek Thought, Oxford University Press 1996

[Glu95] John **Glucker**, Probabile, Veri Simile, and Related Terms, *in:* [Pow95, p.115-143]

[Goa68] Robert J. **Goar**, The Purpose of the *De divinatione*, **Transactions and Proceedings of the American Philological Association** XCIX (1968), p.241-248

[Gui84] François **Guillaumont**, Philosophe et augure: recherches sur la théorie cicéronienne de la divination, Latomus 1984 [Collection Latomus 184]

[Han98a] Robert J. **Hankinson**, Cause and Explanation in Ancient Greek Philosophy, Oxford University Press 1998

[Han98b] Robert J. **Hankinson**, Galen on Antecedent Causes, Cambridge University Press 1998

[Hay69] William H. **Hay**, Stoic Use of Logic, **Archiv für Geschichte der Philosophie** 51 (1969), p.145-157

[Kah69] Charles H. **Kahn**, Stoic Logic and Stoic LOGOS, **Archiv für Geschichte der Philosophie** 51 (1969), p.158-172

[Kne₁Kne₀62] William **Kneale** and Martha **Kneale**, The Development of Logic, Clarendon Press 1962

[Kro00] Brian A. **Krostenko**, Beyond (Dis)belief: Rhetorical Form and Religious Symbol in Cicero's *De divinatione*, **Transactions of the American Philological Association** 130 (2000), p.353-391

[DLa₁DLa₀78] Phillip Howard **De Lacy** and Estelle Allen **De Lacy** (*eds.*), Philodemus on Methods of Inference, Bibliopolis 1978

[Lan00] Marc **Lange**, Natural Laws in Scientific Practice, Oxford University Press 2000

[Leh00] Daryn **Lehoux**, Parapegmata, or, Astrology, Weather, and Calendars in the Ancient World, being an examination of the interplay between the Heavens and the Earth in the Classical and Near-Eastern Cultures of Antiquity, with particular reference to the Regulation of Agricultural Practice, and to the Signs and Causes of Storms, Tempests, &c, University of Toronto 2000, *Ph.D thesis*

[Lon₀88] Anthony A. **Long**, Reply to Jonathan Barnes "Epicurean Signs", **Oxford Studies in Ancient Philosophy**, Supplement 1 (1988), p.135-144

[Lon₀Sed87]	Anthony A. **Long** and David N. **Sedley**, The Hellenistic Philosophers, Cambridge University Press 1987
[Lon₁77]	Francesca **Longo Auricchio** (*ed.*), Ricerche sui papiri Ercolanesi, Naples 1977
[Mac02]	Ian **Maclean**, Logic, Signs, and Nature in the Renaissance: the Case of Learned Medicine, Cambridge University Press 2002
[Man87]	Giovanni **Manetti**, Le teorie del segno nell'antichità classica, Bompiani 1987
[Mat53]	Benson **Mates**, Stoic logic, University of California Press 1953
[Mig96]	Mario **Mignucci**, Ammonius on Future Contingent Propositions, *in:* [Fre₁Str96, p.279-310]
[Mue69]	I. **Mueller**, Stoic and Peripatetic Logic, **Archiv für Geschichte der Philosophie** 51 (1969), p.173-187
[Net99]	Reviel **Netz**, The Shaping of Deduction in Greek Mathematics: a Study in Cognitive History, Cambridge University Press 1999
[Pea79]	Arthur Stanley **Pease** (*ed.*), M. Tulli Ciceronis de divinatione, Arno Press 1979
[dPl79]	Edouard **des Places** (*ed.*), La préparation évangélique / Eusèbe de Césarée, Edition du Cerf 1979
[PlaAx33]	Otto **Plasberg** and Wilhelm **Ax** (*eds.*), Cicero, De natura deorum, Teubner 1933
[Pop96]	Richard H. **Popkin** (*ed.*), Scepticism in the History of Philosophy, Kluwer 1996
[Pow95]	Jonathan G.F. **Powell** (*ed.*), Cicero the Philosopher, Clarendon Press, 1995
[Rac42]	Harris **Rackham** (*ed., trans.*), De oratore, De fato, Paradoxa stoicorum, De partitione oratoria, 2 volumes, Heinemann, 1942 [Loeb Classical Library 348-349]
[Ras98]	Susanne William **Rasmussen**, Cicero's Stand on Prodigies: A Non-Existent Dilemma? (1998), *in:* [WilIsa00, p.9-24]
[Rey98]	Leighton Durham **Reynolds**, M. Tulli Ciceronis De finibus bonorum et malorum libri quinque, Clarendon 1998
[Sam59]	Shmuel **Sambursky**, The Physics of the Stoics, Routledge Kegan Paul 1959
[Sch₀86]	Malcolm **Schofield**, Cicero For and Against Divination, **Journal of Roman Studies** 76 (1986), p.47-64
[Sch₀Bur₁Bar₀80]	Malcolm **Schofield**, Myles F. **Burnyeat**, and Jonathan **Barnes**, Doubt and Dogmatism, Clarendon Press 1980
[Sch₁+69]	H. **Schoene**, K. **Kalbfleisch**, J. **Kollesch**, D. **Nickel**, and G. **Strohmaier** (*eds.*), Galeni De partibus artis medicativae, De causis contentivis, De diaeta in morbis acutis secundum Hippocratem libellorum versiones Arabicae, edidit et in linguam Anglicam vertit M. Lyons; De partibus artis medicativae, De causis contentivis libellorum editiones alterius ab H. Schoene alterius a K. Kalbfleisch curatae, retractaverunt J. Kollesch, D. Nickel, G. Strohmaier, Berlin 1969 [Corpus Medicorum Graecorum, Supplementum Orientale II]
[Sha83]	Robert W. **Sharples** (*ed., trans.*), Alexander of Aphrodisias on Fate, Duckworth 1983

[vFr83]   Bas C. **van Fraassen**, Theory Confirmation: Tension and Conflict, *in:* [WeiCze83, p.319-329]

[WeiCze83]   Paul **Weingartner** and Johannes **Czermak** (*eds.*), Epistemology and philosophy of science: proceedings of the 7th International Wittgenstein Symposium, 22nd to 29th August 1982 in Kirchberg Am Wechsel (Austria), Holder-Pichler-Tempsky 1983

[WhiRus35]   Alfred North **Whitehead** and Bertrand **Russell**, Principia mathematica, Second Edition, Cambridge University Press 1935

[WilIsa00]   Robin Lorsch **Wildfang** and Jacob **Isager** (*eds.*), Divination and portents in the Roman world, Odense University Press 2000 [Odense University Classical Studies 21]

Received: May 15th, 2003;
In revised version: December 1st, 2003;
Accepted by the editors: January 7th, 2004.

Benedikt **Löwe**, Volker **Peckhaus**, Thoralf **Räsch** (*eds.*)
Foundations of the Formal Sciences IV
The History of the Concept of the Formal Sciences

The Logical Background of Pragmatism
C. S. Peirce's Lattice Theory as a Formal Framework for his Pragmatic Account of Meaning

JUSTUS LENTSCH[*]

Institute for Science & Technology Studies (IWT)
Bielefeld University
Universitätsstraße 25
33615 Bielefeld, Germany
E-mail: `lentsch@iwt.uni-bielefeld.de`

> "If thought were not connected with action,
> it is difficult to say what it would be"
> (CSP to Victoria Lady Velby, October 12, 1904)

ABSTRACT. This paper reconstructs the connection between the logic of relations and the pragmatic theory of meaning which the American logician and philosopher Charles Sanders Peirce (1839-1914) tacitly assumes in his epistemology: How can Pierce's formal concepts serve the understanding and improvement of reasoning and empirical knowledge acquisition? This paper shows that the pragmatic maxim achieves the clarification of the meaning of abstract concepts by establishing a relation between the formal and material properties of reasoning and empirical knowledge acquisition processes, implicitly drawing on the formal framework of order- and lattice-theoretical concepts.

1 Introduction

This paper reconstructs the connection between the logic of relations and the pragmatic theory of meaning which the American logician and philosopher

---

[*]The ideas presented in this paper developed with comments and questions from Helmut Pape, Nathan Houser, Karin Hartbecke, Michael Hoffmann, Martin Neumann and the team at the *Peirce Edition Project*. I would also like to thank two anonymous reviewers for providing very helpful comments and suggestions. Further thanks go to Christian Roerecke, the Institute for Science & Technology Studies (IWT) at Bielefeld University and, finally, the *Deutsche Forschungsgemeinschaft* for financial support.

Charles Sanders Peirce (1839-1914) tacitly assumes in his epistemology. It will be shown that this tacit assumption is the link that connects Peirce's pragmatism with the rest of his philosophy. This systematic approach has its intellectual roots[1] in a kind of "logical idealism", that is, the thesis that the most general logical features of reasoning are reflected in the processes of nature itself.[2] Another important move in this idealistic strategy is a kind of process ontology.[3] This amounts to the claim that the basic unities of the universe are not entities but processes, *i.e.*, relations and operations. The logic of relations provides for this reason the logical patterns that constitute the very core of Peirce's philosophy.

Therefore, Peirce conceived logic to be the science of reasoning. Improving on George Boole's "Algebra of Logic" and Augustus De Morgan's relative operations, Peirce has developed a calculus or logic of relations primarily as a tool for the understanding and improvement of reasoning and not in order to secure the foundations of mathematics like the logicists. But how can the formal concepts developed by Peirce in the course of his work on symbolic algebra and the logic of relations really serve the understanding and improvement of reasoning? How is it that we are able to find out from the consideration of what we know, something else, which we do not know? (*Cf.* [*EP*, I.111]; [*PEP*, III.244, 1877].) Those questions are implicitly addressed in the *Illustrations of the Logic of Science* published 1877/78 in the *Popular Science Monthly*. Simultaneously with his work on the logic of relations between 1870 and 1885 Peirce suggests in these essays a methodological rule for the clarification of the meaning of concepts. His general objective in developing this rule is to secure the viability of empirical knowledge acquisition processes. Later on his rule became famous as the "Pragmatic Maxim". Peirce himself emphasizes that his pragmatism is a "logical doctrine" and provides a "principle of logical analysis" [*CP*, 6.490, 1908]. Presenting pragmatism as a branch of logic, Peirce claims that he has first "preached this principle as a sort of logical gospel" [*CP*, 6.482, 1908].[4] Peirce's emphasis on the logical nature of pragmatism as well as the simultaneity in origin raises the question, whether there is indeed a systematic connection between pragmatism and the logic of relations. This question will be addressed by reconstructing the logical background of pragmatism and by making its formative principles explicit. To achieve this objective, the paper proceeds as follows:

Firstly, a brief outline of the essential and most remarkable features of

---

[1] *Cf.* most notably [Pap97, Len$_0$01].
[2] *Cf.* for Peirce and his affinity to a Schellingian objective idealism, *e.g.*, [Rey02].
[3] *Cf.*, *e.g.*, [Res96].
[4] *Cf.* also [*MS*, 137, 1904].

Peirce's logic of relations will be given. Secondly, the logical background of pragmatism and the logical structure of the pragmatic maxim in particular will be reconstructed. It will turn out, that Peirce's pragmatic maxim implicitly employs as formative principles the notion of a partial ordering, which he has invented in his papers on symbolic algebra and the logic of relations from 1870 onwards. Finally, it will be examined how Peirce operationalizes these formal concepts in his Pragmatism as to a methodology for clarifying the meaning of abstract concepts. Moreover, it will be shown, that — given the formal background of order-theoretical concepts — pragmatism to a large extend legitimately claims to describe a viable strategy of knowledge acquisition.

2 Peirce's Logic of Relations and the "Relational Stance"

In his report "The Century's Great Men of Science" for the Smithsonian Institution, 1901, Peirce presents the logic of relations as one of the greatest achievement of 19th century's science:

> [...] In the nineteenth century Boole created a method of miraculous fruitfulness, which aided in the development of the logic of relatives, and threw great light on the doctrine of probability, and thereby upon the theory and rules of inductive reasoning. De Morgan added an entirely new kind of syllogism, and brought the logic of relatives into existence, which revolutionizes general conceptions of reasoning. [Pei01, p.696 sq]

The value of Boole's and De Morgan's work is seen by Peirce not *sui generis* but in their contribution to the overall progress of the development of formal logic, culminating in his, Peirce's own, logic of relations. Peirce considers the logic of relations so important because it revolutionizes, as he writes, "our general conceptions of reasoning" [Pei01, p.696sq]. But what are the features of the logic of relations which predestines it to revolutionize "general conceptions of reasoning"?

Due to the pioneering work in algebraic logic by Boole, Peirce, Schröder, Tarski, Brouwer and others it is nowadays widely acknowledged that many patterns of logical reasoning follow order- and lattice-theoretical principles. Obviously, a partial-ordering is a very general structural feature of every logical system. [5]

---

[5]Following Clarence Irving Lewis, Bertrand Russell, Aron Gurwitsch and others this feature of modern systems of symbolic logic can be traced back to Leibniz and his fundamental inesse-relation (*cf.*, *e.g.*, [Gur74, p.75sq], [Lew60, p.5sq]). Leibniz's inesse-relation denotes that the predicate is contained in the subject term. According to Leibniz, the validity of judgements can be reduced to the inesse-relation. This leads to Leibniz's conviction that all knowledge is analytical. Countering critics, who maintain that Leibniz's

But, as Garrett Birkhoff remarks, it was Peirce who first gave an explicit and axiomatic formulation of a partial ordering and a lattice in his *On the Algebra of Logic* [*PEP*, IV.163-209, 1880] (*cf.* [Bir40, p.ii,5,16]) as mathematical objects in their own right. [6]

Using a partial ordering as a formal framework Peirce formulates in his *Description of a Notation for the Logic of Relatives* (DNLR), 1870, a system of propositional logic based on implication and negation which anticipates some of the main features of modern systems of natural deduction and sequent calculus.[7] Ten years later, in his *On the Algebra of Logic* (1880), he uses a lattice theoretic formulation of Boolean Algebra as a formal framework for his logic of relations. In his 1870 paper Peirce develops a logic of relations of arbitrary valency. In this way he, for the first time in history, formulates the syntax of a complete system of propositional logic, even two years before Frege (*cf.* [Bur05]). And by 1885 Peirce has invented a full syntax of quantificational logic, too.[8] Already in the early 1870 paper Peirce takes a "relational stance" which is of major importance for understanding the logical background and, consequently, the functioning of pragmatism as a methodological strategy for clarifying the meaning of abstract concepts.

The relational stance or point of view corresponds to the following features of Peirce's early 1870 paper: (1.) the introduction of three classes of terms, "absolute", "simple relative" and "conjugative" terms, (2.) a logical relative product[9], complemented by a binary "comma operation" and (3.)

---

system has serious deficits caused by the asymmetry of this relation, Volker Peckhaus shows that already Leibniz's system, comprising one direct operation, the inverse and negation, is complete in the modern, lattice-theoretical sense (*cf.* [Pec97, p.56*sq*] and also [Len$_1$84, p.191*sq*]).

[6]But in the same vein Birkhoff attenuates his appraisal of Peirce's achievements by critically noticing that Peirce lacks a proof of the law of distribution. Therefore, he attributes to Peirce the view that all lattices have to be distributive lattices (*cf.* [Bir40, p74]) and concludes that Peirce has no general conception of a lattice (*cf.* also [Meh79, p.48]). On the contrary, it has been argued that this view is actually mistaken ([Hou91] and [CraRob69]). Nathan Houser, most notably, has reconstructed the missing proof of both parts of the law of distributivity as it is given by Peirce in an earlier version of his 1880 paper (*cf.* [Hou91, p.18*sq*]; [*MS*, 575, 1880]) and from the correspondence between Edward V. Huntington and Peirce in December 1903 [*MS*, L210, 1903]. Following Houser, via C.I. Lewis (*cf.* [Lew60, p.119*sq*]), who received it from Huntigton, "[...] Peirce's proof of the problematic part of the law of distribution came into the mainstream of the development of formal logic" ([Hou91, p.22]).

[7]*Cf.* [Bra00, p.6].

[8]*Cf.* [Pei80], [Pei83], [Pei85b] and [Pei85a].

[9]Peirce also gives an extension of the Boolean Addition to a full fledged logical addition. As the Boolean partial addition or "exclusive or" is applicable only if the alternatives were exclusive, Peirce replaces it by a logical addition or "inclusive or" — a modification William Stanley Jevons had already made in 1864 (*cf.* for the latter point [Pec99]). Thereby logical multiplication becomes the dual of logical addition. This du-

a basic binary relation characterized as a partial order.

2.1 Three Classes of Terms

Peirce builds on a logical interpretation of relational algebra. One of its main features is the introduction of three classes of logical terms, "absolute", "simple relative" and "conjugative" terms. The first class of terms corresponds to the "objects as such" ("absolute terms"), as "horse", "tree" or "man" (*cf.* [*PEP*, II.365, 1870]). The second class of terms "[...] embraces terms whose logical form involves the concept of a relation. They regard an object as over against another, that is as relative" [*PEP*, II.365, 1870] like "father-of" ("simple relative"). The third class of terms are those whose logical form involves the conception "of bringing things into relation" or the "objects as medium or third" ("conjugative" terms), *e.g.* "giver — of — to — ..." or "buyer of — for — from — ...".[10] Peirce claims further that all relations of higher valency can be obtained as a logical product of those three fundamental types of terms.

2.2 The Logical Product

Peirce introduces a logical multiplication, xy, as an associative operation which distributes over addition. Extending Boole, Peirce interprets the multiplication (xy) logically as a relative product or application of relations:

> I shall adopt for the conception of multiplication the application of a relation, in such a way that, for example, lw shall denote whatever is lover of a woman. [*PEP*, II.369, 1870]

But if multiplication is defined as the application of a relative term how, then, can it be applied to absolute terms? In order to define logical multiplication even for absolute terms, Peirce introduces the comma-operation. It converts an absolute term into a relative term and thus increases its valency.

In this way, Peirce demonstrates how the logical multiplication between absolute terms can be defined in terms of the application of relative terms. For example, by the application of the comma-operation the term "man" is converted into "man-that-is- ...". To get for example "man that is black" Peirce applies "man-that-is-..." (m) to the absolute term "black" (b) and thus gets "man that is black", in comma-notation: (m, b).[11] That is, what

---

ality corresponds to the validity of the law of distribution. Using the logical addition Peirce introduces the second half of the law in [Pei68]. In his review of Frege's "*Begriffsschrift*", Schröder refers to Peirce's 1868 paper and emphasizes the importance of Peirce's improvement [Sch80, p.85].

[10]Structurally these three classes of logical terms correspond to Peirce's three fundamental categories, firstness, secondness and thirdness.

[11]The term (m, b) corresponds extensionally to the intersection of the class denoted by "man" and "black", *i.e.* $(m, b) = m \cap b$. Thus (m, b) is an absolute term which denotes all man that are black; *cf.* [Bra00, p.35].

we nowadays would consider as a binary relation of the two relata "man" (m) respectively "black" (b) and what extensionally denotes the intersection of the two classes man and black, is considered by Peirce as an operator (m, b) that yields a relative term m applied to b. Thus, Peirce considers the multiplication of relative terms as the primitive logical operation. Multiplication for absolute terms, then, is explained as a syntactically more complex expression (m, b) involving a comma compared with simple juxtaposition than that of relative terms. Summarizing, when Peirce subsequently defines the multiplication of absolute terms it occurs as a derived concept and the resulting relation as a complex outcome of this operation.

2.3 The Order-Theoretical Characterization of the Implication

Extending Boole's Algebra in his 1870 paper Peirce introduces a special symbol for the inclusion or implication, "—<", as an operator for the material implication. He defines inclusion or implication as a partial order, *i.e.*, when ever it is used "—<" denotes a transitive, reflexive and anti-symmetric relation. The second development worth mentioning is, that Peirce gives priority to inclusion over equality. He emphasizes that if a partial order R is given, the equality relation might be defined as aRb and bRa but not vice versa. Thus, he defines equality in terms of the inclusion or implication relation:

> I use the sign —< in place of ≤. [...] It is universally admitted that a higher conception is logically more simple than a lower one under it. Whence it follows from the relations of extension and comprehension, that in any state of information a broader concept is more simple than a narrower one included under it. Now all equality is inclusion in, but the converse is not true; hence inclusion in is a wider concept than equality, and therefore logically a simpler one. [*PEP*, II.360, 1870]

This is reinforced in his 1880 Algebra of Logic:

> There is a difference of opinion among logicians as to whether —< or = is the simpler relation. But in my paper on the Logic of Relatives, I have strictly demonstrated that the preference must be given to —< in this respect. The term simpler has an exact meaning in logic; it means that whose logical depth is smaller; that is, if one conception implies another, but not the reverse, then the latter is said to be simpler. [Pei80, p.20][12]

---

[12] Peirce also gives primacy to a partial order over a strict order, *i.e.*, a transitive, anti-symmetrical and irreflexive relation, because one can define a strict order from a partial order but not vice versa:

> *Being less than* is being small as with the exclusion of its converse. To say that x ≤ y is to say that $x \Rightarrow y$, and that it is not true that $y \Rightarrow x$.
> [*PEP*, II.360, 1870]

Identity, therefore, is defined in terms of a partial ordering relation by comparing two elements. As it will turn out later, this is a logical feature of the pragmatic maxim, too.

The invention of the notion of a partial order is an important step towards Peirce's lattice theoretical treatment of Boolean Algebra and the development of an inferential calculus in his 1880 paper. Drawing on the conceptual improvements and the material of his 1870 paper in *On the Algebra of Logic* (1880), Peirce uses a lattice theoretical treatment of Boolean Algebra to develop a sequent calculus of implicational propositional logic. In the chapter on *Internal Multiplication and the Addition of Logic* Peirce gives a scheme of two operations which he calls "non-relative addition" and "multiplication" (*cf.* [PEP, IV.182*sq*, 1880]), *i.e.*, disjunction ("+") and conjunction ("×"). He shows that these operations satisfy the conditions of associativity, commutativity and idempotency and hence form a lattice. A lattice can be characterized algebraically[13] as a set with two binary operations which are associative, commutative and idempotent. Peirce's two operations anticipate the elimination and introduction rules of contemporary systems of Gentzen-style sequent calculus of natural deduction. The philosophically important point is that Peirce treats the concepts of a partial ordering and a lattice and, hence, also inference relations as structurally basic.

To sum up, formulating his logic of relations, Peirce uses a partial ordering as a formal framework and adopts a relational stance. On a syntactical level, the relational stance amounts to the claim that expressions denoting relations are more fundamental than atomic or singular terms and their logical composition. On a semantic level the relational point of view amounts to the claim that we can understand a complex expression without having to grasp the meaning of its compounds before and applying the logical vocabulary separately. Thus, the reason why the logic of relations is most appropriate to "revolutionize general conceptions of reasoning" is, according to Peirce, that the fundamental operations of reasoning are relational in character and follow order-theoretic patterns.

3 The Logical Background of Pragmatism

3.1 The Logical Structure of Pragmatism and the Logic of Relations

What is the logical background of pragmatism? And what is the role the formal notions and ideas originating in Peirce's logic of relations play within pragmatism? In the following it is argued that those concepts serve as formative principles. That is, they function as a kind of background theory

---

[13] The other way is to define a lattice as a partially ordered set where every two members have a common least upper and a greatest lower bound.

about the logical structure of experience which pragmatism operationalizes to a methodological rule for the clarification of the meaning of concepts.

In the following chapter, a brief logical analysis of the pragmatic maxim will be given. It will turn out, that the formal concept of a partial ordering constitutes a formal framework for both, the formulation of Peirce's logic of relations and his pragmatic account of meaning. In Peirce's pragmatism (but also in his theory of categories and his semiotics) the notions of a partial order (and, as cannot be shown in further detail here, of a lattice) gain philosophical significance as implicit formative principles for the design of theories about the actual processes connecting reasoning, experience and cognitive actions with our environment.[14] This is argued by demonstrating that Peirce's pragmatism, seen as a methodological strategy for the clarification of the meaning of concepts, is informed by the ideas and notions he has developed simultaneously within the context of his logic of relations. In particular, this thesis holds for the relational point of view and the invention of a partial ordering.

In order to improve our understanding of the process of reasoning it is crucial, Peirce argues, to understand the internal relationship between formal logic, reasoning and action. That is the point where pragmatism comes in. But what is pragmatism? Pragmatism claims the primacy of practice over theory and emphasizes the relation to action even for theoretical purposes. Already the first versions of the pragmatic principle, which can be found in Peirce's review of Fraser's edition of the works of Berkeley, is formulated as a methodological rule to avoid the deceits of language by comparing words according to the differences their application makes: "Do things fulfill the same function practically? Then let them be signified by the same word. Do they not? Then let them be distinguished" [*EP*, I.102, 1871].

A few years later, in the second essay of the series *Illustrations of the Logic of Science* entitled *How to Make our Ideas Clear*, although without using its later name, Peirce introduces the pragmatic maxim as a rule to render our ideas clear:

> (Prag-Max:) Consider what effects, that might conceivably have practical bearings, we conceive the object of our conceptions to have. Then, our conception of these effects is the whole of our conception of the object.
> [*EP*, I.132, 1878]

Now, what is the logical structure of the pragmatic principle? The pragmatic maxim is a methodological rule: It suggests to replace one concept

---

[14]For the sake of brevity two important links have to be omitted: Peirce's theory of categories and his doctrine of signs. But it can be shown that the notions of a partial order and a lattice hold the key to Peirce's system of categories as well to his doctrine of signs, too.

by another in order to achieve the clarification of meaning with regard to the practice within which it is applied, that is, the practice of fixing and revising the beliefs containing the concepts at stake.

If someone wants to clarify the meaning of a concept, Peirce writes,

> [...] a "practical preference" will lead him to translate abstract expressions of concepts into expressions of his conceptions of their essential conditional "practical consequences" [...]   [*MS*, 300, 1905]

The pragmatic maxim suggests how to replace one belief by another in such a way that logically valid inferences are preserved and certain criteria of pragmatic relevance are met. From a relational point of view, to clarify the meaning of a concept means to explicate it in terms of inferential and order relations between the corresponding beliefs.

The pragmatic maxim is a normative rule in two respects:[15] First, it says how to replace one belief by another (*cf.* especially the early 1870 version given in the Berkeley review). Second, it establishes a relation between theoretical and practical beliefs.[16] The pragmatic maxim is a principle which says how to compare concepts about the same object according to the practice within which they are employed and how to explore the differences their application makes. Thus, crucial for an understanding of the function of the pragmatic maxim is the notion of comparability. It is the problem of (pragmatically) comparing two beliefs what Peirce is after in formulating his pragmatic maxim. That is, the pragmatic principle is not primarily a definition of the meaning of a concept in terms of the actions it gives rise to, as it is usually assumed.[17] Instead, it *explicates* the inherent meaning governing the practice of forming and revising beliefs about the same object. Now, the proposed relation of "pragmatic comparability" can be reconstructed in the following way:

> (Prag-Ord:) Given two co-referential concepts a and b of the same object and the corresponding beliefs $P(a)$ and $P(b)$ in which expressions of the two concepts figure as subject or predicate terms. Then we say that b is pragmatically over a with respect to the relation of pragmatic comparison if and only if we are ready to act on $P(b)$ whenever we are ready to act on $P(a)$.

Reconstructed as in (Prag-Ord), the pragmatic maxim induces a relation of a partial order between concepts of the same object by relating our beliefs in which these concepts occur. We can now reformulate Peirce's initial strategy:

---

[15] *Cf.* [Pap02b].

[16] Practical beliefs are beliefs about actions and perceptions. They guide our access to reality and, hence, are decisive for what we take as real, *cf.* [Pap02b].

[17] The only exceptions are [Kuh96], [Ols83], and [Pap98].

> (Prag-Clear:) To clarify the meaning of a concept means to place it in an order structure of other concepts of the same object by pragmatically comparing the beliefs about the same object in which the concept occurs.

The function of the rule (Prag-Max) is nicely illustrated by one of the examples given by Peirce himself: namely, the task of clarifying the meaning of the concept of transubstantiation and, hence, deciding the quarrel between Protestants and Catholics concerning the meaning of wine as one of the elements of the sacrament. Applying the pragmatic principle Peirce comes to the following conclusion:

> [...] we can [...] mean noting by wine but what has certain effects, direct or indirect, upon our senses; and to talk of something as having all the sensible characters of wine, yet being in reality blood, is senseless jargon.
> [EP, I.131, 1878]

What Peirce is saying here is that we can clarify the meaning of a concept by pragmatically comparing beliefs in which they figure according to the consequences they might give rise too. Therefore, it is the notion of comparability which is at the very core of pragmatism. In this way, the pragmatic maxim is a normative principle which says how to structure the inferential relations between our beliefs in such a way that they become order relations. The pragmatic maxim extends the order-theoretic view of the logic of relations to the case where concepts require clarification. Pragmatism is thus concerned with a method of comparing beliefs. Hence, the pragmatic principle, then, has to be conceived as a method to establish a relation of pragmatic comparability between our beliefs in the same way in which order-theory describes the general structure of logical inference.

3.2 Pragmatism and the Formal Structure of Knowledge Acquisition

Concluding, we shall ask on what grounds pragmatism can legitimately claim to describe a viable strategy of knowledge acquisition. Peirce's highly contested suggestion to this problem is the idea of the convergence of the process of inquiry towards reality. Peirce gets there by applying the pragmatic principle to the clarification of the meaning of the fundamental concepts of truth and reality itself:

"The opinion which is fated to be ultimately agreed to by all who investigate, is what we mean by the truth, and the object represented in this opinion is the real" [EP, I.139, 1878]. W. V. O. Quine challenges Peirce to make a "faulty use of numerical analogy [...] since the notion of a limit depends on that of 'nearer than' which is defined for numbers and not for theories" (cf. [Qui60]).

Quine's objection can be met if we consider a broader notion of an ordering: The numerical notions of convergence and of "nearer that" are themselves instances of the much more general structure of a partial ordering. And it is this general conception of a partial ordering which Peirce employs when he speaks of convergence towards truth and reality. But what does that mean for pragmatism considered as a strategy of knowledge acquisition?

Before we can turn to Peirce's answer and hence the function of pragmatism we will have to shed light on another misunderstanding: It is often assumed that the notion of a limit refers to a definite space-time point at which for every question an answer will be settled. But that is not what Peirce had in mind. Pragmatism only proposes that, if the notion of truth should have any meaning, it should make a difference for the practice of inquiry, namely, that further research will not alter the answer.[18] Pragmatism, though it makes the meaning of truth in the practice of inquiry explicit, gives no definition of "truth" in a strict sense. Peirce's suggestion to the problem of knowledge acquisition is, that the best we could do is to organise the transitions between subsequent beliefs according to the pragmatic maxim.[19]

In this way, Pragmatism operationalizes the formal concepts and ideas developed within Peirce's logic of relations as to a methodology for clarifying the meaning of concepts by placing them in the network[20] of already held beliefs about the same object with respect to the conceivable consequences it might give rise to. Pragmatism, thus, clarifies the meanings of concepts by establishing a relation between the formal and material properties of reasoning and empirical knowledge acquisition processes, implicitly drawing on the formal framework of order-theoretical concepts, because, as Peirce writes: "If thought were not connected with action, it is difficult to say what it would be" (Letter to Victoria Lady Velby, 12. October 1904).

Primary Sources.

[CP]   Charles **Hartshorne**, Paul **Weiss**, and Arthur W. **Burks** (eds.), Collected Papers of Charles Sanders Peirce, 8 volumes, Harvard University Press 1931-1958; reprint, manuscripts are referred to by volume, section and publication date.

---

[18] According to [KelGly89] Quines misunderstanding is due to a faulty permutation of quantifiers.

[19] This can be stated in terms of convergence on a lattice of belief contexts; cf. for a similar suggestion [BraRes80]. But this cannot be argued in detail here.

[20] Even a lattice in the formal sense; but this cannot be argued here. Cf., e.g., [BraRes80].

[EP]   Nathan **Houser** and Christian **Kloesel** (*eds.*), The Essential Peirce: Seleted Philosophical Writings Vol. I (1867-1893), Indiana University Press 1992; *reprint, manuscripts are referred to by volume, page number and publication date.*

[MS]   Richard S. **Robin**, Annotated Catalogue of the Papers of Charles S. Peirce, University of Massachusetts Press 1967; *Manuscripts are referred to by microfilm number and publication date.*

[PEP]   **Peirce Edition Project** (*ed.*), The Writings of Charles S. Peirce vol. I-V, Indiana University Press 1982-1989; *Manuscripts are referred to by volume, page number and publication date.*

References.

[Bir40]   Garett **Birkhoff**, Lattice Theory, American Mathematical Society 1940 [American Mathematical Society Colloquium Publications]

[Bra00]   Geraldine **Brady**, From Peirce to Skolem: a neglected chapter in the history of logic, North-Holland 2000

[BraRes80] Robert **Brandom** and Nicholas **Rescher**, The Logic of Inconsistency, Blackwell 1980

[BruFor97] Jacqueline **Brunning** and Paul **Forster** (*eds.*), The Rule of Reason, University of Toronto Press 1997

[Bur05]   Robert W. **Burch**, Charles Sanders Peirce, *in:* Edward N. Zalta (*ed.*), The Stanford Encyclopedia of Philosophy, Winter 2005, http://plato.stanford.edu/archives/win2005/entries/peirce/

[CraRob69] Henry H. **Crapo** and Don D. **Roberts**, Peirce Algebras and the Distributivity Scandal, **Journal of Symbolic Logic** 34 (1969), p.153-154

[Dru91]   Thomas **Drucker** (*ed.*), Perspectives on the History of Mathematical Logic, Birkhäuser 1991

[Gur74]   Aron **Gurwitsch**, Leibniz, Philosophie des Panlogismus, de Gruyter 1974

[Hou91]   Nathan **Houser**, Peirce and the Law of Distribution, *in:* [Dru91]

[Joh02]   P.N. **Johnson-Laird**, Peirce, Logic, Diagrams, and the Elementary Operations of Reasoning, **Thinking and Reasoning** 8 (2002), p.69-95

[KelGly89] Kevin T. **Kelly** and Clark **Glymour**, Convergence to the Truth and Nothing but the Truth, **Philosophy of Science** 56 (1989), p.185-220

[Kuh96]   Friedrich **Kuhn**, Ein anderes Bild des Pragmatismus: Wahrscheinlichkeitstheorie und Begründung der Induktion als maßgebliche Einflußgrößen in den "Illustrations of the logic of science" von Charles Sanders Peirce, Klostermann 1996 [Philosophische Abhandlungen]

[Len$_0$01] Justus **Lentsch**, Wissenschaft als Lebensform: Methodologischer Pragmatismus als Nexus von Logischer Form und Faktizität des Normativen, University of Hannover 2001; *unpublished Ph.D Thesis*

[Len₁84]	Wolfgang **Lenzen**, Leibniz und die Boolesche Algebra, **Studie Leibnitiana** 16 (1984), p.197-203
[Lew60]	Clarence Irving **Lewis**, A Survey of Symbolic Logic, Dover 1960 [University of California Press 1918]
[Meh79]	Herbert **Mehrtens**, Die Entstehung der Verbandstheorie, Gerstenberg 1979 [Arbor scientiarum. Beiträge zur Wissenschaftsgeschichte]
[Ols83]	M. **Olshewski**, Peirce's Pragmatic Maxim, **Transactions of the C.S. Peirce Society** 29 (1983), p.199-210
[Pap97]	Helmut **Pape**, The Logical Structure of Idealism, *in:* [BruFor97]
[Pap98]	Helmut **Pape**, Peirce and his Followers, *in:* [PosSeb98]
[Pap02a]	Helmut **Pape**, Der dramatische Reichtum der konkreten Welt: der Ursprung des Pragmatismus im Denken von Charles S. Peirce und William James, Velbrück Wissenschaft 2002
[Pap02b]	Helmut **Pape**, Pragmatism and the Normativity of Assertion, **Transactions of the C.S. Peirce Society** 58 (2002), p.521-542
[Pec97]	Volker **Peckhaus**, Logik, *mathesis universalis* and allgemeine Wissenschaft: Leibniz und die Wiederentdeckung der formalen Logik im 19. Jahrhundert, Akademie-Verlag 1997
[Pec99]	Volker **Peckhaus**, 19th Century Logic Between Philosophy and Mathematics, **Bulletin of Symbolic Logic** 5 (1999), p.433-450
[Pei68]	Charles S. **Peirce**, On an Improvement in Boole's Calculus of Logic, **Proceedings of the American Academy of Arts and Sciences** 7 (1868), p.250-261
[Pei80]	Charles S. **Peirce**, On the Algebra of Logic, **American Journal of Mathematics** 3 (1880), p.15-57
[Pei83]	Charles S. **Peirce**, The Logic of Relatives, Boston 1883
[Pei85a]	Charles S. **Peirce**, On the Algebra of Logic: A Contribution in the Philosophy of Notation, **American Journal of Mathematics** 7 (1885), p.180-196
[Pei85b]	Charles S. **Peirce**, On the Algebra of Logik (Continued), **American Journal of Mathematics** 7 (1885), p.197-202
[Pei01]	Charles S. **Peirce**, The Century's Great Men of Science, **The Smithsonian Institut** 1302 (1901), p.693-699
[PosSeb98]	Roland **Posner** and Thomas A. **Sebeok** (*eds.*), Ein Handbuch zu den zeichentheoretischen Grundlagen von Natur und Kultur, de Gruyter 1998
[Qui60]	Willard Van Orman **Quine**, Word and Object, Wiley 1960
[Res96]	Nicholas **Rescher**, Process Metaphysics, State University of New York Press 1996
[Rey02]	Andrew **Reynolds**, Peirce's Scientific Metaphysics, Vanderbilt University Press 2002
[Sch80]	Ernst **Schröder**, Review of "Begriffsschrift" by Gottlob Frege, **Zeitschrift für Mathematik und Physik** 25 (1880), p.81-94

Received: May 17th, 2003;
In revised version: March 15th, 2004;
Accepted by the editors: July 7th, 2004.

Benedikt **Löwe**, Volker **Peckhaus**, Thoralf **Räsch** (*eds.*)
Foundations of the Formal Sciences IV
The History of the Concept of the Formal Sciences

The Status of Logic in the Seventeenth Century

JAAP MAAT

Institute for Logic, Language and Computation
University of Amsterdam
Nieuwe Doelenstraat 15
1012 CP Amsterdam, The Netherlands
E-mail: j.maat@uva.nl

> ABSTRACT. In the seventeenth century, logic was a well-established discipline forming a standard part of learning. However, during this period the prestige of logic declined drastically, suffering attacks by influential writers. A general explanation for this can be found in the fact that logic was intimately connected with Aristotelianism. This paper seeks to contribute to a more specific explanation of the decline of logic's reputation by examining some of the arguments put forward by its critics.

1 Introduction

If one is looking for an overview of the development of logic in Western thought and consults some general survey of the subject, the following picture is likely to emerge. In Antiquity, a lot of interesting activity was going on, primarily centered around the towering figure of Aristotle, with some considerable additions made by the Stoics. The medieval period proper is largely uninteresting, but at the end of the Middle Ages, in the scholastic period, another series of important developments took place. Then, during the sixteenth, seventeenth and eighteenth centuries, very little happens that is worth mentioning, with just one notable exception, namely the work of Leibniz. At least a whole century before him, and another century after him, logic seems to be a dying, if not completely dead, subject. It is not until the nineteenth century that it comes to life again, starting with the work of Boole, getting stronger and thriving at the end of the nineteenth century in the work of Frege and others. This formed the start of a new era in which logic reached levels of sophistication and development it had never seen before.

This picture becomes apparent, for instance, from the classic work by Kneale and Kneale [Kne$_1$Kne$_0$62], which devotes 176 pages to Antiquity and 123 to the medieval period, while the fifteenth up to and including the eighteenth centuries are treated on 62 pages, more than one third (25) of which are concerned with Leibniz. The nineteenth century, including the work of Frege, is described on 158 pages, and another 220 pages deal with various developments "after Frege". Most conspicuous for present purposes is the long period between the Middle Ages and the nineteenth century, which receives such a relatively brief treatment. The authors explicitly justify this, stating that logic "no longer attracted the attention of many of the best minds. From the 400 years between the middle of the fifteenth and the middle of the nineteenth century we have in consequence scores of textbooks but very few works that contain anything at once new and good" [Kne$_1$Kne$_0$62, p.298]. A similar evaluation of this period can be found in other general histories of logic, such as Blanché 1970, who characterizes it as one in which logic was "asleep" [Bla70, p.169].

In this paper I do not wish to dispute the correctness of this picture. It is of course true that broad perspectives on history tend to overlook distinctions and events that were considered to be of the greatest significance in some particular period, and the present case is no exception. There were developments and controversies in the history of logic, such as the rise of Ramism in the sixteenth century, that were seen as revolutionary by its advocates and as disastrous by its opponents, but are treated by the broad-brush historian as minor events, unable to disturb a subject that is sound asleep. Furthermore, one might object that our understanding of what happened in the past may be obscured if too much emphasis is put on innovation, if modern standards of what is to be regarded as good logic are applied to a historical period in which these standards were different, and if history is viewed solely in terms of gradual progress and growth towards present-day insights. Nevertheless, I think it is quite legitimate to take a broad perspective and to characterize the period between the Middle Ages and the nineteenth century as one of stagnation if not decay. But I also think that this long period of near-death deserves more attention than it has received thus far. For it is a remarkable fact, which calls for an explanation, that logic as a discipline could be in such a miserable state for such a long time while its main insights and results survived unscathed. My own interest in this period originated from an unwillingness to believe that Leibniz could have pursued his logical interests entirely on his own, without encouragement or stimulation by at least a few of his contemporaries. I must add that thus far I have found little evidence to the contrary.

The present paper focuses on the seventeenth century. This century is the more interesting for our purpose, as this was a period in which revolutionary developments took place in many fields, especially mathematics, philosophy and natural science. The contrast with the lack of significant change in the field of logic is the more apparent. In an important sense logic was very far from dead in the seventeenth century. For every educated man was thoroughly familiar with the subject, as it formed a standard part of training at undergraduate level. Furthermore, a large number of textbooks on the subject were written and widely used. At the same time, logic gradually ceased to be a completely respectable discipline. A number of influential writers voiced vigorous attacks on its usefulness and proposed to discard most or all of it. A very general explanation for this disapproving view of logic is not hard to find, and has in fact often been given in the literature. Due to various broad developments such as Humanism and the rise of a new kind of science which challenged age-old assumptions about the structure of the universe, the once overriding influence of Aristotelianism was getting weaker and weaker. And clearly, logic was closely associated with Aristotelianism. For this reason, logic had a flavour of precisely that backward, authority-bound attitude about it that was detested by those who felt at home in the forward-looking, progress-oriented atmosphere that gained ever more ground.

This paper seeks to contribute to a more specific explanation of the decline of logic's reputation in the seventeenth century, by examining some of the arguments put forward by its critics. Special attention is paid to the assessment of the formal character of logic. It is argued that this was fully acknowledged by most writers on the subject, but that it was precisely this formal character that was viewed as a drawback rather than an advantage.

2 Bacon

One of the most famous and most influential attacks on logic was launched by Francis Bacon (1561-1626). In his *Novum Organum*, indicating by its title that the old, Aristotelian instruments for the advancement of learning were obsolete, he explained that logic is useless, as follows:

> XI As the sciences which we now have do not help us in finding out new works, so neither does the logic which we now have help us in finding out new sciences.
>
> XII The logic now in use serves rather to fix and give stability to the errors which have their foundation in commonly received notions than to help the search after truth. So it does more harm than good.
>
> XIV The syllogism consists of propositions, propositions consist of words, words are symbols of notions. Therefore if the notions themselves (which

> is the root of the matter) are confused and over-hastily abstracted from the facts, there can be no firmness in the superstructure. Our only hope therefore lies in a true induction. [*Bacon, Aphorisms*, p.48-49]

To fully appreciate the strength of this argument, it will be useful to take a brief look at the contents of an average logic textbook. Although there were dozens of different books of that sort, the contents overlapped to a large extent. In spite of considerable differences of emphasis, organization and breadth, they were all concerned with the exposition of Aristotle's logical doctrines. The additions made by scholastic writers to logical theory were not usually treated in them. In the sixteenth century, the writings of Peter Ramus had caused some stir about the organization of logical theory and about its subject matter. Ramus chose to omit various traditional doctrines such as the categories from logic while giving others, such as the topics, more prominence. Subsequent writers could be, and were in fact, distinguished according to their stance on Ramus's proposals. Thus Burgersdijck [*Burgersdijck*] divided logicians into three schools: first, Aristotelians, who adhered to the tradition, secondly, Ramists, and thirdly, those who mixed the Aristotelian canon with Ramus's teaching. The first school is also known as the 'Scholastics' and as the "Philippists" (after Philip Melanchton), and the third school is alternatively called the Philippo-Ramists, or the Systematics. Authors of the latter school restored what Ramus had left out, while giving more prominence to the subjects he had emphasized (*cf.* [How56], [Ris64], [Ash74]). The most widely used textbooks in the seventeenth century (such as [*Blundeville*], [San18], [*Burgersdijck*], [*Jungius*]) were either Aristotelian or Philippo-Ramist, so that the influence of what Ramus had intended to be a reform had largely fizzled out. There was thus an overall uniformity in logic textbooks which provides useful background to an understanding of Bacon's argument against logic.

The standard treatment of logic consisted of three major parts, which were commonly presented in the same fixed order. The first part dealt with terms, or words, or notions. Subjects treated in this part of logic included the so-called predicaments, that is, categories. One could usually find some version of Aristotle's theory of categories here, often enriched with illustrative tables. Related to this was the doctrine of predicables, that is, the definition of, and distinction between the terms genus, species, difference, accident, and property. The second part dealt with propositions, or judgments, that is, with the combination of terms into a more complex whole. The subjects treated here derived from Aristotle's work *De interpretatione*, and the theory of the quantity (universal or particular) and the quality (affirmative or negative) of propositions was a standard part of this. The third part, finally, dealt with discourse, or reasoning, or syllogisms, that is, with the combination of propositions into even more complex structures. This obviously covered the same ground as did Aristotle's Prior Analytics, and explained the various moods and figures of the categorical syllogism, among other things. These three parts were thus presented as forming a coherent structure, each part forming a level of complexity, starting with the theory of terms at the lowest level, the proposition one level up, and finally syllogisms at the top. In addition to these subjects, logic books commonly contained an exposition of the theory of definition, the topics, fallacies and other matters.

This organization of the contents of logic books had a profound influence on the way the entire subject was viewed at the time. Even with writers such as the messieurs of Port Royal who were quite happy to put a lot of the traditional Aristotelian material overboard, these three levels of complexity were of primary importance for their own work [Port Royal]. In their system, as in that of many others both before and after them, each of the three levels is connected with a specific operation of the mind (thus, e.g., [Jungius, Gass. Inst., Wallis]). Terms corresponded with the operation of conceiving, propositions with judgment, and syllogisms with reasoning. This connection between the elements of logical theory and various operations of the mind can be traced back at least as far as Thomas Aquinas [Ash74, p.28]. From a modern point of view, we tend to concentrate on the syllogism as the core of traditional logic, and to view the other doctrines as at best a historical curiosity. But to a seventeenth-century observer this was completely different. Syllogisms were just one element within a larger structure, in which it was firmly embedded. It seems a reasonable conjecture that this structure suggested to many that the validity of the inferences covered by syllogistic logic somehow rested on the soundness of the levels below.

Now this is precisely what Bacon is implying: he likens logic to a building, the firmness of which depends on the strength of its building-blocks. Given the structure of the standard presentation of logical theory, it seemed natural to assume that if the bricks, the terms signifying concepts, were unsound, the whole edifice was. Clearly, Bacon did not so much object to syllogistic logic by itself, as to its uselessness when it comes to the investigation of the natural world. All that syllogisms could offer, were formal patterns, which, without the right matter fed into them, were just empty husks.

One might add that in view of [Bacon, Aphorisms, XII], Bacon's objections went further than that. As he states there, logic hinders the detection of errors in our knowledge by giving them stability. This suggests that in Bacon's view the formal patterns of logic make wobbly things look reliable.

In this was way logic contributes to erroneously taking for granted what is in fact ill-defined and unsubstantiated. A related point is made in Aphorism XX, in which Bacon says that the human mind is tempted to jump to general conclusions while neglecting experiment, an evil that is increased by logic "because of the order and solemnity of its disputations".

The latter phrase illustrates a point which was important for many seventeenth-century criticisms of logic, but which can only be briefly touched upon here. This is that logic was as much regarded as the art of disputation as it was viewed as the art of thinking. Thus Webster and Locke (see below) expressed their distaste for the training in debating skills that formed part of logic teaching.

3 Descartes

Another extremely influential assault on the reputation of logic was made by Descartes, who took a different line of attack. In his *Rules for the Direction of the Mind*, he remarked:

> Some will perhaps be surprised that in this context, where we are searching for ways of making ourselves more skilful at deducing some truths on the basis of others, we make no mention of any of the precepts with which dialecticians [that is, logicians] suppose they govern human reason. They prescribe certain forms of reasoning in which the conclusions follow with such irresistible necessity that if our reason relies on them, even though it takes, as it were, a rest from considering a particular inference clearly and attentively, it can nevertheless draw a conclusion which is certain simply in virtue of the form. (...) Our principal concern here is thus to guard against our reason's taking a holiday while we are investigating the truth about some issue; so we reject the forms of reasoning just described as being inimical to our project. Instead we search carefully for everything which may help our mind to stay alert. [*Regulae*, Rule X (p.4-5)]

This type of criticism is based on an assumption about the way real knowledge is acquired, namely by chains of reasoning in which each step is as small as possible, so that it can be accompanied by a clear and distinct intuition that the step concerned is correct and free from error. From this perspective, the precepts of syllogistic reasoning amounted to quite the opposite. Thanks to their formal character, conclusions can be drawn mechanically, regardless of the contents of either the premises or the conclusion. As long as true premises of the right form are ordered in the right way, the truth of the conclusion can be relied upon without the mind's actively perceiving that the inference is correct. This is what Descartes expresses as our reason taking a rest or going on holiday, and in his view this is precisely what must be avoided at all times. The formalization of reasoning that traditional logic had achieved was viewed by Descartes, not as a remarkable result, but as a threat to clear thinking, which to him is identical with conscious and active thinking.

To this, Descartes added another observation:

> But to make it even clearer that the aforementioned art of reasoning contributes nothing whatever to knowledge of the truth, we should realize that, on the basis of their method, dialecticians are unable to formulate a syllogism with a true conclusion unless they are already in possession of the matter of the conclusion, *i.e.*, unless they have previous knowledge of the very truth deduced in the syllogism. It is obvious therefore that they themselves can learn nothing new from such forms of reasoning, and hence that ordinary dialectic is of no use whatever to those who wish to investigate the truth of things. [*Regulae*, Rule X (p.4-5)][1]

What Descartes apparently means by this is that the truth of the conclusion of a valid syllogism is only guaranteed if the premises are known to be true. In this sense, the conclusion does not yield any new knowledge, for conclusions merely state what was already entailed by the premises. And hence, the inference teaches us nothing new. This argument, sounding like a familiar general objection against the value of deduction, raises broader questions about Descartes's epistemology that must be left aside here.

It may be briefly noted however, that a similar objection against the value of logic, amidst a host of other, more specific ones, was put forward by Pierre Gassendi in his Exercitationes paradoxicae adversus Aristoteleos. The first book of this work appeared in 1624, while the second, which is wholly devoted to a detailed criticism of Aristotelian logic, existed in manuscript at that time, not to be published until 1659. Gassendi was influenced by the work of the second-century sceptic Sextus Empiricus, who argued extensively "against the logicians" [*Adv. Log.*]. Gassendi argued, following Sextus, that syllogistic reasoning is both circular and superfluous [*Gass. Exerc.*, p.422-428]; Sextus Empiricus [*Pyrr. Hyp.*, p.144-203].

4 Webster

Both Bacon's and Descartes's criticism of logic, however differently focused, took the internal correctness of the deductive procedures of Aristotelian logic for granted. There was yet another way in which logic was criticized, and this criticism was directed against the procedures themselves. It can be found in a booklet which was published in 1654 in England, entitled *Academiarum Examen*. It contains a critical review of the university curriculum at Oxford and Cambridge. The author was John Webster, and his censure of university education provoked immediate and rather irritated rebuttal by various authors from academic circles. One of the objects of Webster's scorn was logic. His chapter on the subject is informative as in it

---

[1] Changing their rendering of "*materia*" as "substance" to "matter".

he put together all the various objections against logic he had found with other writers, such as Bacon, Gassendi and van Helmont. Thus he repeated Bacon's remarks on the lack of a proper foundation for the logical building:

> The main defect of Logick is, that it teacheth no certain rules, by which either notions may be truly abstracted and gathered from things, nor that due and fit words may be appropriated to notions, without which it fails in the very fundamentals, and falls as an house built upon sand.
> [*Webster*, p.34]

To this he added an argument he had found in a tract by Joan Baptista van Helmont (1579-1644), entitled *Logica inutilis* [*van Helmont*, p.39-43]. Webster argued that there was something wrong with the logician's claim that the conclusion of a valid syllogism necessarily follows from the premises. Copying van Helmont's example, he offered the following instance of a valid syllogism from false premisses:

> *Nullum adorabile est Creator.*
> *Omne simulachrum est adorabile.*
> *Ergo, Nullum simulachrum est Creator.*[2]
>
> Which is a true conclusion. From whence it cannot be judged that the Conclusion of Syllogisms doth of necessity compel assent, nor that the Conclusion doth necessarily depend upon the Premisses. Therefore as the truth is not contained or hid in a ly, nor the knowledge of it: so the consequent is, that the knowledge of the conclusion is not necessarily included in the Premisses. [*Webster*, p.37]

Both premisses are taken to be false, and the conclusion is regarded as true. Yet, so the argument runs, the conclusion follows from the premisses, according to the logicians. So there must be something wrong with the logicians' formalism.

This argument is clearly based upon a misunderstanding of what the logicians' notion of validity was, and still is. As every student of logic was supposed to know, the (semantic) notion of validity boils down to the requirement that if both premisses are true, the conclusion cannot fail to be true as well. The example used by Webster meets this requirement: it is one of the so-called perfect syllogisms of the first figure, and its mood was mnemotechnically expressed as Celarent.

Consequently, this mistake of Webster's was an easy target for those defending the university curriculum, and logic along with it. Seth Ward, for example, did not even take the trouble of refuting Webster's argument, but dismissed it in a ironic remark, which at the same time pointed out where Webster went astray:

---

[2] Nothing to be worshipped is the Creator. Every image is to be worshipped. Therefore, no image is the Creator.

> Their Conclusions doe not necessarily compell Assent, *viz.* M. Webster is one who can grant the premises in a true Syllogisme, and yet deny the conclusion. I Answere this is by a speciall gift. [*Ward*, p.25-26]

Webster's little book gave rise to other reactions as well, among which was a short treatise especially devoted to the defence of logic, written by an anonymous author, referred to as "a very learned pen". This author went into detail in order to refute every single argument put forward by Webster, including this one. He corrected Webster's mistake regarding the notion of validity, clearly distinguishing formal correctness and truth, as follows:

> We say not that in syllogisms which of necessity compels assent is the conclusion itself, but the premises, when out of them it is rightly proved (i) when the premises both are true and well ordered in Mood and figure, Assent to the conclusion is made necessary. [...] The conclusion indeed doth necessarily depend upon the premises, in respect of the forme at least, as the conclusion of a true syllogisme, and so doth that of his syllogisme before mentioned. [Deb70, p.302]

As is clear from the context, the expression "true syllogisme" must be taken to mean a syntactically well-formed syllogism, "true" being used here to refer to a formal as opposed to a semantic property.

One of the advantages of Webster's eclectic criticisms for the historian is that it gives us an impression of what Ward and "the very learned pen" had to say against the arguments Webster had borrowed from Bacon: that if the terms on which the formal building was supposed to stand were ill-defined, the whole building was shaky. Their line of defence consisted mainly in saying that there was nothing wrong with the way in which logic treated the theory of terms and that if notions were not rightly abstracted, one could not blame logic for it. This strategy could only lead to defeat. For it must have reinforced the impression that logic was completely tied to the lore of Aristotelian metaphysics and science, which was rapidly getting obsolete.

5 Conclusion

John Locke, towards the end of the seventeenth century, pointed to logic as one of the sources of the terminological confusion that characterized the writings of the schooolmen', as the representatives of traditional learning were called:

> To this abuse, and the mischiefs of confounding the Signification of Words, Logick, and the Liberal Sciences, as they have been handled in the Schools, have given Reputation; and the admired Art of Disputing, hath added much to the natural imperfection of Languages, whilst it has been made use of, and fitted, to perplex the signification of Words, more than to discover the Knowledge and Truth of Things.   [*Locke, Essay*, book III, chapter X, §6]

To Locke, and many of his contemporaries, logic was one of those remnants of Aristotelian pretended learning that had to be abolished as soon as possible.

As we have seen, various writers put forward more specific arguments against the merits of logical theory as well. Most writers on the subject, adversaries and defenders alike, perceived that logic, or some bits of it at least, qualified as a formal science. But this by itself was no reason for esteem to most scholars, and indeed it was seen as a disadvantage by many.

Primary Sources.

[Port Royal] Antoine **Arnauld** and Pierre **Nicole**, La Logique ou l'Art de Penser, Paris 1662

[Bacon, Aphorisms] Francis **Bacon**, Aphorisms concerning the Interpretation of Nature and the Kingdom of Man, in: [SpeEllHea57, Volume IV]

[Blundeville] Thomas **Blundeville**, The Art of Logike, John Windet 1599

[Burgersdijck] Franco **Burgersdijck**, Institutionum Logicarum Libri duo, Lugdunum Batavorum (Leiden) 1626

[Regulae] René **Descartes**, Regulae ad directionem ingenii, *incomplete; english translation in:* [CotStoMur85]

[Gass. Exerc.] Pierre **Gassendi**, Exercitationes paradoxicae adversus Aristoteleos (1659), *french translation in:* [Roc59]

[Gass. Inst.] Pierre **Gassendi**, Institutio Logica (1658), *english translation in:* [Jon81]

[van Helmont] Joan Baptista **van Helmont**, Opera Omnia, Christianus Paulli 1707

[Jungius] Joachim **Jungius**, Logica Hamburgensis (1638/1681), *german translation in:* [Mey57]

[Locke, Essay] John **Locke**, An Essay concerning Human Understanding (1689), in: [Nid75]

[Adv. Log.] **Sextus Empiricus**, Adversus logicos, in: [Bur33-49]

[Pyrr. Hyp.] **Sextus Empiricus**, Pyrroneioi Hypotyposeis, in: [Mat96]

[Wallis] Johannus **Wallis**, Institutio Logicae, ad communes usus accommodata, Leonard Lichfield 1686

[Ward] Seth **Ward**, Vindiciae Academiarum, containing, some briefe animadversions upon Mr Websters Book, stiled, The Examination of the Academies, Leonard Lichfield 1654

[Webster] John **Webster**, Academiarum Examen, or the Examination of Academies, Giles Calvert 1654

References.

[Ash74]   Earline J. **Ashworth**, Language and Logic in the Post-Medieval Period, Reidel 1974

[Bla70]   Robert **Blanché**, La Logique et son Histoire, d'Aristote à Russell, Armand Colin 1970

[Bur33-49]   Robert Gregg **Bury** (ed., trans.), Sextus Empiricus, 4 volumes, Heinemann 1933-36 [The Loeb Classical Library 273, 291, 311, 382]

[CotStoMur85]   John **Cottingham**, Robert **Stoothoff**, and Dugald **Murdoch** (eds.), Rules for the Direction of the Mind, The Philosophical Writings of René Descartes, Cambridge University Press 1985

[Deb70]   Allen G. **Debus**, Science and Education in the Seventeenth Century, The Webster–Ward Debate, MacDonald 1970

[How56]   Wilbur Samuel **Howell**, Logic and Rhetoric in England, 1500-1700, Princeton University Press 1956

[Jon81]   Howard **Jones** (ed., trans.), Pierre Gassendi, Institutio Logica, A Critical Edition with Translation and Introduction, van Gorcum 1981

[Kne$_1$Kne$_0$62]   William **Kneale** and Martha **Kneale**, The Development of Logic, Clarendon Press 1962

[Mat96]   Benson **Mates** (ed., trans.), Sextus Empiricus, The Skeptic Way, Sextus Empiricus's Outlines of Pyrrhonism, Oxford University Press 1996

[Mey57]   Rudolf W. **Meyer** (ed., trans.), Joachim Jungius, Logica Hamburgensis, J.J. Augustin 1957

[Nid75]   Peter H. **Nidditch** (ed.), John Locke, An Essay concerning Human Understanding, Clarendon Press 1975

[Ris64]   Wilhelm **Risse**, Die Logik der Neuzeit, Frommann 1964-1970

[Roc59]   Bernard **Rochot** (ed., trans.), Pierre Gassendi, Exercitationes paradoxicae adversus Aristoteleos, Vrin 1959

[San18]   Robert **Sanderson**, Logicae Artis Compendium, Oxford 1618

[SpeEllHea57]   James **Spedding**, Robert Leslie **Ellis**, and Douglas Denon **Heath** (eds.), The Works of Francis Bacon, Philosophical Works, London 1857-1874

Received: May 17th, 2003;
In revised version: March 15th, 2004;
Accepted by the editors: April 19th, 2004.

Benedikt **Löwe**, Volker **Peckhaus**, Thoralf **Räsch** (eds.)
Foundations of the Formal Sciences IV
The History of the Concept of the Formal Sciences

A formal bridge between epistemic cultures
Objective Possibility in the times of the second German empire

MARTIN NEUMANN[*]

Universität Osnabrück
Fachbereich Sozialwissenschaften
Seminarstraße 33
49074 Osnabrück, Germany
E-mail: `martneum@freenet.de`

> ABSTRACT. The physiologist Johannes von Kries elaborated a concept of probability, called *Spielräume* (*i.e.*, scope, range), to reconcile probability with deterministic causality. Its difference with regard to statistical laws can be characterized by two aspects. First a different conceptualization of a probabilistic explanation, and second a different field of application: Originally motivated to fill the explanatory gap between the reversible laws of mechanics and the irreversible 2nd law of thermodynamics, the concept of *Spielräume* has extended into the field of the humanities.

1 The problem of probabilistic explanations

Ian Hacking [Hac90] has shown the importance of the emergence of a statistical style of scientific reasoning in the 19th century. Hacking calls this *taming of chance*. The main work on this *probabilistic revolution* was carried out in western Europe. In Germany, however, there was a lot of doubt about statistical reasoning. The following quote of Christoph Sigwart is typical:

> Es hatte ein Interesse, zu zählen, wie viele Monds- und Sonnenfinsternisse Jahr für Jahr sich ereigneten, so lange sie als unvorhergesehene und unbegriffene Ereignisse eintraten; seit die Regel gefunden ist, nach der sie sich ereignen und auf Jahrhunderte rückwärts und vorwärts berechnen lassen, ist jenes Interesse verschwunden. [Sig11, p.696]

---

[*]I would like to thank two anonymous referees for fruitful comments.

But this doubt did not lead to an abandonment of probabilistic reasoning at all. On the contrary, it led to some quite sophisticated investigations on the meaning of probability.

The most important work on probability theory was written in 1886 by the physiologist Johannes von Kries (1853-1928). It is entitled the theory of *Spielräume* (*i.e.*, range, scope). In a recent article, Michael Heidelberger has shown its contribution to the logical foundation of probability [Hei01]. A detailed analysis of some measure theoretic aspects of this theory was given by Kamlah ([Kam83, Kam87a]), who also investigated its function as a critique of Laplacian probability theory [Kam87b]. Fioretti [Fio01] has investigated von Kries' contribution to the concept of probability developed by Keynes. This paper, however, will concentrate more on the scientific rather than the philosophical impact of this theory.

First, it is quite surprising to find a physiologist working on probability theory. But even more surprising, von Kries became an honorary doctor of law for his theory. This theory applied probabilistic explanations in a significantly different way as is common today. Von Kries rejected the link to error theory that Stephen Stigler called the "Gauß-Laplacian synthesis" [Sti86, chapter 4]. As a result, his theory transformed concepts of science into those of the humanities.

This method of scientific reasoning therefore has to be understood against the backdrop of the German philosophy of science.

In the mid 19th century, the core of science was seen in a combination of mechanism and determinism (*cf.*, *e.g.*, [Sch97]). The principles of mechanics were seen as the ultimate source of scientific reasoning. Within this framework, determinism was a leading principle for scientific research: a law of nature is characterized by a deterministic relation. Otherwise one has to look for deeper –and even more hidden– causes of the events. This was the standard method of a scientific explanation. But at around the same time, in the mid 19th century, something new was developing: The emergence of a statistical style of reasoning. This way of reasoning developed in both social statistics and in the kinetic theory of gases. Hence, both sciences seemed to contradict the principles of science. It is well documented that, in fact, both theoretical developments were intertwined. Porter ([Por81, Por94]) has shown how the development of the kinetic theory of gases was influenced by social statistics. The interaction between Physics and Economic Theory is highlighted by Mirowski [Mir00].

Johannes von Kries was seeking a probabilistic way of reasoning which was nevertheless in accordance with the deterministic picture of science. To reach this goal, he followed the example of Ludwig Boltzmann's work on thermodynamics from the year 1877 [Bol77].

In order to provide a source for an understanding of the style of reasoning developed by von Kries, some elements of this essay that were essential for him will first be highlighted. The paper then continues with an investigation of the epistemological foundation of probability theory elaborated by von Kries. Finally, the fields of application are examined.

2 The influence of the kinetic theory of gases

The research programme of the kinetic theory of gases was to reduce the phenomenological laws of thermodynamics to the laws of mechanics. A gas was thought of as a collection of many identical particles. These "atomic" elements where thought to have no inner structure and to move in a chaotic manner through the gas container. Nothing more could be said about them and the way in which they move. But it is clear that they will collide and this collision will follow the mechanical laws of elastic collision. In this respect, Rudolf Clausius spoke of the motion we call heat [Cla57]. Thus, the kinetic theory of gases followed the mechanistic picture of science.

But in the case of the second law of thermodynamics, this programme proved problematic. The second law states: if there is a body which is hot on one side and cold on the other then gradually the temperature will become equal on both sides. This is an irreversible process. But the laws of mechanics are reversible. Thus, the second law of thermodynamics cannot be reduced solely to the laws of mechanics. This argument, which was developed by J.J. Loschmidt in 1876 [Los76], caused Ludwig Boltzmann to investigate the starting conditions in his article of 1877: If nothing is known about the starting conditions every starting condition is of equal possibility. Then Boltzmann calculated in a purely combinatorial way how many of these conditions would cause an unequal distribution of heat and how many would cause an equal distribution of heat. The result was: The overwhelming majority of the starting conditions would result in the latter. Thus, no purely mechanical explanation of the second law of thermodynamics is possible without regarding the starting conditions. The second law demands an explicit consideration of the starting conditions.

Let us now take a look at von Kries interpretation of Ludwig Boltzmann's probabilistic reasoning.

3 Epistemological foundations of probability

The main idea of the theory of *Spielräume* was that even if there is perfect knowledge of the mechanical laws governing a process there might still be a certain range of possibilities for the concrete outcome. Under certain circumstances, this range can even be measured (*cf.* [Kam83, p.242-248]; [Kam87a, p.318-320]; [Neu02, p.151-179]). In this respect, von Kries followed Ludwig Boltzmann exactly. In accordance with Boltzmann, von Kries considered this to be the source of a probabilistic explanation:

To generalize Boltzmann's main idea he developed the example of a meteorite going down to the earth. Even if it is known that the meteorite will hit the earth there is still left a range on which point exactly the meteorite will smash on the earth. The degree of our expectation that the meteorite will hit a certain district can be measured if the size of this area is compared with the size of the whole. Von Kries states:

> Bei der Bildung von Erwartungen mit Bezug auf das Niedergehen eines Meteors schien es naheliegend, die Wahrscheinlichkeit, dass dasselbe auf irgend einen Teil der Erdoberfläche auftreffe, dem Flächeninhalt dieses Teiles proportional zu setzen. Wir gelangen ohne Schwierigkeiten zu dem Gedanken, dass etwas Ähnliches immer der Fall sein möchte, wenn unsere Annahmen einen Gegenstand betreffen, für welchen, unserem Wissen gemäß ein messbarer und in Teile zu zerlegender Spielraum des Verhaltens möglich erscheint. [vKr86, p.24]

A brief look at this example shows that the behavior of the meteorite can certainly be described by the laws of mechanics. But even though these laws are perfectly deterministic, the range of starting conditions produces an uncertainty in our knowledge of where exactly the meteorite will strike earth. Notice that principally he was following the ideas of Ludwig Boltzmann. He simply replaced the small atomic particles of the kinetic theory of gases by the larger body of a meteorite. Of course, a meteorite striking earth does not follow the laws of elastic collision. But like the collision of atomic particles it is dependent on small variations of the starting conditions. Then von Kries –like Boltzmann– argued that the source of probability cannot be found in the mechanical laws. Instead, he claimed, that it has to be found in the singular conditions of one special event. Therefore von Kries invented a distinction between so-called nomological and the ontological aspects of nature. (*Cf.* [vKr86, p.86]; [Hei01, p.177-178].)

> 'Die Erkenntnis der Wirklichkeit ist eine Aufgabe, an welcher wir zwei wesentlich verschiedene Teile unterscheiden können. [...] Die Kenntnis des Gravitations-Gesetzes, um ein Beispiel anzuführen, lehrt uns noch nichts über die wirklich Statt findende Bewegung der Planeten; sondern um sie zu verwerten, müssen wir noch wissen, welche Massen existieren und in welchem Zustande der räumlichen Verteilung und der Bewegung sie sich irgendwann befunden haben. [vKr86, p.85]

The law of gravitation is the nomological aspect, whereas the existence and distribution of masses are the ontological aspects of the description of nature in this case. Statements about ontological aspects refer to the contingent circumstances of single events; "*das rein Tatsächliche*" [vKr86, p.86], as von Kries writes. Thus, uncertainty in the ontological part of the description says nothing about the deterministic character of the nomological parts. They represent kinds of knowledge which are independent of one another. Following von Kries, probabilistic reasoning can only be found in the ontological part. As a consequence, he is able to reconcile probabilistic reasoning with the deterministic character of the nomological part of the description of nature, *i.e.*, the laws of mechanics:

> Die objektive Möglichkeit bedeutet Konformität mit den Wirklichkeits-Gesetzen und irgend etwas nicht verwirklichtes in einem objektiven Sinne möglich zu nennen hat dann, aber auch nur dann eine völlig bestimmte und klar angebbare Bedeutung, wenn wir annehmen, daß die Wirklichkeits-Gesetze neben dem Verwirklichten einen bestimmten Umfang anderer Verhaltensweisen zulassen. [vKr16, p.54]

This is a brief sketch of the general idea of what von Kries thought as the source of probabilistic reasoning. He has drawn two consequences out of this which have shown to be essential for the style of reasoning developed out of this theory. First, probabilistic arguments can only be applied to singular circumstances and second, he embedded this theory of probability into a framework of a theory of reference.

The first point can be explained by the notion of ontological aspects of the description of nature: They are only singular circumstances of a specific event. For example, the factual distribution of masses at some time point is just one singular event in the history of the universe. At another time point, we find a different distribution. Thus, if probabilistic arguments have to be applied to those ontological aspects, they also can only be applied to singular events.

> Ich möchte dies umso stärker betonen, als von verschiedenen Seiten die Meinung vorgebracht worden ist, dass jede numerische Wahrscheinlichkeit durchaus eine allgemein sein müsse und sich auf ein einzelnes individuelles Ereignis ganz und gar nicht beziehe. Im Gegensatz hierzu muss ich behaupten, dass streng genommen jede Wahrscheinlichkeit eine singuläre, auf ein einzelnes bestimmtes Verhalten sich beziehende ist. [vKr86, p.129-130]

In this respect, this style of reasoning is significantly different to that of statistical mechanics. Instead of dealing with statistics von Kries is considering something similar to propensities (*cf.* [Kam87a, p.316-320]). With respect to the way in which Boltzmann uses probabilistic arguments, he states:

> Mir scheint eine derartige Ungenauigkeit des Ausdrucks wirklich an einer ganz bestimmten Stelle der Untersuchung immer vorzuliegen: Die Kennzeichnung eines Zustandes geschieht nämlich in der Weise, dass die Z a h l derjenigen Moleküle angegeben wird [...] [vKr86, p.203]

This seemed to him to be incorrect. Instead one should reduce the statistical argument to the range of possible behavior of one single molecule. However, the question remains of how to apply probability theory to a single case. This leads to the second point:

Following the psychologism of his day, von Kries defines reference as a judgement. Applied to general terms, however, this judgement can be ambiguous. For example, an object might be described by the word "red". You then have a continuous range of possibilities from the one case, that this notion is clearly false, to the other, that it is clearly true:

> Als einfachstes Beispiel erörtere ich die Subsumption eines Einzelnen unter eine Allgemein-Vorstellung; und zwar wollen wir uns noch vorerst an die einfachste Art von solchen halten, die etwa eine Art sinnliche Empfindung bezeichnen (Süß, Rot etc.) Schon bei diesen ist zu bemerken, daß von denjenigen Fällen, in denen die einzelne Empfindung sogleich und mit Sicherheit der betreffenden Allgemein-Vorstellung subsumiert wird, eine kontinuierliche Abstufung zu denjenigen führt, in denen die Subsumption mehr oder weniger zweifelhaft erscheint und schließlich zu denjenigen, in denen sie verneint wird. [vKr99, p.9]

Note, that von Kries is considering a single object in this example. Now von Kries draws a link to probability theory: he sees a chance experiment as an experiment that can only be described in general terms.

> An den Umständen, von welchen das Eintreten irgendwelcher Ereignisse abhängt, unterscheidet man nämlich einerseits gewisse allgemeine Bestimmungen, andererseits die besonderen Gestaltungen, welche innerhalb des Rahmens dieser allgemeinen Bestimmungen Statt finden können. 'Allgemein' heissen diese hier, wie nachdrücklich hervorgehoben werden muß, zunächst lediglich deswegen, weil sie eine Anzahl verschiedener Verhaltungsweisen als besondere Fälle in sich schließen. So ist es z.B. eine allgemeine Bestimmung, dass "irgend jemand eine Münze aufwirft", weil eine unübersehbare Menge verschiedenen Geschehens unter diesen Begriff fällt.
> [vKr86, p.102]

Following von Kries, this is the reason why a chance experiment leads to probabilistic statements. In a chance experiment like tossing a coin, an infinite range of rotation, velocity, *etc.*, remains. The specific rotation, velocity, *etc.*, in one concrete toss of a coin are singular circumstances of this event. They can vary from case to case. Thus, the ontological aspects of a chance experiment can only be described in general terms. Because all these cases are covered by the one instruction "toss a coin", one can say that this is not a precise but a fuzzy instruction. Nevertheless, we gain exact knowledge of the size of the *Spielraum* left open by it. Following von Kries this is the reason why probability theory can be applied to a chance experiment:

> Denken wir uns z.B. beim Würfeln die Gesammtheit aller Gestaltungen der bedingenden Umstände zusammengefasst, welche den Wurf 1 herbeiführen, ebenso die Gesammtheit aller, welche die Würfe 2, 3, etc. bewirken würden: so lässt sich behaupten, dass diese sechs Complexe alle von gleicher Grösse sind. [vKr86, p.VII]

Note that we know this exactly because we do not know the starting conditions in a concrete case. This is the kind of positive knowledge probabilistic statements refer to. Therefore von Kries stresses the intersubjective validity of this kind of analysis:

> Die Wahrscheinlichkeit, mit einem Würfel 6 zu werfen, setzten wir gleich 1/6, und haben damit etwas ausgesprochen, was für Jeden und unter allen Umständen gültig ist. [vKr86, p.129]

Consequently, he regarded the construction of a model as the core problem of a probabilistic explanation. It has to consist out of general terms that determine exactly the size of a *Spielraum* left open by them. The kinetic theory of gases served him as a paradigmatic case of such an explanation:

> Sagen wir, der Körper K befinde sich zur Zeit t innerhalb des Raumes R, so ist diese Angabe in der Weise ungenau, daß dabei die Lokalisierung in jedem beliebigen Punkte von R zugelassen wird. Die Gesamtheit des Raumes R stellt den von dem Urteil umfaßten Spielraum dar [...] [vKr16, p.412-413]

Boltzmann himself used the term *"Inbegriff"* of a mechanical system. In a contemporary philosophical dictionary one can find the definition: *"Ein in einer logischen Synthese Zusammengefaßtes Ganzes"*, along with references to set, number, and term [Eis27, p.729]. Statistical mechanics deals with sets. This leads to a frequency theory of probability. But it has to be remarked, that within the framework of the kinetic theory of gases atomic particles where not observable entities. Therefore von Kries thought of the statistical approach as an inaccuracy in speech: He was talking about the construction of terms, in this case about the linguistic construction of an atomic particle.[1]

Therefore he reconstructed Boltzmann's approach as a theory of single case probabilities.

This result can be formulated in the following conclusion regarding the theory of applicability of probability theory:

Within this interpretation, indeterminism is to be found on a completely different part of the world as within statistics. Statistics are searching for

---

[1] Without going into details, it might be interesting to note that a further elaboration of von Kries' idea can be found in the framework of Alexius Meinong's theory of objects. In 1890 he wrote a review of von Kries' *Principien der Wahrscheinlichkeitsrechnung* and again in the preface of his book on *Möglichkeit und Wahrscheinlichkeit*, he writes that the origins of his ideas reach back to his reading of von Kries [Mei15, p.VI].

correlations. There is therefore a probabilistic relation between well-defined objects. Within the framework of the theory of *Spielräume*, however, indeterminism is hidden in the problem of reference. Speaking in the words of von Kries, one has to say that in statistics indeterminism is found in the nomological while in this theory it is found in the ontological aspects of the description of nature.

4 *Spielräume* in scientific practice

A look at the scientific context shows us that von Kries had a catalytic function for something that could be called the *German way of taming chance*. He was not very successful in physics but his distinction between nomological and ontological aspects was quite successful in the invention of a new style of scientific reasoning. In the following, some of these fields are investigated. Namely, the theory of law, and the methodology of Max Weber's sociology.

4.1 Fuzziness in law

The problem of reference was highly urgent in the theory of law. Von Kries points out that a complete exclusion of fuzzy terms is neither justified nor practicable (*cf.* [vKr88, p.424]). He states:

> Die Beschaffenheit der sozialen Erscheinungen bringt es mit sich, dass der Umfang derjenigen Kategorie, unter welche ein Einzelfall zu subsumieren ist, keineswegs immer selbstverständlich bestimmt erscheint.
> [vKr88, p.218]

Here both elements we pointed out to be central to the theory of *Spielräume* can be found: the single event probability and the problem of reference.

Note that it was not von Kries himself who raised the problem. It was the *Reichsgericht*, which claimed that one cannot give a strict demarcation of the notion of *danger* [*RGSt (1884)*, p.176]. With the onset of industrialization, however, risk and danger where highly urgent topics. One simply had to deal in some way with the emergence of previously unknown risks (*cf., e.g.*, [Lüb93, Ewa93]; [Ros88, p.291-362]).

The link to probability theory was given by the definition of the notion of danger: danger can be defined as the possibility of damage. Possibility, however, was the central term in Laplacian probability theory. Thus, there seemed to be a natural bridge between a theory of danger and a theory of probability. The canonical point of view, combining philosophical determinism with the notion of subjective possibility, could not handle with this problem in a reasonable way: Within this framework, danger was reduced to fear (*cf.* [Her80], [Fin89]). This was of no practical use for a court. Von

Kries, however, was able to elaborate a concept of danger by making use of conditional probabilities. Let us take a look at the example he used (*cf.* [vKr88, p.201]):

Imagine a coachman who is asleep or drunk while the coach carries a passenger. Imagine further that the coach takes a wrong turn and, because of this, the coach is struck by lightning. Of course, this would not have happened if the coach had been on the right course. So the cause of this event is that the coach had lost its way. But the conditional probability that the coach will be struck by lightning is the same under both conditions— that the coach is on the right or the wrong way. Von Kries calls this a cause by chance. If, on the other hand, the coach crashes while the coachman is sleeping, then the fact that the coachman was asleep or drunk had an effect on the probability of this event. In this case, the drunkenness of the coachman is what von Kries called an adequate cause of the accident. The coachman only is liable for the accident in the latter case.

Since the objective of this theory is not to deal with causes by chance but with adequate causes, there is a remarkable shift of interest from a theory of pure chance to this juridical theory: As indicated by the discussion of the legal notion of danger, it established a framework to handle problems that are not due to pure chance. Nevertheless, they cannot be subsumed under the notion of deterministic causality. This realm strictly between pure chance and deterministic causality was key for applications in the social sphere (*cf.* also [Mir00]), and consequently also Max Weber discussed the theory of adequate cause.

Nevertheless, one has to pose the question as to how this conditional probability can be recognized. The answer was most relevant to the methodology of Max Weber's sociology. Von Kries did not think of accident statistics but of counterfactual conditional sentences: One element X of a chain of causes is replaced by the element X_1 and then follows an investigation of the consequences of this replacement under the conditions of the usual way of life. "*Es wird zunächst das als regelwidrig anzusehende Verhalten [...] weg-, und das regelmäßige an seiner Stelle gedacht*" [Rad02, p.401]. For example, the drunk coachman is replaced by a sober one. Under usual circumstances the coach would not have crashed. This idea was adopted by Weber.

4.2 The range of understanding

Last but not least, Max Weber adopted the theory of *Spielräume* from the theory of law. For him it served as a methodological tool for the foundation of a new science: sociology, namely for the operation called *verstehen* (*cf.* [WagZip85, TurFac81]). Of course, it can not be the aim of this paper to discuss the pitfalls of Weber's methodology. What shall be emphasized is the fact, that even Max Weber's qualitative sociology relied on a concept of probability.

As a historian, Weber wanted to defend the idea of historical causality. To achieve this goal he developed the concept of historical ideal types, which have to be understood as the idea of a historical phenomena. Weber described them as *"ein Komplex von Zusammenhängen in der geschichtlichen Wirklichkeit, die wir unter dem Gesichtspunkt ihrer Kulturbedeutung begrifflich zu einem Ganzen zusammenschließen"* [Web47a, p.30]. With an ideal type the essentials out of the historical material are united into one appropriate notion. Contrary to the statistical notion of the correlation of mass phenomena, an ideal type is intended to cover one single instance of a complex historical process:

> Je mehr es sich um einfache Klassifikationen von Vorgängen handelt, die als Massenerscheinungen in der Wirklichkeit auftreten, desto mehr handelt es sich um Gattungsbegriffe, je mehr dagegen komplizierte historische Zusammenhänge in denjenigen ihrer Bestandteile, auf welchen ihre spezifische Kulturbedeutung ruht, begrifflich geformt werden, desto mehr wird der Begriff –oder das Begriffssystem– den Charakter des Idealtyps an sich tragen. [Web68b, p.202]

As it is the case in von Kries' reconstruction of Boltzmann's arguments, Weber was interested in a methodology for the construction of terms. For this, the question has to be answered as to what is essential within the historical data. Following Weber, something is essential if it has a causal meaning [Web68c, p.86]. To recognize this meaning, Weber adopted the methodology of constructing counterfactual conditional sentences. At this point, von Kries was relevant to him. Weber says: "[...] *als Gegensatz von 'zufällig' (ist) nicht* [...] *zu setzen: 'notwendig', sondern: 'adäquat' in dem* [...] *im Anschluß an von Kries entwickelten Sinn"* [Web68a, p.287]. Note, that he did not make use of the model of deterministic causality, although he wanted to defend the notion of historical causality. Yet, he did not intend to make use of probability theory. He was only interested in constructing counterfactual conditionals. He illustrated this by a variation of the example von Kries had developed for the introduction of his concept of adequance (*cf.* [Web68a, p.266]):

Weber makes a distinction between Bismarck's decisions in 1866 that led straightforward to a war and an exchange of pistol shots in Berlin that caused street fights. The former he says is an adequate cause while the latter is a cause by chance: If you think Bismarck had decided in another way then it is most probable the war had not taken place. This is different in the case of the exchange of pistol shots: The circumstances had been such that any other event could have caused the street fights. Thus, if you think there had been no shots it is most probable that there would still have been street fights.

This is the method Weber used to identify historical causality. Of course he was not interested in the examples themselves but in historical ideal types. Remember that the causal relevance is the criterion that they are correctly composed. The most prominent example developed by Max Weber is the so-called protestant ethics:

According to him, it is an adequate cause for the spirit of capitalism because he was convinced to have identified an influence of Protestantism in the spirit of capitalism. Thus, Weber regarded it as a correctly composed ideal type.

5 Conclusion

Now the catalytic function of von Kries in the *German way of taming chance* is obvious: Although his name is not very well known within this scientific community, he contributed to the creation of the phenomenological tradition of the social sciences, also called *verstehende Soziologie* (as opposed to statistical social research). The function of statistics was to extend the style of reasoning in the natural sciences to the social sphere. Nowadays, within some research communities, the terms "Social Sciences" and "statistical social research" are seen as synonymous. On the other hand, the style of reasoning developed out of his theory of *Spielräume* serves as a tool for the contrary: to defend the humanities as a style of reasoning in its own right. Thus, this theory served as a tool for the transformation of a concept of science into those of the humanities.

Primary Sources.

[RGSt (1884)] Entscheidungen des Reichsgerichts in Strafsachen, Veit & Comp. 1884

References.

[Bol77] Ludwig **Boltzmann**, Über die Beziehung zwischen dem zweiten Hauptsatze der mechanischen Wärmetheorie und der Wahrscheinlichkeitsrechnung respektive den Sätzen über das Wäremgleichgewicht, **Sitzungsberichte der mathematisch-naturwissenschaftlichen Classe der Kaiserlichen Akademie der Wissenschaften** 76 (1877), p.373-435

[Cla57]   Rudolf **Clausius**, Über die Art der Bewegung, welche wir Wärme nennen, **Annalen der Physik und Chemie** 100 (1857), p.353-380

[Coh94]   I. Bernhard **Cohen** (*ed.*), The Natural Sciences and the Social Sciences, Kluwer 1994

[DasHeiKrü87] Lorraine J. **Daston**, Michael **Heidelberger**, and Lorentz **Krüger** (*eds.*), The Probabilistic Revolution, volume 1, MIT Press 1987

[Eis27]   Rudolf **Eisler**, Wörterbuch der philosophischen Begriffe, Mittler 1927

[Ewa93]   François **Ewald**, Der Vorsorgestaat, Surkamp 1993

[Fin89]   August **Finger**, Der Begriff der Gefahr und seine Anwendung im Strafrecht, Prag 1889

[Fio01]   Guido **Fioretti**, Von Kries and the Other "German Logicians": Non-numerical probabilities before Keynes, **Economics and Philosophy** 17 (2001), p.245-273

[Gib02]   J. Willard **Gibbs**, Elementary principles in statistical mechanics, Scribner's 1902

[Hac90]   Ian **Hacking**, Taming of Chance, Cambridge University Press 1990

[Hei01]   Michael **Heidelberger**, Origins of the logical theory of probability, **International Studies in the Philosophy of Science** 15 (2001), p.177-188

[Her80]   Eduard **Hertz**, Das Unrecht und die allgemeinen Lehren des Strafrechts, Hoffmann & Campe 1880

[Kam83]   Andreas **Kamlah**, Probability as a quasi-theoretical concept - J.v. Kries sophisticated account after a century, **Erkenntnis** 19 (1983), p.239-251

[Kam87a]   Andreas **Kamlah**, What can Methodologists learn from the history of Probability, **Erkenntnis** 26 (1987), p.305-325

[Kam87b]   Andreas **Kamlah**, The decline of the Laplacian Theory of Probability: A Study of Stumpf, von Kries, and Meinong, *in*: [DasHeiKrü87, p.91-116]

[vKr86]   Johannes **von Kries**, Principien der Wahrscheinlichkeitsrechnung, Mohr 1886

[vKr88]   Johannes **von Kries**, Über den Begriff der objectiven Möglichkeit und einige Anwendungen desselben, **Vierteljahrsschrift für wissenschaftliche Philosophie** 12 (1888), p.179-240, 287-323, 393-428

[vKr99]   Johannes **von Kries**, Zur Psychologie der Urteile, **Vierteljahrsschrift für wissenschaftliche Philosophie** 23 (1899), p.1-48

[vKr16]   Johannes **von Kries**, Logik: Grundzüge einer kritischen und formalen Urteilslehre, Mohr 1916

[Los76]   Josef J. **Loschmid**, Über das Wärmegleichgewicht eines Systems von Körpern mit Rücksicht auf die Schwere, **Sitzungsberichte der mathematisch-naturwissenschaftlichen Classe der Kaiserlichen Akademie der Wissenschaften Wien** 73 (1876), p.128-142, 366-372

[Lüb93]   Weyma **Lübbe**, Die Theorie der adäquaten Verursachung, **Zeitschrift für allgemeine Wissenschaftstheorie** 24 (1993), p.87-102

[Mei90] Alexius **Meinong**, Anzeige, **Göttingische gelehrte Anzeigen** (1890), p.56*sq*

[Mei15] Alexius **Meinong**, Über Möglichkeit und Wahrscheinlichkeit, Johann Ambrosius Barth 1915

[Mir00] Philip **Mirowski**, More Heat than Light, Economics as Social Physics, Physics as Nature's Economics, Cambridge University Press 2000

[Neu02] Martin **Neumann**, Die Messung des Unbestimmten, Hänsel-Hohenhausen 2002

[Por81] Theodore M. **Porter**, A Statistical Survey of Gases: Maxwell's Social Physics, **Historical Studies in the Physical Sciences** 12 (1981), p.77- 114

[Por94] Theodore M. **Porter**, From Quetelet to Maxwell: Social Statistics and the origins of Statistical Physics, *in:* [Coh94, p.345-362]

[Rad02] Gustav **Radbruch**, Die Lehre von der adäquaten Verursachung, **Abhandlungen des kriminalistischen Seminars Berlin** 1 (1902), p.325-408

[Ros88] Heinrich **Rosin**, Der Begriff des Betriebsunfalls als Grundlage des Entschädigungsanspruches nach den Reichsgesetzen über die Unfallversicherung, **Archiv für öffentliches Recht** 3 (1888), p.291-362

[Sch97] Gregor **Schiemann**, Wahrheitsgewissheitsverlust, Wissenschaftliche Buchgesellschaft 1997

[Sig11] Christoph **Sigwart**, Logik, volume 2, Mohr 1911

[Sti86] Stephen M. **Stigler**, History of Statistics, Harvard University Press 1986

[TurFac81] Stephen **Turner** and Regis A. **Factor**, Objective possibility and adequate causation in Weber's methodological writings, **The Sociological Review** 29 (1981), p.5-28

[WagZip85] Gerhard **Wagner** and Heinz **Zipprian**, Methodologie und Ontologie: Zum Problem kausaler Erklärungen bei Max Weber, **Zeitschrift für Soziologie** 14 (1985), p.115-130

[Web47a] Max **Weber**, Die protestantische Ethik und der Geist des Kapitalismus, *in:* [Web47b, p.17-206]

[Web47b] Max **Weber**, Gesammelte Aufsätze zur Religionssoziologie, Volume 1, Mohr 1947

[Web68a] Max **Weber**, Objektive Möglichkeit und adäquate Verursachung in der historischen Kausalbetrachtung, *in:* [Web68d, p.266-290]

[Web68b] Max **Weber**, Idealtypus, Handlungsstruktur und Verhaltensinterpretation, *in:* [Web68c, p.65-167]

[Web68c] Max **Weber**, Methodologische Schriften, Fischer 1968

[Web68d] Max **Weber**, Gesammelte Aufsätze zur Wissenschaftslehre, Mohr 1968

[Win94]   Wilhelm **Windelband**, Geschichte und Naturwissenschaft, Heitz & Mündel 1894

Received: April 24th, 2003;
In revised version: November 19th, 2003; February 19th, 2004;
Accepted by the editors: March 19th, 2004.

Benedikt **Löwe**, Volker **Peckhaus**, Thoralf **Räsch** (*eds.*)
Foundations of the Formal Sciences IV
The History of the Concept of the Formal Sciences

On Formal Objects

RAINER OSSWALD

Praktische Informatik VII
Informatikzentrum
FernUniversität in Hagen
Universitätsstraße 1
58084 Hagen, Germany
E-mail: `rainer.osswald@fernuni-hagen.de`

> ABSTRACT. One of the most elaborate approaches to specifying formal systems and objects has been given by H. B. Curry. Although Curry is mainly known as a proponent of the formalist viewpoint on mathematics, his conception of a formal system turns out to show constructivist and structuralist aspects as well. In particular, Curry emphasizes that the exact nature of mathematical objects is irrelevant with respect to the truth of mathematical statements. This view is in accordance with a Quinean conception of structuralism, which comes along with a relative notion of ontology.
>
> On the other hand, there is a crucial difference in attitude between Curry and Quine concerning their ontological commitments in specifying formal objects; for example, Curry regards inductively generated classes of objects as intuitively given by means of inductive specifications, whereas Quine abstains from drawing on the intuitively given when it comes to ontology. This conflict is closely related to differing conceptions of logic. For Quine, in contrast to Curry, elementary logic is an indispensable part of any serious science — including the science of formal systems.

1 Introduction

By a formal object we mean things like numbers, strings, tuples, lists, trees, or the abstract data types used in computer science. The starting point of our discussion will be Haskell B. Curry's view of formal objects. Curry is generally considered to be the most prominent (if not the only) proponent of a formalist philosophy of mathematics in the second half of the twentieth century. Without doubt, Curry's elaboration of the formalist position is the most detailed one up to date. The standard reference is his 1951 booklet *Outlines of a Formalist Philosophy of Mathematics* [Cur51], which was already written in 1939.[1] In subsequent years, Curry further developed and modified his approach, culminated in his *Foundations of Mathematical Logic* [Cur63], which we take as our primary reference.

Curry's formalist philosophy has its origins in the ideas of the Hilbert school, which he got acquainted with during a stay in Göttingen.[2] W. V. Quine, on the other hand, is often associated with "logicism" in that he stands in the tradition of Frege, Russell, and Carnap. Although such a classification may be more irritable than useful, it correctly hints at the central role of logic in Quine's philosophy. He regards logic as the grammar of science, which includes of course the formal sciences as well. In what follows, we will try to spell out the different positions of Curry and Quine with respect to logic and ontology and their consequences concerning the nature of formal objects.

2 Formal Systems and Objects

This section briefly introduces the basic notions of Curry's conception of a formal system. A discussion of the underlying assumptions will follow below when Curry's approach is analyzed from a Quinean viewpoint.

2.1 Classes and Processes

Since Curry aims at foundations, he abstains from presupposing anything like axiomatic set theory, which he regards as a higher part of logic. Nevertheless, in specifying formal systems he needs to refer to certain "totalities":

> We shall often have to formulate [...] properties (or relations) which define, in a strictly intuitive (or contensive) way, a totality of elements of notions. In order to distinguish such intuitive totalities from the "sets" or "classes" formed later (and conceived rather as objects of some theoretical study than as intuitive notions), we shall call them *conceptual classes* (or relations).
> [Cur63, p.38]

A conceptual class is thus a totality of elements defined in a strictly intuitive or contensive way.[3] Such a class is said to be *definite* if the question of membership can be decided by an effective process.

An *effective process* for attaining a certain goal for a given element is a sequence of transformations to be applied successively to the element such that the goal is reached after a finite number of steps (*cf.* [Cur63, p.37]).

---

[1] As an invited paper for an International Congress for the Unity of Science in the same year; see [SelHin80a, p.v].

[2] He wrote his dissertation *Grundlagen der kombinatorischen Logik* (1929) under Hilbert, although he did most of his work with Paul Bernays; *cf.* [SelHin80a, p.viii].

[3] "Contensive" is Curry's translation of *"inhaltlich"*.

In particular, one needs to know which elements are *admissible* for which transformations and what the results of the latter will be. Contrasting his notion of an effective process with other constructivist programs, like that of the inuitionists, Curry states that his notion "does not depend on any idealistic intuition, temporal of otherwise." Although a critical analysis is postponed to later sections, we can at least note that if not on idealistic intuitions, his proposal heavily relies on non-idealistic ones.

A conceptual class is called *inductive* if it "is generated from certain initial elements by certain specified modes of combination" [Cur63, p.38]. More precisely, an inductive class X is given by *initial* and *generating specifications*. The initial specifications determine a definite class of *initial objects*, the *basis* of X; the generating specifications define a definite class of *modes of combination* of finite degree, each of which when applied to a tuple of elements of X produces an element of X. In addition, an inductive class is assumed to satisfy the *closure specification*, which says that "every element [of the class] can be reached by an effective process [...] which starts with certain initial elements and at each later step applies a mode of combination to arguments already constructed." [Cur63, p.39]

A *construction* of an element of an inductive class is a process for reaching that element by iterated application of the modes of combination. An inductive class is called *monotectonic*, if every element has a unique construction, and *polytectonic*, otherwise. The prototypical example for the latter type is given by the inductive class of finite strings over some alphabet, with concatenation of strings as the single mode of combination.

2.2 Formal Systems

In short, a formal system is a theory about formal objects.[4] More exactly, a *formal system* is given by a conceptual class of *formal objects*, a conceptual class of *basic predicates* of finite degree, and a conceptual class of *elementary statements*, the *elementary theorems*, which assert that certain basic predicates hold of certain formal objects [Cur63, p.50]. The system is called *deductive*, if the class of theorems is inductive, in which case the initial elements are referred to as *axioms* and the modes of combination as *deductive rules* [Cur63, p.46]. In case the class of basic predicates is empty, Curry speaks of the formal system as *pure morphology*.

Curry distinguishes between *syntactical systems* and *ob systems* [Cur63, p.51 *sq*]. In syntactical systems, the formal objects are the *finite strings* over the *letters* of some *alphabet*. There are two principal ways of conceiving these formal objects as an inductive class. The first option is to use, for each letter,

---

[4]For a comparison of Curry's notion of a formal system with others discussed in the literature *cf.* [Cur63, Section 2S].

the *affixation* of that letter to the right as a (unary) mode of combination, and the letters as initial elements. The second option employs concatenation as a single (binary) mode of combination, where again the letters serve as initial elements. Notice that affixation leads to a monotectonic inductive class whereas concatenation gives rise to a polytectonic one.

An *ob system* is based on a monotectonic inductive class of formal objects, called *obs*; the initial elements are called *atoms*, the modes of combination *operations*. Observe that an affixative syntactical system is also an ob system. Since the focus of this paper is on formal objects, we are primarily interested in the "pure morphology" of ob systems. To give a simple example, consider a single atom a and a binary operation $\langle\,,\,\rangle$. The obs are a, $\langle a,a\rangle$, $\langle\langle a,a\rangle,a\rangle$, $\langle\langle a,a\rangle,\langle a,a\rangle\rangle$, etc.

For an even simpler example take an atom 0 and a unary operation S. The resulting inductive class of formal objects, which consists of 0, $S0$, $SS0$, $SSS0$, and so on, can be employed as a basis for arithmetic. To this end, consider the formal system consisting of these obs, a two-place basic predicate =, the axiom $0 = 0$, and the deductive rule that if $x = y$ then $Sx = Sy$. Here x and y are unspecified obs, *i.e.*, the rule is actually a *scheme* that gives rise to infinitely many rules (*cf.* [Cur63, p.55]). Peano's third axiom, for instance, which corresponds to the rule that if $Sx = Sy$ then $x = y$, is *admissible* in the sense that adjoining it to the system does not affect the class of elementary theorems [Cur63, p.256].[5] The rule is thus an *epitheorem* of the system, *i.e.*, a provable statement about elementary statements.[6] Without going into details let us note that addition and multiplication can be introduced into the system by *definitional extension* [Cur63, p.107]. For a full account of this formal system as a basis for arithmetic, the reader is referred to [Sel75].

We close our brief exposition of formal systems by introducing the notion of a *representation*, which is essentially a structure preserving one-to-one correspondence between the formal objects of a formal system and certain objects "given from experience":[7]

> Any way of regarding the formal objects as specified objects given from experience will be called a *representation* of the system, provided the contensive objects retain the structure of the formal objects. [...] It means that there is a separate contensive object for each formal object, and [...] that the operations must be reflected in some way as modes of combination of the contensive objects. [...] In technical terms there must be a one-to-one correspondence, isomorphic with respect to the operations and modes

---

[5] *Cf.* also [Lor69].
[6] Curry uses the prefix "epi-" instead of the more common "meta-".
[7] For the sake of completeness, we should also mention Curry's notion of an *interpretation* of a formal system, which resembles to some degree the standard concepts of model theory; *cf.* [Cur63, p.59*sq*].

> of combination, between the formal objects [...] and the contensive objects
> of the representation. [Cur63, p.57]

For instance, the formal objects can be represented by their *names* that are introduced by the specification of the formal system; this is called the *autonymous* representation [Cur63, p.57]. So, every formal system has a syntactical representation. Another example is the *Gödel representation*, where each formal object is assigned its Gödel number [Cur63, p.58]. Curry emphasizes that it is "possible to present a system without having any specific representation in mind" [Cur63, p.58]. He calls such a system *abstract*.

3 Logic and Language

Let us now look more closely at Curry's conception of logic and contrast it with Quine's views on logic and language. Since Curry tries to provide mathematical logic with a foundation by means of formal systems, it is not surprising that in his opinion his formalist approach does not hinge on logic at all:

> [...], from the standpoint of formalism [...] one can characterize a mathematical system objectively without presupposing anything which would be natural to call "logic". [Cur63, p.18]

Presumably this does not include *philosophical logic*, which, according to Curry, is concerned with the "principles of valid reasoning" [Cur63, p.1]. For we may assume that he considers his reasoning valid too.

However, it is hard to draw a line between philosophical and *symbolic* logic if the latter is taken as concerned with the regimentation and formalization of the logical structure of ordinary language and the principles of valid reasoning. According to Quine, "[t]he effect of the regimentation is to reduce grammatical structure to logical structure" [Qui87, p.158] and "to paraphrase a sentence of ordinary language into logical symbols is virtually to paraphrase it into a special part still of ordinary or semi-ordinary language" [Qui60, p.159]. Since one of the main purposes of regimentation is to resolve ambiguities in ordinary language and thereby to increase conceptual clarity, Curry as a scientist should welcome such a move — whether or not logical symbols are used instead of ordinary language expressions.

At least, Curry is quite aware of the importance of language in describing and communicating formal systems:

> The construction of a formal system has to be explained in a communicative language understood by both the speaker and the hearer. We call this language the *U-language* (the language being used). [...] It is well determined but not rigidly fixed; new locutions may be introduced in it by way of definition, old locutions may be made more precise, etc. Everything we do depends on the U-language; we can never transcend it; whatever we study

> we study by means of it. Of course, there is always vagueness inherent in the U-language; but we can, by skillful use, obtain any degree of precision by a process of successive approximation. [CurFey58, p.25]

Although at first glance, Curry's goal of a sufficiently precise language seems to be in accordance with Quine's program of regimentation, there is a big difference in attitude. For Quine, regimentation means reformulation in a restricted language that allows a direct symbolization within the language of quantificational logic, because he regards formulability in that language more or less as equivalent to full intelligibility.[8]

Curry, in contrast, does not give an explicit criterion of precision. For instance, he regards the notions of effective process and class or totality as sufficiently explicated in the form presented in Section 2.1 above. In particular, he does not address the question of ontological commitment, which for Quine is intimately connected to properly analyzing and regimenting the discourse in question.

Before we will take up this issue in more detail in Section 4.1, let us dwell a bit on the status of logic and language in Quine's overall picture of scientific inquiry. Quine thinks of "logic as the grammar of strictly scientific theory" [Qui01, p.219]. So, in a sense, language and logic are prerequisite to science.[9] But if language is regarded as essential for science –including the science of formal systems– and if the language in question, though ordinary, is regimented in certain ways, then it is fair to ask for a precise definition of that very language. At this point, a problem arises: we cannot devise a theory of regimentation that meets Quine's scientific standards without getting into an infinite regress.

The same problem is virulent when logical reasoning is at issue. In his *Methods of Logic* (as well as in other writings), Quine uses so-called *schematic letters* as a notational device for specifying logically valid sentence and inference schemata. Schematic letters, say "p" and "q" in "$p \,\&\, q \to p$", are to be treated as "placeholders" for expressions, here sentences, and not as variables ranging over the members of some universe. With Alex Orenstein we can question the status of these placeholders:

> We are told that schematic letters [...] are neither object language expressions nor metalinguistic variables. This is only a negative characterization and out of keeping with Quine's requirement for being precise. Worse still, the introduction of schemas involves positing additional types of expressions and additional rules determining their wellformdness. [Ore02, p.116]

Quine might respond that schematic letters are a convenient but dispensable device. He could refer us to his method of *quasi-quotation*, which allows

---

[8] *Cf.*, *e.g.*, [Qui60].

[9] This observation has to be qualified insofar that Quine considers logic a part of his holistic *web of belief* and thus open to revision, at least in principle.

to quantify over parts of expressions, and which is definable within a fully formalized theory of expressions, called *protosyntax*, as demonstrated in [Qui51, Chap.7]. This response, however, can be criticized on two grounds: First, any protosyntactical theory already makes use of elementary logic. Second, defining concatenation in terms of writing and inscriptions, as done in [Qui51, p.288], is insufficient for protosyntax, as Quine himself points out, for example, in [Qui69a, p.42], where he proposes to employ finite sequences instead, whose definition in turn employs the natural numbers.[10] So, there can be no first theory of regimentation and logical inference as a scientific theory about the expressions of a natural or formalized language.

The alternative is to take up an ontogenetic point of view. Quine endorses the assumption that "the basic laws of logic [...] are internalized in childhood, in acquiring the use of the logical particles 'not', 'and', 'or', 'some', 'every'" [Qui95a, p.51]. Learning elementary logic is thus on a par with learning to master your mother-tongue.[11]

To sum up, for Quine, elementary logic (the principles of valid reasoning) is an indispensable part of any serious science. Moreover, he regards regimentation into the (externalist) language of predicate logic (his canonical notation) as a strong requirement for full intelligibility. On the other hand, and Quine is everything but explicit on this point, elementary logic cannot be described in a way meeting these standards without giving rise to an infinite regress. This negative conclusion should not be too surprising since Quine repudiates first philosophy anyway. So Quine's attitude is probably better seen as normative rather than as foundational.

Curry, in contrast, does not address the logical structure underlying his presentation of formal systems. He seems to take "the principles of valid reasoning" as obvious.[12] But there is every reason to make logical structure and reasoning explicit in order to detect hidden assumptions and to avoid errors. In the words of Donald Davidson:

> By prompting us to decide on the logical form of sentences [the program of formalizing a natural language] can reveal our basic ontological commitments, it can tell us where inferences are truly logical and where they are not, and it can reveal problems we had barely appreciated. [Dav99, p.715]

---

[10]Noticeably, Quine [Qui46] also espouses the idea of reducing arithmetic to protosyntax.

[11]But notice that logic, according to Quine and contra Carnap, cannot be learned by convention, because conventionalism already presupposes language and logic; *cf.* [Qui76a].

[12]There are some exceptions; see, for instance, the preliminary remarks about the logical connectives in the context of epitheoretic reasoning; *cf.* [Cur63, pp.96*sq*].

4 Structure and Ontology

4.1 Ontological Commitment and Individuation

One of the hallmarks of Quine's philosophy is "his insistence upon being scrupulously clear and consistent about one's ontological commitments" [Ore02, p.24]. These commitments are manifest in (regimented) discourse:

> We can very easily involve ourselves in ontological commitments by saying, for example, that *there is something* (bound variable) [...] which is a prime number larger than a million. But this is, essentially, the *only* way we can involve ourselves in ontological commitments: by the use of bound variables. [Qui61, p.12]

Hence the famous slogan "to be is to be the value of a bound variable." Notice that this slogan only indicates how to reveal ontological commitments and not whether they are acceptable in discourse, scientific or otherwise:

> We look to bound variables in connection with ontology not in order to know what there is, but in order to know what a given remark or doctrine, ours or someone else's, *says* there is; and this much is quite properly a problem involving language. [Qui61, p.15sq]

For Quine, a necessary requirement for accepting an ontological commitment is to be able to formulate criteria as to whether any two of the postulated entities are identical or not; in short: no entity without identity. For "[w]e cannot know what something is without knowing how it can be marked off from other things. Identity is thus of a piece with ontology" [Qui69a, p.55]. Intersubjective science is hardly possible if two scientists would not be able to make sure that they are talking about the same thing.

Let us reconsider the key notions underlying Curry's definition of formal systems under this perspective; *cf.* Section 2.1 above. First of all, he commits himself to the existence of *conceptual classes* as "totalities of elements defined in a strictly intuitive way." So, when are two conceptual classes X and Y identical? To say that they are identical if everybody has the intuition that they are should surely not count an acceptable criterion. Happily, there is a rather straightforward alternative: X and Y are identical if they have the same elements, that is, with "$x \in X$" for "x is an element of X", if $\forall x(x \in X \leftrightarrow x \in Y)$. In other words, conceptual classes are identical if they are *coextensive*. Curry would presumably agree. The more pressing problem is to pin down the conceptual classes Curry commits himself to exist. His reference to a "strictly intuitive" way of definition is of minor use to anybody lacking Curry's intuitions. Moreover, he agrees that (naïve) intuition gives rise to *Russell's paradox* [Cur63, p.4]. Admittedly, Curry's primary interest is in *inductive* classes.

4.2 Inductive Constructions

Recall from Section 2.1 that an inductive class is generated from certain initial elements by certain modes of combination. To individuate classes, one needs to individuate their elements. So we need to state identity conditions for the elements of an inductive class. Let us confine our discussion to monotectonic classes. Then, by definition, two elements are identical if and only if they have the same construction. This leaves us with identity conditions for processes that consist in iterated applications of the modes of combination. We can take for granted that two modes of combination are identical if they take the same arguments to the same values; hence we can as well speak of *functions* or *functional relations* instead. But how to individuate processes of iterated applications of them? Certainly, they are not meant to be processes in space and time nor mental processes of a particular person. Apparently, we are hopelessly thrown upon Curry's appeal to intuition.[13]

From a Quinean perspective, we better dispense with processes altogether and explicate the notion of iteration in the first place. Let us assume for the moment the natural numbers as given. Then the *iterate* of a function or relation can be defined in terms of *finite sequences*.[14] Consider, for example, the monotectonic inductive class given by a class A of initial elements and a single binary function f such that $f(x, y) = f(u, v)$ only if $x = u$ and $y = v$. Let g be the function that takes a given class X to $X \cup f(X \times X)$. Then x is an element of the generated inductive class if and only if there is a natural number n such that $x \in g^n(A)$. As for the ontological commitments underlying this explication, a careful analysis shows that if we are only inclined to posit the generated formal objects and not the whole class of them, then the only postulates needed are the existence of the classes $\{0, 1, \ldots, n\}$ and *replacement* on them; *cf.* [Qui69c] for details.[15] In other words, by defining a formal object we need to refer to the class of objects it is "built of", which is a rather modest assumption if anything is.

Of course, if the question is how to individuate natural numbers we should not take them as already individuated. Curry does not share such scruples, for he freely utilizes numbers and finite sequences in his specification of inductive classes. Interestingly, when it comes to explaining his informal use of the natural numbers, Curry [Cur63, p.42] refers to the following

---

[13] It is worth mentioning that even Charles Parsons, who concedes intuition a certain role in grasping mathematical objects, is "inclined to deny that even very simple inductive conclusions are intuitive knowledge" [Par80, Sec. VIII].

[14] *Cf.* [Qui69c, §14].

[15] In particular, there is no need to postulate the existence of the union or the product of classes; for to say that $x \in g(X)$ is just a convenient way of saying that $x \in X \vee \exists y \exists z (y \in X \,\&\, z \in X \,\&\, x = f(y, z))$.

characterization:

> Any system of objects, no matter what, which is generated from a certain initial object by a certain unary operation in such a way that each newly generated object is distinct from all those previously formed and that the process can be continued indefinitely, will do as a set of natural numbers.
> [Cur63, p.12]

But this is essentially a specification of a monotectonic inductive class, with the notions of process and generation as unexplicated as ever.

Quine [Qui69c, Chap.IV] offers a definition of the natural numbers along the following lines: Take 0 as anything you like, and take as S any function such that $S(x) \neq 0$, for every x, and $S(x) = S(y)$ only if $x = y$. Then x is a natural number if

$$\forall X(x \in X \ \& \ \forall y(S(y) \in X \to y \in X) \to 0 \in X).$$

Quine's idea is that his "inverted" definition circumvents the need of infinite classes because the outer variable is required to range only over finite sets — in contrast to the classical definition of the closure specification by Frege, Dedekind, and Peano.[16] In order to make sure that there are enough finite sets, we can follow von Neumann and define 0 as $\{\}$ and $S(x)$ as $x \cup \{x\}$. Notice that assuming the existence of these sets involves genuine ontological commitments since we are not presupposing any axiomatic set theory. The existence of a natural number thus hinges on the existence of the class of all predecessors of that number.

4.3 "The Ordered Pair as a Philosophical Paradigm"

Although the characterization of inductive constructions is an important issue for the science of formal systems, one can study the nature of formal objects also by looking at such elementary constructs as the *ordered pair*. In his presentation of formal systems, Curry does not mention ordered pairs at all. Quine, on the other hand, devotes a whole section of his *Word and Object* to the ordered pair.[17] To motivate that the notion of an ordered pair calls for explication, Quine cites the following characterization by Peirce:

> The Dyad is a mental Diagram consisting of two images of two objects, one existentially connected with one member of the pair, the other with the other; the one having attached to it, as representing it, a Symbol whose meaning is "First", and the other a Symbol whose meaning is "Second".[18]

---

[16] See also the discussion in [GeoVel98, p.321 sq]. An alternative approach within weak second-order logic, with separate variables for finite sets, is proposed by [FefHel95].
[17] *Cf.* [Qui60, §53], from which the present section borrows its title.
[18] Quoted from [Qui60, p.257].

Ordered pairs are in charge when binary relations are taken as classes (of ordered pairs). They are then typically used as values of variables of quantification and are thus to be treated as entities. As to the question what ordered pairs are, Quine points out that two ordered pairs $\langle x, y \rangle$ and $\langle u, v \rangle$ are identical if and only if $x = u$ and $y = v$. This identity condition is the only thing that matters when referring to ordered pairs. In *Word and Object*, Quine puts forward the slogan that "explication is elimination", which means to systematically choose already-recognized objects as ordered pairs subject to the restriction that they satisfy the identity condition. Since he aims at ontological economy, his preferred candidates are set theoretic constructs like Kuratowski's $\{x, \{x, y\}\}$. For him, "the question 'What is an ordered pair?' is dissolved by showing how we can dispense with ordered pairs in any problematic sense in favor of certain clearer notions" [Qui60, p.260].

In later writings, Quine shifts emphasis more towards ontological relativity and indifference. The point is now not so much one of reduction or elimination but one of positing entities that satisfy such and such conditions. In order "to affirm something about a pair $\langle u, v \rangle$, and to do so without choosing any one of the various ways of constructing ordered pairs" [Qui95a, p.74], Quine considers to employ the technique of *Ramsey sentences*. A sentence about $\langle u, v \rangle$, say "$P\langle u, v\rangle$", is then replaced by

$$\exists f (\forall x \forall y \forall z \forall w (fxy = fyw \leftrightarrow x = z \,\&\, y = w) \,\&\, P(fuv)).$$

As Quine observes, "Ramsey's treatment [...] brings out indeterminacy of reference not by reinterpretation, but by waiving the choice of interpretation." He furthermore observes that "each Ramsey sentence is a fresh existential quantification; consequently there is no assurance of sameness of objects from sentence to sentence", which he sees as unproblematic in the case of formal objects because "they can be happily dismissed after each application and introduced anew for the next." We should add, however, that the existential quantification has to take scope over the whole discourse of the application in question.

The ontological scrupulous might hesitate to posit functions in order to cope with ordered pairs.[19] A more modest solution could run as follows: Take the locution "the order pair of x and y" as a definite description, i.e., if "$Fxyz$" stands for "x is the ordered pair of y and z", then "$\langle y, z \rangle$"

---

[19] Quine, in fact, speaks of an "unrealistic" assumption. What he presumably has in mind is that functions are usually defined in terms of ordered pairs (or triples, *etc.*). However, though ordered pairs are technically useful to define relations and functions as sets of ordered pairs, one can do without them, as already done so in *Principia Mathematica*.

is short for "$\iota x Fxyz$". Now eliminate the definite description in favor of a uniqueness and an existence assumption. The exact treatment of the latter is open to various options. Russell, for instance, regards the existence assumption as part of the sentence under consideration, whereas Peano and Hilbert take it as a presupposition. Although these differences are of interest to our discussion, especially the scope considerations in the case of Russell's contextual definition, lack of space prevents us from going into details.

It is tempting to suggest that there is nothing more to say about the nature of ordered pairs than to require existence and uniqueness and the condition that ordered pairs are identical only if their components are identical. This view seems to be in full accordance with Hilbert's axiomatic method. The question "What are ordered pairs?" is then answered by saying that ordered pairs are something we assume to exist and to have such and such identity conditions. Quine's above argument that things become clearer if ordered pairs are identified with sets, say with Kuratowski's representation, is not convincing. For what are sets? The only legitimate answer can be: entities we assume to exist with such and such identity conditions.

4.4 Structuralism

Structuralism is, as Charles Parsons puts it, "the view that reference to mathematical objects is always in the context of some background structure, and that the objects involved have no more to them than can be expressed in terms of the basic relations of the structure" [Par90, p.303]. The slogan is that formal objects are nothing but "positions" in structures.

Quine is a confessing structuralist—not only with respect to *formal* objects. He holds the view of a global structuralism, which is closely connected to his doctrine of *ontological relativity* or the *inscrutability of reference*:

> [...] if we transform the range of objects of our science in any one-to-one fashion, by reinterpreting our terms and predicates as applying to the new objects instead of the old ones, the entire evidential support of our science will remain undisturbed.
>
> [...] there can be no evidence for one ontology as over against another, so long anyway as we can express a one-to-one correlation between them. Save the structure and you save all. [Qui92, p.8]

For Quine, considerations of this sort "belong not to ontology but to the methodology of ontology, and thus to epistemology" [Qui81a, p.21].

Michael Resnik, who is also inclined to prefer an epistemic interpretation of structuralism, points out that one can adhere to a structuralist ontology in mathematics without committing oneself to ontological structuralism "all the way down", *i.e.*, without saying that all objects are literally positions in structures.[20] However, Resnik is only prepared to accept an ontologi-

---
[20] *Cf.* [Res97, Sect. 12.8].

cal reading of structuralism that posits "positions" as objects but not the structures themselves.

The version of structuralism favored by Stewart Shapiro, in contrast, treats structures as genuine objects. Shapiro maintains that "[Quine's] thesis of inscrutability blocks the final ratification of structuralism" [Sha97, p.141]. At the same time, Shapiro is well aware of the epistemic link between ontology and language:

> [...] grasping a structure and understanding the language of its theory amount to the same thing. There is no more to understanding a structure and having the ability to refer to its places than having an ability to use the language correctly. [Sha97, p.137]

Quine could surely agree. But Shapiro seems to be after a true ontology which is independent of our epistemic grasp, whereas for Quine, "to ask what reality is *really* like [...] apart from human categories, is self-stultifying" [Qui92, p.9].

Quine's dictum that explication is elimination gives his structuralism an *eliminative* connotation. Explication of the natural number structure, for instance, means for him to eliminate it in favor of any progression, preferably of sets:[21]

> I prefer to say with Benacerraf simply that there are no natural numbers, and there is no need of them, since whatever purposes we might have used them for can be served by any progression, and set theory affords progressions in generous supply.
> [Qui98, p.403]

Here, we can argue as in the case of ordered pairs that aside from ontological economy there is no reason not to grant natural numbers (or other formal objects) the same ontological status as sets. One can nevertheless be scrupulous with respect to explicitly positing structures in addition to positions in structures, whereas the unscrupulous may employ the technique of Ramsey sentences as indicated in our discussion of ordered pairs. Shapiro in essence takes the latter route, via second order logic and implicit definitions.

After this brief overview of structuralist positions,[22] let us seek for traces of structuralism in Curry's formalist framework. According to the above characterization of structuralism by Parsons, we need to address the following two questions: does it make sense to refer to formal objects, *i.e.*, to Curry's obs, without having a formal system in background, and is there more to a formal object than its relation to the other objects of the system?

---

[21] *Cf.* also [Qui69a, p.44*sq*].
[22] It should be noticed that there are further variants of structuralism; *cf.*, for instance, Geoffrey Hellman's *modal* structuralism [Hel89].

The negative answer to the first question is immediate since any ob belongs to an inductive class defined by some ob system. As to the second question, recall that an ob is uniquely determined by its construction; in fact, "an ob can be identified with [...] a construction, objectified, if you will, by means of a tree diagram (or a normal construction sequence)" [Cur63, p.54]. An ob is thus fully characterized by two things: a certain mode of combination and a certain tuple of obs, where the ob in question is the result of applying the mode of combination to the given tuple. Under the reanalysis given in Section 4.2 this comes down to saying that an ob is fully characterized by a certain functional relation the ob in question bears to certain other obs of the system — and there is apparently nothing more to say of an ob.

We can conclude that the structuralist viewpoint is well suited for inductive classes and hence for ob systems.[23] Furthermore notice that Curry's characterization of the natural numbers cited above in Section 4.2 cannot deny a rather strong structuralist flavor.

At the close of Section 2.2, we saw that for Curry, formal systems can be *abstract* in that it is "possible to present a system without having any specific representation in mind." In spite of Curry's reluctance concerning explicit ontological commitments, it is tempting to read the foregoing statement to the effect that formal systems or, better, the structures determined by formal systems, are posited as genuine (abstract) objects.

5 Conclusion

In sum, Curry's position concerning the nature of formal objects appears to be compatible with the basic assumptions of structuralism — and thus in some respect with a Quinean viewpoint. We can also concede a considerable agreement between Curry and Quine about the importance of language in accessing formal objects.

There is a stark contrast between their attitudes towards the role of logic and its relation to language. While Curry draws a sharp line between mathematical and philosophical logic, which goes along with a distinction between different levels of language, Quine prefers "the fiction of an all-purpose scientific language" [Qui91, p.243], with logic as its grammar. For Quine, ontological commitments are intimately connected to the logical syntax of language, whereas Curry does not address such questions at all.

Curry aims at a foundation of mathematical logic on the basis of formal systems. What makes his approach particularly interesting to the study of formal objects is his departure from Hilbert in that he does not restrict

---

[23] *Cf.* also the discussion in [Sha97, Chap.6] on structuralist interpretations of constructivism.

himself to syntactical systems. All in all, it might be worthwhile to continue the reanalysis of Curry's approach to formal objects under a Quinean perspective –as sketched in this essay– thereby retaining its constructivist appeal without falling back on the intuitively given.

References.

[Cur51]	Haskell B. **Curry**, Outlines of a Formalist Philosophy of Mathematics, North-Holland 1951
[Cur63]	Haskell B. **Curry**, Foundations of Mathematical Logic, McGraw-Hill 1963
[CurFey58]	Haskell B. **Curry** and Robert **Feys**, Combinatory Logic, volume 1, North-Holland 1958
[DalOli98]	Garth **Dale** and Gianluigi **Oliveri** (*eds.*), Truth in Mathematics, Oxford University Press 1998
[Dav99]	Donald **Davidson**, Reply to Ernie Lepore, *in:* [Hah99, p.715-717]
[FefHel95]	Solomon **Feferman** and Geoffrey **Hellman**, Predicative foundations of arithmetic, **Journal of Philosophical Logic** 24 (1995), p.1-17
[FloShi01]	Juliet **Floyd** and Sanford **Shieh** (*eds.*), Future Pasts: The Analytic Tradition in Twentieth-Century Philosophy, Oxford University Press 2001
[GeoVel98]	Alexander **George** and Daniel J. **Velleman**, Two conceptions of natural numbers, *in:* [DalOli98, p.311-327]
[Hah99]	Lewis Edwin **Hahn** (*ed.*), The Philosophy of Donald Davidson, Open Court 1999 [The Library of Living Philosophers XXVII]
[HahSch98]	Lewis Edwin **Hahn** and Paul Arthur **Schilpp** (*eds.*), The Philosophy of W. V. Quine, 2nd edition, Open Court 1998 [The Library of Living Philosophers XVIII]
[Hel89]	Geoffrey **Hellman**, Mathematics without Numbers, Oxford University Press 1989
[Lor69]	Paul **Lorenzen**, Einführung in die operative Logik und Mathematik, 2nd edition, Springer 1969
[Ore02]	Alex **Orenstein**, W.V. Quine, Princeton University Press 2002
[Par80]	Charles **Parsons**, Mathematical intuition, **Proceedings of the Aristotelian Society** 80 (1979-1980), p.145-168
[Par90]	Charles **Parsons**, The structuralist view of mathematical objects, **Synthese** 84 (1990), p.303-346
[Qui46]	Willard Van Orman **Quine**, Concatenation as a basis for arithmetic, *in:* [Qui95b, p.70-82]
[Qui51]	Willard Van Orman **Quine**, Mathematical Logic, 2nd edition, Harvard University Press 1951
[Qui60]	Willard Van Orman **Quine**, Word and Object, MIT Press 1960
[Qui61]	Willard Van Orman **Quine**, On what there is, *in:* [Qui61a, p.1-19]

[Qui61a]	Willard Van Orman **Quine**, From a Logical Point of View, 9 logico-philosophical essays, 2nd edition, Harvard University Press 1961
[Qui69a]	Willard Van Orman **Quine**, Ontological relativity, *in:* [Qui69b, p.26-68]
[Qui69b]	Willard Van Orman **Quine**, Ontological Relativity and Other Essays, Columbia University Press 1969
[Qui69c]	Willard Van Orman **Quine**, Set Theory and its Logic, 2nd edition, Harvard University Press 1969
[Qui76a]	Willard Van Orman **Quine**, Carnap on logical truth, *in:* [Qui76b, p.107-132]
[Qui76b]	Willard Van Orman **Quine**, The Ways of Paradox and Other Essays, 2nd edition, Harvard University Press 1976
[Qui81a]	Willard Van Orman **Quine**, Things and their places in theories, *in:* [Qui81b, p.1-23]
[Qui81b]	Willard Van Orman **Quine**, Theories and Things, Harvard University Press 1981
[Qui87]	Willard Van Orman **Quine**, Quiddities: An Intermittently Philosophical Dictionary, Harvard University Press 1987
[Qui91]	Willard Van Orman **Quine**, Immanence and validity, *in:* [Qui95b, p.242-250]
[Qui92]	Willard Van Orman **Quine**, Structure and nature, **Journal of Philosophy** 89 (1992), p.5-9
[Qui95a]	Willard Van Orman **Quine**, From Stimulus to Science, Harvard University Press 1995
[Qui95b]	Willard Van Orman **Quine**, Selected Logic Papers, 2nd edition, Harvard University Press 1995
[Qui98]	Willard Van Orman **Quine**, Reply to Charles Parsons, *in:* [HahSch98, p.396-403]
[Qui01]	Willard Van Orman **Quine**, Confessions of a confirmed extionalist, *in:* [FloShi01, p.215-221]
[Res97]	Michael D. **Resnik**, Mathematics as a Science of Patterns, Oxford University Press 1997
[Sel75]	Jonathan P. **Seldin**, Arithmetic as a study of formal systems, **Notre Dame Journal of Formal Logic** 16 (1975), p.449-464
[SelHin80a]	Jonathan P. **Seldin** and J. Roger **Hindley**, A short biography of Haskell B. Curry, *in:* [SelHin80b, p.vii-xi]
[SelHin80b]	Jonathan P. **Seldin** and J. Roger **Hindley** (*eds.*), To H. B. Curry: Essays on Combinatory Logic, Lambda Calculus and Formalism, Academic Press 1980
[Sha97]	Stewart **Shapiro**, Philosophy of Mathematics, Oxford University Press 1997

Received: July 2nd, 2003;
In revised version: July 31st, 2004;
Accepted by the editors: August 19th, 2004.

Benedikt **Löwe**, Volker **Peckhaus**, Thoralf **Räsch** (eds.)
Foundations of the Formal Sciences IV
The History of the Concept of the Formal Sciences

The Equals Sign: a Peircean View

Michael Otte

Institut für Didaktik der Mathematik
Universitätsstraße 25
33615 Bielefeld, Germany
E-mail: michael.otte@uni-bielefeld.de

ABSTRACT. The equals sign seems to contain a paradox or ambiguity as it might indicate the equal within the different or the different within the equal, that is it could be interpreted as a contingent fact or as as an association of ideas and thus represent either synthetic or analytic knowledge. The paper will discuss these ambiguities in the light of Peirce's semiotic pragmatism, which enables one to conceive of the matter in terms of (semiotic) activity itself, rather than merely in terms of symbolic language.

— I —

Equality is beset with a lot of riddles and gives rise to challenging questions since Robert Recorde introduced its symbol $A = B$ into mathematics in 1557, on account of the view that "no 2 thynges can be more equal" [Recorde].

The symbol $A = B$ contains, however, something equal and something different as well: therefore it differs from $A = A$. This implies that equality has to be distinguished from numerical identity. Depending on where one places one's priorities, one may see such an equation in two different ways. One may begin with a distinction and would then have to search and think of a possible relationship or similarity, between the different entities. Or one might conceive of A and B as of different properties, or representations, of one and the same object. Each of these decisions would produce philosophical and logical consequences of a different kind.

The symbol $A = B$ nowadays is, following Frege, commonly interpreted by saying that A and B are different intensions of the same extension. Both terms A and B have the same reference, while the sense or the mode of presentation is different. In Frege's famous essay on *"Sinn und Bedeutung"*, the author quotes some examples from elementary geometry. Frege writes:

> Let a, b, c be the lines connecting the vertices of a triangle with the midpoints of the opposite sides. The point of intersection of a and b is then the same as the point of intersection of b and c. So we have different designations for the same point, and these names ("point of intersection of a and b"; "point of intersection of b and c") likewise indicate the mode of presentation, and hence the statement contains actual knowledge.
> [*Sinn & Bed.*, p.41][1]

To adopt the other view of $A = B$, beginning with difference or with a distinction and then looking for some common property or some relationship might be appropriate in cases where the common extension is not given as such as in the examples of mathematical entities or in case of theoretical terms, like energy –of which heat and motion are different representations, for instance,– or of the electro-magnetic field. Even in case of Frege's now famous example "the Evening star = the Morning star" one might doubt Frege's interpretation. If I say, writes Shwayder,

> that the Morning star is the same as the Evening star what I say is not that the "the Evening star" and "the Morning star" have the same meaning or even refer to the same object; what I do say is that the brightest heavenly body different from the moon which is sometimes seen to precede the rising sun in the east is the same as that heavenly body which is at other times brightly seen in the west after the setting of the sun. [Shw65, p.18]

What we have here is an empirical fact, rather than an assertion about language. The people who have invented these names quite certainly did not know that the Evening star is the Morning star. The same is true when after some calculation we find out that $x = 25$ or that $2 + 2 = 4$; or that $(5 - 2) \cdot (5 + 2) = 21$. With respect to these equations Shwayder contests, however, that the number expressions are descriptions. This seems to be a misunderstanding, as the properties of theoretical entities like numbers are given by the axioms system and each of these equations may be verified by means of Peano's axioms.

According to Frege, "$2 + 2$" and "4" have the same meaning but distinct senses. This interpretation presupposes the existence of numbers as objects in a quasi-Platonic sense. Such views met with strong objections by constructively minded mathematicians, for whom mathematical existence claims make sense only relative to a language or an axiomatic system. Lebesgue, for example, favors a treatment of arithmetic completely within the boundaries of the decimal system of numeration. What could be the reasons for declining such an approach? — he asks.

---

[1] Author's translation.

First of all our metaphysical habits. Is it not a blasphemy to call a number a symbol since numbers once constituted the very essence of things? Here we have a fear manifesting itself in the most varied forms. For instance, let us say we may certainly use interchangeably the English word chair or the French word *chaise* because they both refer to the same object, but what is the analogue of the object chair in the use of the symbols *101* in the binary system and *5* in the decimal system? Since there is no chair hidden under 5, we can avoid the difficulty by a verbal pirouette and speak of the metaphysical entity 5, which will replace the physical reality chair. This amounts to refusing to answer the question. [Lebesgue, Mesure, p.16]

Lebesgue is able to distinguish between true and false propositions only because he considers arithmetic as an applied discipline. Mathematics is, according to him, nothing but an instrument for other sciences. And arithmetic is not a theory of its own applicability, or, as Lebesgue said: "*L'arithmétique s'applique quand elle s'applique*" (Arithmetic applies when it applies, [Fél74, p.31]).

Hilbert, like Lebesgue, considered pure mathematics to be incomplete in so far as its growth depends on applications of various kinds. This makes the intensional or instrumental view popular even in school. Frege adopts a completely extensional view of $A = B$, even saying that a statement which has meaning but no reference "belongs to poetry, not, however, to science" [Gab01, p.154]. Frege does also not admit the hypostatization of predicates and rules, or what is commonly called "definition by abstraction" by means of an equivalence relation $A = B$, because the resulting equality could not be verified with respect to the whole universe [*Grundlagen*, §66*sqq*]. Such a verification would, in fact, require that A and B were "complete concepts" in Leibniz's sense. The universality of language and logic and the fact that we cannot vary the interpretations of our logical language lead to the "ineffability of semantics" [Hin97a, p.24], and this implies that we cannot in general decide whether $A = B$ holds or not. Thus if one starts with a distinction then searching for a relation of equivalence, this makes sense only relatively to some context or theory. Two objects are considered equal if they are functionally equivalent in a certain way, specified by theory. This principle has been much emphasized by constructively minded mathematicians and in AI research. In order to establish equality in the context of an axiomatic theory, we would have to single out those functions and predicates that make up the substitution axioms which distinguish equality from other equivalence relations, those n-ary functions f or predicates p for which it holds that:

$x_i = y_i$ for $i = 1, ..., n$ implies $f(x_1, ..., x_n) = f(y_1, ..., y_n)$ and $p(x_1, ..., x_n) = p(y_1, ..., y_n)$.

At the same time Frege adopts Leibniz's definition of identity [*Grundlagen*,

§ 65], which contradicts his own arguments of *"Sinn und Bedeutung"* quoted above, according to which intensions would make a cognitive difference even if extensions were identical (*cf.* also [Ang67]). Frege's extensional view, and the rather rigid distinction between general and singular, or concept and objects associated with it, deprives meanings of their dynamical qualities. Frege claims that mathematicians do neither define concepts, nor their contents, but rather their extensions. Equality of functions or concepts is then established via the axiom of extensionality.

> For the mathematician, it is no more correct and no more incorrect to define a conic section as the circumference of the intersection of a plane and the surface of a right circular cone than as a place curve whose equation with respect to rectangular co-ordinates is of degree 2. Which of these two definitions he chooses, or whether he chooses another again, is guided solely by grounds of convenience, although these expressions neither have the same sense nor evoke the same ideas.   [Fre94]

With respect to the growth of knowledge, it seems very relevant indeed which definition is chosen, which perspective is taken, or how a problem situation is represented. Two concepts could be extensionally equivalent and yet could be different and might function differently within a certain cognitive context. Heuristically or cognitively, there are generals, like the famous general triangle of school geometry for example, which are not predicative and which cannot be described extensionally in terms of collections of particular instantiations. Such generals are objects of some representation in their own right, they are concrete universals, as they are called sometimes. A general triangle is a free variable, and not a collection of determinate triangles. Which properties are essential to a "general triangle" depends on context, on the activity and its goals. If the task, for instance, is to prove the theorem that the medians of a triangle intersect in one point, the triangle on which the proof is to be based can be assumed to be equilateral, without loss of generality — because the theorem in case is a theorem of affine geometry and any triangle is equivalent to an equilateral triangle under affine transformations. This fact considerably facilitates conducting the proof because of such a triangle's high symmetry. The truth of a proposition about the "general triangle" then means nothing but that this proposition is provable in a certain way, *i.e.*, by means of construction and analysis.

Another argument against the relational interpretation of the form $A = B$ is sometimes furnished by examples like "Socrates is rare", which seems to make no sense as long as we consider Socrates as a particular. The statement could however reasonably be rephrased as meaning "Men like Socrates are rare". Socrates would then have to be considered as a type, and as something, which Peirce has characterized by the term "subjective generality" [*CP*, 5.429].

Mathematics is neither completely intensional nor merely extensional knowledge. Both views do not really represent alternatives. Borrowing the notions of function and argument to replace the traditional logical notions of predicate and subject, one may make this obvious. In an intensional theory objects are identified by their properties. This can be expressed as follows, taking into account our convention:

$x = y$ if and only if $f(x) = f(y)$; for every function f.

This is nothing but a version of Leibniz' principle of the identity of indiscernibles (PI), which consists in the thesis that there are no two substances which resemble one another entirely, differing only numerically. And we see that Leibniz' principle is nothing but the dual of the "Axiom of Extensionality", by which concepts or functions are identified:

$f = g$ if and only if $f(x) = g(x)$; for every argument x.

This complementarity is well known from the duality principles of projective geometry and linear algebra. It makes sense only if functions or concepts on the one hand, and arguments or objects on the other are considered to be of equal ontological status, as well as, that we do not aspire to speak about "all" concepts functions or objects, but confine ourselves to the context of axiomatized theory and its models or intended applications.

These considerations lead us to consider the extensions and intensions of mathematical terms as relatively independent and as complementary to each other. We assume therefore that mathematical meaning is to be conceived in terms of the complementarity of extension and intension. Meaning has two objective components, one of which refers to objects or indicates them; the other relating to linguistic expressions or diagrammatic representations, which show the characteristics of the object of activity (which in general is not the object named) and which express how the characteristics hang together. The complementarity is established by processes of generalization and verification. Castonguay, in a similar vein, tried

> [...] to lay bare two objective components of meaning, one of which refers to objects, and which it is appropriate to name the extensional, or correspondence component of meaning; the other relating to concepts or linguistic expressions, and which it is suitable to call the intensional, or coherence, component, in that it expresses, how a given concept or expression coheres, or hangs together, with its fellows through relations of consequence.
> [Cas72, p.3]

— II —

Let us look how this complementarity is dealt with in the works of Russell and Peirce respectively.

The interpretations of $A = B$ are countless and dependent of what A and B are meant to signify, the difference between $A = A$ and $A = B$ being, however, the invariant focus of all these interpretations. This difference is captured in Russell's famous theory of description by the distinction he draws between names and descriptions. We have, Russell writes, "two things to compare: A *name*, which is a simple symbol, directly designating an individual which is its meaning (or referent), and having this meaning in its own right independently of the meanings of all other words. A *description*, which consists of several words, whose meanings are already fixed, and from which results whatever is taken as the "meaning" of the description".

> A proposition containing a description is not identical with what that proposition becomes when a name is substituted, even if the name names the same object as the description describes. "Scott is the author of *Waverley*" is obviously a different proposition from "Scott is Scott": the first is a fact in literary history, the second a trivial truism. [Rus19, p.174]

But it would not be different if "the author of *Waverley*" were just a name. "Thus so long as names are used as names, 'Scott is Sir Walter' is the same trivial proposition as 'Scott is Scott'" [Rus19, p.175]. We cannot, Russell believes, gain knowledge by just giving things new names. This view, which in fact implies that the "interpretation of a name is initially, a matter of choice" Abraham Robinson, [Rob$_2$79], is very ambiguous, and would be acceptable only if we believed in an absolute difference of objects and concepts, which we don't.

"Unicorn" then would be an abbreviated description, and "$\sqrt{-1}$" as well. For these descriptions, the affirmation "x exists" makes sense, whereas, according to Russell, "x exists" is meaningless if "x" is a name, because "x exists" means just "$x = x$" and this latter expression is hardly to be distinguished from "$x = y$", where y is another name. A name is an index, that is, an existence claim. A description in contrast has a sense but not necessarily a meaning or reference.

> We may even go so far as to say that, in all such knowledge as can be expressed in words –with the exception of "this" and "that" and a few other words of which the meaning varies on different occasions– no names, in the strict sense, occur, but what seem like names are really descriptions. [...] And so, when we ask whether Homer existed, we are using the word "Homer" as an abbreviated description: we may replace it by (say) "the author of the *Iliad* and the *Odyssey*". The same considerations apply to almost all uses of what look like proper names. [Rus19, p.178-9]

But the essential point is that indices as well as icons (predicates or descriptions) are essential, although we may never be able to separate them completely, as we always use our linguistic terms both referentially as well as attributively. To illustrate the latter point let us discuss the following example. Let us assume an English tourist visiting Amazonia sees a larger animal near the shore of a lake and asks what kind of animal this is. He receives the answer that what is seen is a *Capivara*. As the tourist does not know Brazilian Portuguese this is only an indexical or referential designation which leaves him without any representation for the moment. If he is offered, to relieve his frown, an anglification by the term "water hog", his face lights up and he says "aha", actually believing to have understood what it is, the fact being that he is able to link something meaningful with the words of "water" and "hog". This is thus a case of some kind of descriptive designation, which has the disadvantage, however, of creating completely false notions. For the *Capivara* is no swine at all, but a grass-eating rodent. The Amazonian, against that, is in the opposite situation, as for him the Indian name of *Capivara* meaning "grass-eater", while the designation "water hog" tells him absolutely nothing.

Now such a referential use sometimes serves the starting point of further observations if a motive or curiosity results. After some time, the tourist may observe some characteristics and habits of the *Capivara*, and then will be able to say "*Capivaras* are good swimmers and divers", or "the *Capivara* lives in family groups", *etc*. Gradually use of the term changes and is transformed into a description. And indeed theories *in statu nascendi* are mainly used "referentially" by their exponents as well as by their opponents while being at their zenith they are used "attributively" until a new theory emerges and ascends to its zenith, when the former theory is used "referentially" again."

The key thing about a name or an index is that it has a direct connection with its object. In case of the present example this connection is being established by concrete ostentation. It indicates its objects without giving any information about it. Therefore we are able to understand an index as a *sign* only by means of some "collateral experience" or contextual acquaintance with what the sign denotes to make the interpretation work.

> For instance, I point my finger to what I mean, but I can't make my companion know what I mean, if he can't see it, or if seeing it, it does not, to his mind, separate itself from the surrounding objects in the field of vision. It is useless to attempt to discuss the genuineness and possession of a personality beneath the histrionic presentation of Theodore Roosevelt with a person who recently has come from Mars and never heard of Theodore before. [*CP*, 8.314]

In everyday knowledge names are given on the basis of some arbitrarily chosen characteristics. The cuckoo got his name from the sound he makes. In scientific taxonomy, in contrast, the name is chosen to locate the object named within some theoretical context and the choice is thus constrained. Within axiomatized mathematical theories the names are constrained by the axiom system, that is they apply to those entities and only to those, which comply with the axioms. But the names are general names, they refer to universal, not completely determined objects and they may thus not have a proper denotation at all (if the axiom system is inconsistent, for example). These names are to be understood in intension, or as Russell might say, they are just descriptions. Theory as a whole determines the intensions of its terms and the intensions determine the extensions. The theory becomes a world of its own totally incommensurable with any other theory. Theory development thus is completely discontinuous, or revolutionary and without reason. The growth of knowledge becomes incomprehensible as soon as theories are exclusively taken in intension and are identified with their languages.

For several years after the publication of Kuhn's *The Structure of Scientific Revolutions* lively discussions were held concerning the problem of continuity and discontinuity of scientific development. The extreme positions were Feyerabend's incommensurability thesis of historically successive theories, on the one hand, and the traditional cumulative model of scientific development on the other. The conception of meaning as a dual entity might be helpful here. Marta Feher, for instance, writes:

> We shall say that scientific terms and descriptive phrases, the senses or intensions of which are given by the systems of laws and lawlike statements belonging to the theories, can be (and are) used "attributively" as well as "referentially" in scientific discourse. That is, the terms occurring in the laws of a theory can be regarded, on the one hand as giving "descriptions" of their referents, to be applied to those and only those entities with reference to which they are true and so referring to those objects, which they are denoting [...]
>
> Let us now turn to the second interpretation according to which the laws and the terms contained in them can be used "referentially" too. In this case we do not regard the expressions of the theory as referring to those objects which satisfy the given denotation, but as saying something (may be falsely) about objects, *i.e.*, about referents fixed independently of the given description [...] Summing up: in our opinion the thesis of meaning variance does not lead to such severe consequences as Feyerabend assumes, provided one is ready to accept the possibility as well as the actuality of the distinction given above. [Feh81, p.342*sq*]

And the structural theory of science, following Suppes and Sneed, had reminded us, a theory remains incomplete without an area of intended applications, presented independently from the theory itself.

— III —

> The decisive insight both for Frege and for Peirce was that a judgment is not an aggregate of terms which represent concepts or classes but that its elements have different kinds of roles in their contexts. Two of those basic roles were that of representing relations and that of denoting individuals. My suggestion is that this idea occurred to them because they first confessed the priority of judgments over their constitutive concepts and then started to analyze judgments following Kant's epistemological lessons.
> [Haa93, p.115sq]

In what consisted these Kantian lessons then? The essential point seems to refer to Kant's affirmation that no description of existence is possible. Existence cannot be defined nor proved but must, according to Kant be given in intuition, or as one might say to-day, must be indicated or postulated. "Being is not a real predicate", says Kant [Kr. d. r. V., B 625].

We cannot establish existents by means of language, nor fix the referents of our linguistic terms by descriptions. This Kantian insight is a result of his criticism of the Ontological Argument for the existence of God. The kernel of this argument of the Rationalists of the 17th century was to claim that the notion of the nonexistence of God is a contradiction; for God is perfect and existence is perfection, hence God must exist.

Rationalism depended on this argument, because Gods mind was supposed to provide stability, generality and truth for human intuition and knowledge (cf. [Hac80]). The immense dynamics, which resulted in the *Scientific Revolution* of the 17th century was based on bold imaginations or anticipations, fostered and expressed by new representational systems and operative ideas, as their meanings. The new mathematics and logic of the time was conceived of as purely intensional. The mathematical ideas were the objects themselves, capable of being represented formally and existing only in the mind (of God). Leibniz argued that in reality there are no indivisible atoms and that the simple indivisible and enduring substances, on which everything real is to be based, must be spiritual entities or monads. And he claimed that even sensory perceptions are more accurately interpreted as the properties of unextended monads. Therefore the ideal has to found the existent, and for this God's mind was indispensible.

Kant objected to the Ontological Argument, and one of his reasons for objecting to it was that, in his view, existence is not a property. Since to say of some x that it exists adds nothing to the concept of x, "exists" is not a predicate. Kant concluded that there are two sources of knowledge –concepts and intuitions– and he made the latter the basis of possible existence claims.

Grayling believes

> What has been said so far is not quite right as it stands, for, in a sense, to say of tigers that they exist does add something; it says that the concept of a tiger has instances in reality — that is, that *there are* tigers to be met with in the world. [Gra97, p.89/90]

A statement of existence is, according to this view, a higher-order statement involving reference to a propositional function. It becomes a second order predicate. This is the orthodox view, of which Russell was the main proponent: "Existence is essentially a property of a propositional function. It means that the propositional function is true in at least one instance" [*Phil. Log. Atom.*, p.232]. McGinn illustrates this traditional conception as follows:

> When you think that tigers exist you do not think of certain feline objects that each has the property of existence, rather you think of the property of tigerhood, that it has instances. [...] The concept of an object existing simply is the concept of a property having instances. [McG00, p.18]

Or stated in Russell's terms: To say that tigers exist is to say, that "x is a tiger" is sometimes true.

It is, however, necessary to postulate *existent* things as instantiating the property. Russell wanted to furnish an objective possibility for this by means of his axiom of infinity, when trying to reduce the number concept to logical and set theoretic terms (*cf.* [Rus19]). The minimal requirements to justify the concept of number, or the propositional function "x is a number" are to show that it has instantiations, that it is not "empty" in the Kantian sense. We have to understand "number as the number of a quantity" and to provide an application for the concept thus defined by demonstrating the existence of sets of arbitrary cardinality. This can obviously only be done axiomatically. In doing so, however, the notion of axiom must not be understood in the Peano-Hilbertian sense; the term must rather be conceived of according to the classical Euclidean tradition, that is as an intuitively evident truth and as a precondition of mathematics. Mathematical axioms, in the modern, Hilbertian sense, make no existence claims, with respect to the objects, so described. This stands in definite contrast to Euclidean axiomatics. This is why Russell introduces an "axiom of infinity".

Kant assumes that all our knowledge extending cognitions are synthetical. Kant's synthesis springs from the function of cognizant consciousness itself which this way becomes aware of itself. The synthetic unity of consciousness, according to Kant, is "an objective condition of all knowledge. [...] For in the absence of this synthesis, the manifold would not be united in one consciousness" [*Kr. d. r. V.*, B 138].

For Kant this synthesis was primarily intended to provide an objective content for our thoughts. Kant admits, however, that pure mathematics is not concerned with objective existence at all [*Kr. d. r. V.*, B 747]. Peirce now stresses that this very unity is due to the objective reality of the representation. Thirdness, or representation, replaces Kant's so-called "highest point", that is, synthetic unity of consciousness. Under the perspective of Thirdness the human subject is to be characterized primarily by its capacity to grow, or to learn and evolve, rather than by mere constructivity.

> The highest kind of synthesis is what the mind is compelled to make neither by the inward attractions of the feelings or representations themselves, nor by a transcendental force of necessity, but in the interest of intelligibility, [...]
>
> Kant gives the erroneous view that ideas are presented separated and then thought together by the mind. This is his doctrine that a mental synthesis precedes every analysis. What really happens is that something is presented which in itself has no parts, but which nevertheless is analyzed by the mind, that is to say, its having parts consists in this, that the mind afterward recognizes those parts in it. Those partial ideas are really not in the first idea, in itself, though they are separated out from it. It is a case of destructive distillation. When, having thus separated them, we think over them, we are carried in spite of ourselves from one thought to another, and therein lies the first real synthesis. An earlier synthesis than that is a fiction.
> [*CP*, 1.383-84]

Which resembles closely Marx's characterization of the dialectical method. Still Peirce had philosophically been educated by Kant and the conclusions he had drawn from Kant's "epistemological lessons" he summarized in a review, published in 1885, of a book of the Hegelian philosopher J. Royce. Royce, Peirce writes,

> seems to think that the real subject of a proposition can be denoted by a general term of the proposition; that is, that precisely what it is that you are talking about can be distinguished from other things by giving a general description of it. Kant already showed, in a celebrated passage of his cataclysmic work, that this is not so; and recent studies in formal logic have put it in a clearer light. We now find that, besides general terms, two other kinds of signs are perfectly indispensable in all reasoning. One of these kinds is the *index*, which like a pointing finger exercises a real physiological force over the attention, like the power of a mesmerizer, and directs it to a particular object of sense. One such index at least must enter into every proposition, its function being to designate the subject of discourse. [...] But the index, which in point of fact alone can designate the subject of a proposition, designates it without implying any characters at all [...]
>
> One instant of time is, in itself, exactly like any other instant, one point of space like any other point; nevertheless dates and positions can be approximately distinguished. And how are they so distinguished? By *intuition* says Kant; perhaps not in so many words; but it is because of this property that he distinguishes Space and Time from the general conceptions of the understanding and sets them off by themselves under the head of intuition.

> But I should prefer to say that it is by volitional acts that dates and positions are distinguished. The element of feeling is so prominent in sensations, that we do not observe that something like Will enters into them, too.
> [CP, 8.41]

Peirce claims that intuition or feeling with its tendencies to see analogies, on the one hand, and volition or the activity of drawing distinctions, on the other hand, normally occur together, or semiotically speaking, that icons and indices can rarely be completely separated. It is impossible to find a proposition so simple as not to have reference to indices as well as icons.

> Take, for instance, "it rains". Here the icon is the mental composite photograph of all the rainy days the thinker has experienced. The index, is all whereby he distinguishes that day, as it is placed in his experience.
> [CP, 2.438]

Once more we come to the conclusion that the complementarity of attributive and referential uses of concepts provides both an essential orientation and a fundamental problem.

"What I call volition", Peirce says, still in the passage quoted from the review of Royce,

> is the consciousness of the discharge of nerve-cells, either into the muscles, etc., or into other nerve-cells; it does not involve the sense of time (*i.e.*, not of a continuum) but it does involve the sense of action and reaction, resistance, externality, otherness, pairedness. It is the sense that something has hit me or that I am hitting something; it might be called the sense of collision or clash. It has an outward and an inward variety, corresponding to Kant's outer and inner sense, to will and self-control, to nerve-action and inhibition, to the two logical types $A : B$ and $A : A$.
>
> The capital error of Hegel, which permeates his whole system in every part of it, is that he almost altogether ignores the Outward Clash. Besides the lower consciousness of feeling and the higher consciousness of nutrition, this direct consciousness of hitting and of getting hit enters into all cognition and serves to make it mean something real. It is formal logic which teaches us this. [CP, 8.41]

The conceptions of logics Peirce has in mind here are those developed by DeMorgan, Grassmann and Schröder, alongside with his own. Van Heijenoort had, as was said already, characterized them by the term "logic as calculus". He writes:

> Answering Schröder's criticism of his *Begriffsschrift*, Frege states that, unlike Boole's, his logic is not a calculus ratiocinator, or not merely a calculus ratiocinator, but a lingua characteristica. If we come to understand what Frege means by this opposition, we shall gain useful insight into the history of logic. [vHe67, p.324]

Now the logics of Boole, Grassmann, or Schröder evolved from an observation of the development of mathematical thought during the 19th century. And the mathematician uses, in fact, "exists" as a predicate, but employs it relatively to a model or an intended universe of discourse. The mathematician uses what Peirce had called "degenerate indices". Does the real number x, which makes the equation $x^2 = -1$, or written differently, $x = -\frac{1}{x}$ true, exist? If so, it must be equal to 1 or to -1 and this yields $1 = -1$; a contradiction. But the mathematician enlarges his universe and finds a new system of numbers, complex numbers. Truth seems to depend on consistency. But consistency is given relative to a possible world or model. And model theory vice versa "presupposes the view of logic as a calculus... The development of the notion of model and the emergence of the idea of truth have gone largely hand in hand in our century", writes Hintikka [Hin97a, p.25-29].

Neither existence nor identity can be defined absolutely, but must be stated or affirmed and this can be done only relatively to some universe of discourse. Truth thus becomes a relative term and we would have to search for some intended applications to get a sense of truth according to common sense understanding.

— IV —

Mathematical thought, as Aristotle already says, begins with the Pythagoreans, with *"theoremata"* like: "The product of two odd numbers is odd". Or: "If an odd number divides an even number without remainder, it also divides half that number without remainder". These are theorems which, as one says, go beyond what can be experienced concretely, because they state something about infinitely many objects. Actually, they do not state anything at all about objects (*e.g.*, about numbers), but they are analytic sentences, which unfold the meaning of certain concepts or hypostatic abstractions. This kind of conceptual inference finds its most exalted expression in modern axiomatics, a method not at all confined to mathematics and logic.

How precisely do we prove, however, those seemingly analytic propositions, like the already quoted "the product of two odd numbers is odd"? We intuitively represent certain activities. We will say, for instance, if an odd number is divided by 2, there will, by definition, be left a remainder of one. The concept's meaning is represented as a hypothetico-deductive statement or operation. From this we infer that there is for each odd number x another number n such as that $x = (2n+1)$ holds. If we now have two odd numbers represented in this way before us, and if we multiply these, the said theorem will result quasi automatically by applying the distributive and commutative laws. Mathematics typically proceeds by constructing (algebraic or geometric) diagrams and by observing them and analyzing the effects certain activities have on them. Mathematical judgments thus become apodictic and intuitive and the diagrammatization serves the purpose to show this (as Kant had already affirmed so strongly).

We understand a mathematical diagram in the sense of Wittgenstein's famous "picture theory" of language — the diagram or picture "represents reality by exhibiting the possibility of the existence or non-existence of a relation" [*Tractatus*, 2.201], it has the "logical form" in common with what is represented [*Tractatus*, 2.2].

Observe, that the diagram makes no absolute existence claims with respect to definite matters of fact or factual things. The indices employed serve to fix reference in the first place. One could, however, claim that mathematical concepts, like the number concept in the present example are themselves to be represented in the form $A = B$, where the "equality" perhaps might sometimes be better understood as function or representation according to which a certain property A implies another one B: $A \to B$. Mathematical reasoning consists, according to Peirce, as was quoted above already,

> in constructing a diagram according to a general precept, in observing certain relations between parts of that diagram not explicitly required by the precept, showing that these relations will hold for all such diagrams, and in formulating this conclusion in general terms. [*CP*, 1.54]

In the example $A = B$ is established by means of definitions in a rather straightforward manner. Very often things are, however, much more complicated and involved. Bernays has exemplified this problem using the example of the rule of *modus ponens*.

> Suppose that in a formal inference this rule has to be employed and suppose that neither A nor "A implies B" are among the formulas to begin with. Rather, we have a chain of deductions S resulting in A, and another one T, which leads to "A implies ". Then these formulas together provide the formula B according to the rule of modus ponens. If we want analyze what is going on here, we have to be careful not to anticipate the essential point by our symbolization already. The formula which stands at the end of the chain of reasoning T is given to us only by this very chain of reasoning, and to see that this formula is identical with the other, namely "A implies B" represents new knowledge. The verification of an identity is by no means always a tautological act. [Ber76, p.26]

This verification depends in principle on symbolizing all possible multiple relationships, or rather on their conjoint effects. The consequences of a set of premises are not always to be known by means of a simple analysis of meaning relations, because they depend, as Bernays emphasizes, on a combination of assumptions and rules. It cannot be explicitly described or defined in general which combination of assumptions and rules has to be taken into account. The set of presuppositions does not form a well defined set, but is rather, as Peirce would express it, a sort of continuum. The situation is similar to that in the empirical sciences, where the results do not depend exclusively on the natural laws, but also on the initial conditions.

Leibniz already had stated that the infinite proofs of the contingent truths of matter demand the use of the continuity principle and cannot be established by logical analysis alone. Today one might add that Leibniz' insight has to be applied also on the problem of the proof of the necessary truths of pure mathematics. Gödel's incompleteness results could, in fact, be interpreted in this direction.

> Deductive completeness can fail either because the non-logical theory is descriptively incomplete or because the underlying logic is semantically incomplete. Gödel's result does not as such say which of these happens, although in fact he did assume that the logic he was using was first-order logic which he had himself proved complete. [Hin97a, p.40]

Peirce affirms that in the more involved cases a "theorematic deduction" is required, which

> performs an ingenious experiment upon the diagram,[2] and by the observation of the diagram, so modified, ascertains the truth of the conclusion.
> [CP, 2.267]

This modification depends on the abductive introduction of a new idea according to which the diagram is then modified to render the conclusion more or less obvious.

> What I call the theorematic reasoning of mathematics consists in introducing a foreign idea, using it, and finally deducing a conclusion from which it is eliminated. ... The principal result of my closer studies of it has been the very great part, which an operation plays in it, which throughout modern times has been taken for nothing better than a proper butt of ridicule. It is the operation of abstraction, in the proper sense of the term, which, for example, converts the proposition "Opium puts people to sleep" into "Opium has a dormitive virtue" [...]
>
> I am able to prove that the most practically important results of mathematics could not in any way be attained without this operation of abstraction.
> [Eis76, IV.42-49]

One might thus summarize things by saying that deduction requires a generalization, that is, the introduction of an ideal object or relation, resp. a law to mediate between the premises and the conclusion. Every law, or general rule, or symbol is categorized by Peirce as *Thirdness*; whereas icons

---

[2]That is, on the image of the premisses.

represent *Firstness* and indices *Secondness*. "In Deduction, then, *Firstness* by the operation of *Thirdness* brings forth *Secondness*" (Peirce), that is, a mathematical fact.

With respect to our problem of understanding the meaning of "=" or of $A = B$ we come to the conclusion that it is either a formal fact, to be verified by syntactical operations, or that there is given a relation, which is mediated by some sort of idea, by some universal, as in what Peirce had called "theorematic reasoning". In this case, $A = B$ represents, in fact, not a dyadic relation, but rather a triadic one. All relations of meaning, all meaningful mediations, are triadic, according to Peirce. In his "*A Guess at the Riddle*" (1890), as well as in "*One, Two, Three: Fundamental Catagories of Thought and of Nature*" (1885) or in his letter to Lady Welby of 1904, October 12, and at various other places in his collected writings[3] Peirce describes the fundamental importance of triadic relationships to his philophy and logic and he claims

> that all plural facts can be reduced to triple facts [...] any number of termini can be connected by roads which nowhere have a knot of more than three ways. [*PEP*, V.244]

> A tetradic, pentadic, etc. relationship is of no higher nature than a triadic relationship; in the sense that it consists of triadic relationships and is constituted of them. But a triadic relationship is of an essentially higher nature than a dyadic relationship, in the sense that while it involves three dyadic relationships, it is not constituted by them. If A gives B to C, he, A, acts upon B, and acts upon C; and B acts upon C. Perhaps, for example, he lays down B, whereupon C takes B up, and is benefited by A. But these three acts might take place without that essentially intellectual operation of transferring the legal right of possession, which axiomatically cannot be brought about by any pure dyadic relationships whatsoever. [*CP*, VI.323]

Recently logicians have interpreted $A = B$ in a manner similar to Peirce and have provided new solutions to the semantic paradoxes on this basis (*cf.* [Wen01]).

Primary Sources.

[*Log. Unters.*]	Gottlob **Frege**, Logische Untersuchungen, *in:* [Pat03]
[*Grundlagen*]	Gottlob **Frege**, Die Grundlagen der Arithmetik: eine logisch-mathematische Untersuchung über den Begriff der Zahl, Koebner 1884; *english translation in:* [Aus68]
[*Sinn & Bed.*]	Gottlob **Frege**, Über Sinn und Bedeutung, **Zeitschrift für Philosophie und philosophische Kritik** NF 100 (1892), p.25-50

---

[3]See for instance: [*CP*, 1.474, 1.520, 2.86, 3.424, 4.572, 5.469, 5.473, 6.323].

[Phil. Log. Atom.]	Bertrand **Russell**, The Philosophy of Logical Atomism (1918), *in:* [Mar71, p.177-281]
[Kr. d. r. V.]	Immanuel **Kant**, Critik der reinen Vernunft, *in:* [Wei83]
[Lebesgue, Mesure]	Henri Léon **Lebesgue**, Sur la mesure des grandeurs, 8 parts, **Enseignement Mathématique** 31 (1932), p.173-206; 32 (1932), p.23-51; 33 (1934), p.22-48, p.177-213; 33 (1935) p.270-284; 34 (1936), p.176-219; *english translation in:* [May66]
[CP]	Charles **Hartshorne**, Paul **Weiss**, and Arthur W. **Burks** (*eds.*), Collected Papers of Charles Sanders Peirce, 8 volumes, Harvard University Press 1931-1958; *Manuscripts are referred to by volume and paragraph*
[PEP]	**Peirce Edition Project** (*ed.*), The Writings of Charles S. Peirce vol. I-V, Indiana University Press 1982-1989; *Manuscripts are referred to by volume and page number.*
[Recorde]	Robert **Recorde**, The Whetstone of Witte, London 1557; *facsimile:* Theatrum Orbis Terrarum 1969 [The English experience 142]
[Tractatus]	Ludwig **Wittgenstein**, Tractatus logico-philosophicus, *in:* [Vos01]

References.

[Ang67]	Ignacio **Angelelli**, On Identity and Interchangeability in Leibniz and Frege, **Notre Dame Journal of Formal Logic** 8 (1967), p.94-100
[Aus68]	John L. **Austin** (*ed., trans.*), Gottlob Frege, The Foundations of Arithmetic: A Logico-Mathematical Enquiry into the Concept of Number, Northwestern University Press 1968
[Ber76]	Paul **Bernays**, Abhandlungen zur Philosophie der Mathematik, Wissenschaftliche Buchgesellschaft Darmstadt 1976
[Cas72]	Charles **Castonguay**, Meaning and Existence in Mathematics, Springer 1972
[Eis76]	Carolyn **Eisele** (*ed.*), The New Elements of Mathematics by Charles S. Peirce, vol. I-IV, Humanities Press 1976
[Feh81]	Márta **Fehér**, Some Remarks on Meaning Invariance and Incommensurability, **Science of Science** 2 (1981), p.339-344
[Fél74]	Lucienne **Félix**, Henri Lebesgue, Message d'un mathématicien, pour le centenaire de sa naissance, Introductions et extraits choisis, A. Blanchard 1974
[Fre94]	Gottlob **Frege**, Rezension von Dr. E. G. Husserl: Philosophie der Arithmetik, Psychologische und logische Untersuchung, Erster Band, **Zeitschrift für Philosophie und philosophische Kritik** 103 (1894), p.313-332
[Gab01]	Gottfried **Gabriel** (*ed.*), Gottlob Frege, Schriften zur Logik und Sprachphilosophie, Aus dem Nachlaß, Meiner 2001 [Philosophische Bibliothek 277]
[Gra97]	Anthony C. **Grayling**, An introduction to philosophical logic, 3rd edition, Blackwell Publishers 1997

[Haa93]	Leila **Haaparanta**, Charles Peirce and the Logic of Logical Discovery, *in:* [Moo93, p.158-179]
[Hac80]	Ian **Hacking**, The Theory of Probable Inference: Neyman, Peirce and Braithwaite, *in:* [Mel80, p.141-160]
[vHe67]	Jean van **Heijenoort**, Logic as Calculus and Logic as Language, **Synthese** 17 (1967), p.324-330
[Hin97a]	Jaakko **Hintikka**, *Lingua Universalis* vs. *Calculus Ratiocinator*, *in:* [Hin97b]
[Hin97b]	Jaakko **Hintikka**, An Ultimate Presupposition of Twentieth-Century Philosophy (Selected papers 2), Kluwer 1997
[Mar71]	Robert C. **Marsh** (*ed.*), Logic and Knowledge, G.P. Putnam's Sons 1971
[May66]	Kenneth O. **May** (*ed., trans.*), Henri Léon Lebesgue, Measure and the integral, Holden-Day 1966 [The Mathesis series]
[McG00]	Colin **McGinn**, Logical properties: identity, existence, predication, necessity, truth, Oxford University Press 2000
[Mel80]	David H. **Mellor** (*ed.*), Science, Belief and Behavior: Essays in Honour of R.B. Braithwaite, Cambridge University Press 1980
[Moo93]	Edward C. **Moore** (*ed.*), Charles S. Peirce and the Philosophy of Science, The University of Alabama Press 1993
[Pat03]	Günther **Patzig** (*ed.*), Gottlob Frege, Logische Untersuchungen, Vandenhoek & Ruprecht, 2003 [Kleine Reihe V&R 4031]
[Rob$_0$79]	Georg W. **Roberts** (*ed.*), Russell Memorial Volume, Allen+Unwin 1979
[Rob$_1$67]	Richard S. **Robin**, Annotated Catalogue of the Papers of Charles S. Peirce, University of Massachusetts Press 1967
[Rob$_2$79]	Abraham **Robinson**, On Constrained Denotation, *in:* [Rob$_0$79, ch.6]
[Rus19]	Bertrand **Russell**, Introduction to Mathematical Philosophy, Routledge 1919
[Shw65]	David S. **Shwayder**, "=", **Mind** 105 (1965), p.16-37
[Tex02]	Mark **Textor** (*ed.*), Gottlob Frege, Funktion – Begrif – Bedeutung, Vandenhoeck & Ruprecht 2002 [Sammlung Philosophie 4]
[Vos01]	Wilhelm **Vossenkuhl** (*ed.*), Ludwig Wittgenstein, Tractatus logico-philosophicus, Akademie Verlag 2001
[Wei83]	Wilhelm **Weischedel** (*ed.*), Immanuel Kant, Kritik der reinen Vernunft, 2 volumes, Wissenschaftliche Buchgesellschaft Darmstadt 1983 [Immanuel Kant Werke in zehn Bänden 3-4]
[Wen01]	Lan **Wen**, Semantic Paradoxes as Equations, **Mathematical Intelligencer** 23 (2001), p.43-49

Received: May 19th, 2003;
In revised version: February 16th, 2004;
Accepted by the editors: February 24th, 2004.

Benedikt **Löwe**, Volker **Peckhaus**, Thoralf **Räsch** (*eds.*)
Foundations of the Formal Sciences IV
The History of the Concept of the Formal Sciences

Mathematics — Cultural Product or Epistemic Exception?

SUSANNE PREDIGER

Institut für Entwicklung und Erforschung des Mathematikunterrichts
Fachbereich Mathematik
Universität Dortmund
44227 Dortmund, Germany
E-mail: prediger@math.uni-dortmund.de

> ABSTRACT. How does mathematics differ from the natural sciences and the arts? Many differences can be stated but one of the most important deals with the epistemic status of the disciplines: Whereas the arts and even the natural sciences are often seen to be invented by humans and culturally influenced in their developments, mathematics is still regarded as an epistemic exception, a culture-independent discipline without any contingency. This paper emphasizes a cultural but not relativist view on mathematics searching for an explanation for the high level of coherence of mathematical theories and concepts and the wide-spread consensus among mathematicians.

1 A duality in experiencing mathematics as a culture

> It is characteristic for mathematics as a scientific domain that it has disconnected from everyday life and the socio-cultural foundation which it originally came from. Scarcely any other subject regards itself that definitively as being independent of time, values and culture. The exclusive reference on the formal and the abstractable [...] makes it difficult to discuss the relation between mathematics and cultural or social elements. Mathematics is [...] widely seen as the paradigm of the formal, the structural, or the algorithmical and contrasted to culture — *i.e.*, the historical, the dynamical, the informal, or the intuitive or social: thus mathematics and culture are conceived to be extremes, which are not reconcilable.
>
> [Sch$_1$00, p.452]

This quote provides us with a concise characterization of mathematics as it is conceived in the minds of many non-mathematicians, but also mathematicians and mathematics teachers. However, in the disciplines that systematically reflect on mathematics, this characterization is questioned more and more. Rejecting the old, absolutist image of mathematics, many philosophers of mathematics have established humanistic or social constructivist positions, in which mathematics is understood as a *cultural product* (*cf.* [Ern98, Tym85, ResvBeFis93] for the social-constructivist view, or [Whi$_0$93] for the humanistic position). Emphasis is put on both aspects: "product" accounts for the fact that mathematics cannot only be discovered but must be created by humans. These creations always take place in a specific cultural setting, thus it is a "cultural" product.

One of the most prominent proponents of this position is Reuben Hersh. In his book *What is mathematics, really?* [Her97], he describes mathematics as a human activity:

> From the viewpoint of philosophy mathematics must be understood as a human activity, a social phenomenon, part of human culture, historically evolved, and intelligible only in a social context. [Her97, p.11]

To understand mathematics as a cultural product means to acknowledge the human influence on mathematics. From this perspective there is no principal difference in the epistemic status between mathematics and the arts or natural sciences.

Nevertheless, Schroeder has pointed out an important difference in the way individuals experience mathematics: whereas in other disciplines, the human and cultural roots of the discipline are still visible, in mathematics every individual is confronted with an "objectively accessible" theory, an apparently unchangeable corpus of ideas, notions, and theorems. Mathematical realities seem to have an existence independent of the human mind.

This experience has led to long discussions about the ontological status of mathematical objects. A convincing explanation for this experience has been given by Leslie White. He explains this phenomenon by locating the mathematical reality in the intersubjective world of culture:

> Mathematics does have objective reality. And this reality, as Hardy insists, is not the reality of the physical world. But there is no mystery about it. Its reality is cultural: the sort of reality possessed by a code of etiquette, traffic regulations, the rules of baseball, the English language or rules of grammar. [Whi$_1$47, p.302f]

The dual character of mathematical experience has recently been described by Jessica Carter also in the case of working mathematicians: In her historical case study about the development of K-theory, she puts emphasis on the phenomenon that on the one hand, mathematical objects are invented by individual mathematicians, but on the other hand, their objects gain an autonomous existence, a reality that is outside the individual's range

[Car02]. The term "constructivist realism", by which she names her position, fits well with Whites cultural approach if we locate the "reality" of this "realism" in the intersubjective world of the mathematicians' culture.

> The concept of culture clarifies the entire situation. Mathematical formulas, like other aspects of culture, do have in a sense an 'independent existence and intelligence of their own'. The English language has, in a sense, 'an independent existence of its own'. Not independent of the human species, of course, but independent of any individual or group of individuals, race or nation. [Whi$_1$47, p.295]

For White, mathematics is not only a cultural *product*, but a *living culture*. This shift of emphasis has various implications, especially concerning the dynamical character of mathematics, the importance of implicit knowledge, and the sociological dimensions of the discipline.

Even if the duality of mathematical realities can partly be explained by locating them in the intersubjective world of culture, this explanation is not completely satisfying, because it cannot give any account for the *difference* between mathematics and other disciplines. As long as the similarities between mathematics and other disciplines concerning their cultural character are emphasized, we cannot explain why mathematics is in fact experienced in a different why than other disciplines, as it is described in the first quotation of this section.

2 Coherence and consensus in mathematics — evidence for the epistemic exception?

Why is mathematics experienced as being more disconnected from its cultural roots than other disciplines? Why do mathematical objects seem to have an existence being more autonomous than the products of other scientific disciplines?

The most obvious answer, appealing to the formalization of argumentation, is prominent and often given: the axiomatic-deductive constitution of mathematical reasoning eliminates human influences to the highest possible degree. It is the achievement of Bettina Heintz to have specified two other important phenomena: the high coherence of mathematical concepts and theories, and the wide consensus among mathematicians [Hei00]. Since they often serve as evidence for claiming a special epistemic status for mathematics, they shall be considered in detail here. Heintz describes *coherence* in mathematics as follows:

> In contrast to other domains that decompose into separate and partly contradictory theories, mathematics is still a connected ensemble. In view of the enormous specialization [...] this coherence is not natural by any means. Mathematics is a collective product but not coordinated centrally. There

> is no authority which would ensure that the individual results match one another. But although mathematicians operate relatively isolated and restrict themselves to a small domain of work, connections can be discovered again and again between areas which were developed independently.
> [Hei00, p.19]

Coherence, in Heintz's sense, refers on the logical level to the fact that there are only very few (famous) contradictions within mathematical theories. On the conceptual level, it refers to unexpected connections between concepts from different branches of mathematics. The logical and the conceptual aspect have a counterpart on the discursive level, the high *consensus* among mathematicians. For Heintz, this aspect is even more important:

> Ludwig Wittgenstein says in a famous passage that in mathematics, there is hardly a controversy, and if there is one, 'it is safe to decide' (Wittgenstein 1983: 571). In contrast to other sciences, mathematics does not provide any flexibility for interpretation. The conclusions of mathematics are mandatory. Whoever follows the rules of the mathematical method will inevitably arrive at the same result. [Hei00, p.20]

Starting from this observation, even the sociologist Heintz (certainly not a supporter of an absolutist view of mathematics) comes to the conclusion that mathematics somehow is an epistemic exception:

> Modern mathematics is characterized by features that hardly leave a scope for a sociological analysis. [...] A sociological perspective is legitimate and appropriate where it concerns the reconstruction of the development which led to that epistemic structure being typical for modern mathematics and singular in its coherence and argumentative rationality. [Hei00, p.274-275]

If Heintz were right, mathematics would not only be an inappropriate domain for a sociological analysis of social factors of influence on the scientific development, as she discusses, but also immune against human influence. Hence, it would be useless to consider mathematics as a cultural product. A human factor could only be detected in the historical development of mathematics, during the long phases in which humans made decisions, *e.g.*, about the style or the rigor of formal proofs. Due to its mandatory conclusions, contemporary mathematicians could only be creative in their ways of discovering theorems and proofs. In its consequence, Heintz's thesis of the "special epistemic status" claims that contingency in mathematics is only located in the *ways* to mathematical contents not in the *contents* themselves.

We do not need to follow David Bloor in his strong programme for the sociology of mathematics with its emphasis on contingency or even relativity [Blo91] in order to reject the thesis of the epistemic exception that is outlined by Bettina Heintz. Emphasizing the character of the "proving discipline" (subtitle of [Hei00]), Heintz ignores crucial areas of mathematical activities in her theoretical conclusions. She neglects the entire process of mathematization (*i.e.*, the question of how initial non-mathematical problems are to be translated into mathematics), concept formation, the development of theories, as well as the criteria of relevance for research questions:

- How are mathematical concepts found?

- What influences the process of concept formation?

- How does the community decide whether a problem is adequately mathematized?

- Which factors affect the development of a theory?

- Who decides about the relevance of questions or theorems?

In all these questions, the contingent character of mathematics is much more evident than when one restricts oneself to the process of proving theorems.

Following Hersh in his distinction between the "front" and the "back" of mathematics (*i.e.*, the way of presenting finished mathematics and the creating of mathematics, resp.), we can see that consensus is essentially restricted to the "front":

> There's amazing consensus in mathematics as to what's correct or accepted. But just as important is what's interesting, important, deep or elegant. Unlike correctness, these criteria vary from person to person, speciality to speciality, decade to decade. They're no more objective than esthetic judgments in art or music. [Her97, p.39]

Just as many philosophers of mathematics, the sociologist Heintz has taken the easy way out and concentrated her epistemological considerations exclusively on proofs. In contrast, in the empirical part of her study she describes that proofs only appear at the end of the mathematicians' working process. And it is exactly in this "back" of mathematics, in the mathematics in the making, where the contingent, humanly influenced aspects of mathematics can be found.

Rejecting the thesis of the "special epistemic status" of mathematics, a further analysis of the phenomena of coherence and consensus is needed. If we insist on the cultural character of mathematics and the contingency of parts of mathematical knowledge, then the coherence and the absence of real conflicts or revolutions in mathematics cannot be explained easily. In order to find an account for it in the cultural framework, Fleck's philosophy of science and a suited theory of mathematical development are presented in the following section.

3 Approaches to an alternative explanation for coherence and consensus

In order to explain the phenomena of coherence and consensus, let us first take a look at Ludwig Fleck's philosophy of science, developed in his book *Entstehung und Entwicklung einer wissenschaftlichen Tatsache: Einführung in die Lehre vom Denkstil und vom Denkkollektiv* (English title: 'Genesis and development of a scientific fact') [Fle35]. Today Fleck is widely recognized as a pioneer of the constructivist-relativist tendencies in philosophy of science and of the sociologically-oriented approach to the study of the evolution of scientific and medical knowledge. He deserves this recognition and respect all the more as during his lifetime his philosophical achievement passed completely unnoticed, until the well-known philosopher of science Thomas Kuhn recognized Fleck's main work as a source of inspiration of his *The Structure of Scientific Revolutions* [Kuh70]. As Fleck is not well known in philosophy of mathematics, his work shall be described in some detail.

3.1 Fleck's theory of the thought-collectives and thought-styles

Fleck may be called a pioneer of cultural epistemologies, because he did not only consider the subject and the object of perception, but he added the conditions of perception as a third important component of epistemology. According to his theory, the conditions of perception are determined by the existing standards of knowledge, which are not located in the individual, but in the collective. He describes an

> interaction between the perceived and the perception: the already perceived influences the way of new perceiving; perception enhances, regenerates, reinvents the perceived. Thus, perception is not an individual process of theoretical consciousness, it is the result of a social activity, because the particular standard of perception exceeds the individual's limits.
>
> [Fle35, p.54]

Fleck outlines this idea in a case study of changing concepts of syphilis in medical history. He demonstrates, how the development of a scientific fact is influenced by the culturally determined ways of thinking.

In order to describe the intersubjective character of perception and science in general, Fleck developed the notions of thought-style (*Denkstil*) and thought-collective (*Denkkollektiv*). The thought-collective is defined as a community possessing a common thought-style. This style develops successively, and is at every stage connected to its own history. It creates a certain definite readiness and dictates what and how the members of the thought-collective can observe. Thought-style is defined as directed perceiving. It is characterized by

- common attributes of the problems of interest for the collective,
- common judgments of what is considered to be evident;
- common methods as media of perceiving.
- Eventually, it is accompanied by a technical and literary style of a system of knowledge. [Fle35, p.130]

The thought-styles in which individuals think are the results of their theoretical and practical education. Passing from teachers to students, they contain certain traditional values, which are subjected to a specific historical development and specific sociological laws.

If a certain thought-style is sufficiently elaborated, it does not only determine the perception, but also what is considered to be true. Therefore, truth is located in the intersubjective dimension:

> The notion of truth in its classical significance, as a value independent of the subject of cognition and of social forces, compels one to accept truth as an unattainable ideal. Besides, the history of science teaches us that we do not approach that ideal, even asymptotically, for the development of science is not unidirectional and does not consist only in accumulating new pieces of information, but also in overthrowing the old ones. Thus, classical theories of cognition ought to distinguish between:
>
> (1) the ideal, unattainable truth,
>
> (2) the official "truths" which "should" somehow approach it,
>
> (3) illusions and mistakes. At the same time they have to admit that there is no general criterion of truth. [...]
>
> The epistemology which is the science of thought-styles, of their historic and sociological development, considers truth as the up-to-date stage of changes of thought-styles. [Fle36, p.111]

We can learn a lot from Fleck's theory of thought-collectives and thought-styles for mathematics: In Fleck's view, the sciences are specific thought-collectives that are especially stable. If we consider mathematics to be such a thought-style, Fleck gives us interesting answers to our question why there is this wide consensus and this high coherence. According to Fleck, these phenomena give no evidence for a "special epistemic status," but they are to be understood in correlation to the standard of a discipline:

> The more a field of knowledge is elaborated, the more it is developed, the smaller are the differences [...] It is, as if the scope for development was shortened with the growth of nodes, as if more resistance appeared, as if the room for free thinking was restricted. [Fle35, p.110]

This idea is followed by the notion of "active and passive linking" (*Kopplung*): "every active part of knowledge corresponds to a passive linking, which results mandatorily" [Fle35, p.110]. The more active parts of knowledge belong to a thought-style, the more passive linkings evolve as more or less mechanical consequences. Thus, according to Fleck, we can understand mathematics as a field of knowledge that is elaborated to a high degree. His short thesis,

> The deeper we go into a field of knowledge, the stronger it is bound to a thought-style (*Denkstilgebundenheit*), [Fle35, p.109]

gives a good explanation for the high consensus among mathematicians: Mathematics is a thought-style that is well elaborated and has a long tradition. This enforces constraints on thought (*Denkzwang*). This is most obvious in the field of deductive reasoning, a very important aspect of the mathematical way of generating and justifying knowledge. (Although the discussions about proofs have shown that the myth of strict deductivist criteria for proofs must be questioned since proofs always have logical gaps, see, *e.g.*, [Her97]).

But constraints on thought in Mathematics do not only concern the logical reasoning. Fleck's conception of constraints on thought also comprises common attributes of the problems that are interesting to mathematicians, common assessment of values, and common methods used for mathematical cognition.

To sum this up, we do not need the "special epistemic status" of mathematics as an explanation for consensus of mathematicians. The degree of elaboration of the mathematical thought-style supplies an alternative and more convincing explanation. More than in other fields of knowledge, the active elements of mathematical knowledge produce passive linkings (mainly but not only due to deductive reasoning). Thus, the evolution of the mathematical thought-style has indeed superseded contingency to a high degree. Nevertheless, this process can never result in the complete elimination of contingency in mathematics.

In addition to these explanations given by Fleck's theory, historical investigations can help us understanding the phenomena of coherence and consensus in mathematics. Although the authors cited in the following passages do not all have homogenous philosophical positions, some of their ideas can serve for the paper's argumentation.

3.2 Changes of thought-styles in mathematics: historical investigations

An important contribution to our discussion has been made by Philip Kitcher in specifying some interesting characteristics of the development of mathematics. In his book *The nature of mathematical knowledge* [Kit84], Philip Kitcher compares mathematical change and scientific change. In analogy to Fleck's concept of thought-style, Kitcher defines the notion *mathematical practice*:

> We view a mathematical practice as consisting of five components: a language, a set of accepted statements, a set of accepted reasonings, a set of questions selected as important, and a set of meta-mathematical views (including standards for proof and definition and claims about the scope and structure of mathematics). [Kit84, p.229]

Similarly to Fleck's investigation of scientific progress as transitions of thought-styles, Kitcher describes mathematical change as transition from one mathematical practice to the next:

> The problem of accounting for the growth of mathematical knowledge becomes that of understanding what makes a transition from a practice $\langle L, M, Q, R, S \rangle$ to an immediately succeeding practice $\langle L', M', Q', R', S' \rangle$ a rational transition. [Kit84, p.229]

He shows in various historical examples that these transitions are often initiated by discrepancies between the components of the mathematical practices. By changing one or more components, they can by re-equilibrated. For example, theorems are retained valid by changing the language: Instead of rejecting a theorem when counter-examples are found, mathematicians often restrict the concerned notions in such a way that the theorem becomes again valid (this mechanism has been described in detail in Lakatos' book *Proofs and Refutations* [Lak76]).

> So, where in the case of science we find the replacement of one theory by another [...], in the mathematical case there is the adjustment of language and a distinction of questions, so that the erstwhile "rivals" can coexist with each other. Mathematical change is cumulative in a way that scientific change is not, because of the existence of a special kind of interpractice transition. [Kit84, p.229]

Kitcher considers this mechanism for producing consensus and coherence to be characteristic for mathematics. It helps to avoid explicit discontinuities. Although singular components of the mathematical practices must be revised in order to face inconsistencies, mathematical practices are rarely abandoned completely.

In short: coherence in mathematics emerges, because mathematicians immediately search for solutions to level inconsistencies whenever they appear. In other words, inconsistencies do not exist in mathematics, because they are not tolerated.

This thesis is supported by the work of Raymond Wilder who has analyzed mathematics as a developing cultural system [Wil81]. He emphasizes the cultural relativity of mathematics:

> Because of its cultural basis, there is no such thing as the absolute in mathematics; there is only the relative. [Wil81, p.148]

Anyhow, mathematics is not arbitrary and real discontinuities can only be found on the meta-level, as he postulates in agreement with Crowe [Cro75]:

> Revolutions may occur in the metaphysics, symbolism and methodology of mathematics, but not in the core of mathematics. [Wil81, p.142]

On this matter, Heinz's and Wilder's positions coincide. But for Wilder, the absence of revolutions does not imply that mathematical knowledge grows cumulatively, since the patterns of development are more complicated. When he describes these patterns, he does not focus on standards of rigor for proofs nor on other aspects on the meta-level, but he concentrates on central elements within mathematics: mathematical objects, concepts, and theories. Over the centuries, mathematical objects and concepts undergo radical changes in their meaning and their role within the theories. On the basis of historical case studies, Wilder tries to specify "laws" of this evolutionary process and figures out different characteristic mechanisms.

Besides the mechanisms *abstraction* and *generalization* which have often been described, he attaches importance to *consolidation* by which he means the unification of theories or concepts [Wil81, p.87]. By connecting mathematical concepts, these processes of consolidation play an important role for establishing coherence not only on the logical, but also on the conceptual level. I will come back to this aspect in section 3.3.

Using the notion *hereditary stress*, Wilder characterizes the culturally determined phenomena that initiate the evolution of theory and concepts, such as mathematical or non-mathematical problems, a changing conception of nature, discovered inconsistencies or paradoxes, growing demands for rigor etc. In addition, there is the mechanism of *diffusion* of ideas and methods, by which mathematical thoughts are transferred from one domain to another. This is an important condition for processes of consolidation.

On the whole, Wilder considers these evolutionary processes to be embedded into their cultural background. Therefore, his patterns of change put the singular achievements of individual mathematicians into a cultural perspective (instead of celebrating single mathematicians as genius discoverers, as it was usual in the former historiography of mathematics). Starting from the observation of multiple discoveries and the "before his time" — phenomenon (*i.e.*, concepts or ideas which fail to attract attention at their time but are rediscovered and appreciated later), Wilder describes to what degree individual thinkers rely on their cultural environment. He concludes the following:

> The individual mathematician cannot do otherwise than preserve his contact with the mathematical culture stream; he is not only limited by the state of its development and the tools which it has devised, but he must accommodate to those concepts which have reached a state where they are ready for synthesis. [Wil81, p.145]

Thus, according to Wilder, the cultural influence on every individual thinker provides another explanation for the phenomena of consensus and coherence: If all further developments in mathematics are based on the same cultural background, they coincide significantly in most cases. And when inconsistencies appear, they initiate processes of consolidation, which ensure consistency again.

Paul Ernest has described these patterns of mathematical change in his "generalized logic of discovery" (built on Lakatos' "logic of discovery" [Ern98]). He considers the process of discoveries to be a dialectical cyclic process in which definitions, proposals, and relations are discussed in the community. Along this social process, the proposals are accepted or rejected. Rejecting them initiates modifications of the original proposal [Ern98, p.149-160]. The community always acts in a scientific and epistemic cultural context, "including problems, concepts, methods, informal theories, proof criteria and paradigms, language, and metamathematical views" [Ern98, p.151].

Just as Kitcher, also Wilder and Ernest emphasize the important role of well-working mechanisms that re-establish coherence in mathematics. Thus, these authors do not consider coherence to be a surprising phenomenon that legitimizes the hypothesis of the "special epistemic status", but to be an aim for which mathematicians consequently strike again and again.

3.3 Changing the question: importance of and conditions for coherence

Against the background of these historical investigations, we must pose the question of coherence in a different way: From a cultural perspective, we do not need to ask why mathematical theories are coherent, but why they are always (re)made coherent and how this is possible. Following this question we find the major difference between mathematics and other disciplines: it is the role of the value coherence that makes mathematics exceptional.

Values in the mathematical culture. The most important answer for the question why mathematical theories are always (re)made coherent can be found in the prevalent view of mathematics. When a community (of platonists or others) is convinced that inconsistencies *cannot* appear, the participants will make great efforts to remove them whenever they *do* appear.

It is similar on the conceptual level: Sociological studies have shown that integrating different mathematical theories, *i.e.*, establishing coherence between theories, is highly valued among working mathematicians. Leone Burton describes the "drive to establish connectivities" by "tremendous satisfaction when two apparently different areas are found to connect" [Bur99, p.137]. She illustrates this thesis by various quotations of interviewed mathematicians, *e.g.*, "I am certainly impressed by links and in my own work I feel very happy if I can tie things up." (cited in [Bur99, p.137]). A typical example for the value which is attached to such connections is Euler's formular $e^{i\pi} = -1$: It has won the competition of being the "most beautiful" theorem in an election among readers of the mathematical journal Mathematical Intelligencer [Wel90]. Euler's formular is said to be the "most beautiful" theorem because it shows unexpected connections between originally unconnected mathematical domains [Hei00, p.145-150].

Following this line, the prevalent view of mathematics and its values has proved to be a "self-fulfilling prophecy" again and again: When inconsistencies are not tolerated and coherence is highly estimated within the community, mathematician's efforts are oriented at these values. Many things were changed in order to leave this central value unchanged. It would be an interesting question for historical investigations whether the value of coherence itself has ever been seriously questioned.

Ontological status of mathematical objects as precondition. Why can coherence be re-established more easily in mathematics than in other sciences?

Obviously, one important reason is the ontological status of mathematics. Whenever necessary, mathematical theories have been detached from the physical reality. In this way, refutations of theorems can be answered by changing (mostly restricting) the concerned mathematical concepts. Hence, one reason for the possibility of coherence is the convertibility of mathematical concepts which is the opposite of their often presumed a priori status. Whereas scientific concepts must correspond to reality, mathematical concepts can be built in an explicitly constructed mathematical reality in order to avoid complete refutations.

This has been made possible by an important decision about the conception of truth in mathematics: Mathematicians prefer consistency within the theory to conformity with reality. This preference has even been formalized in Hilbert's notion of truth as absence of contradiction, in short: consistency. To sum up, it is the special ontological status which makes the difference between mathematics and other sciences or arts possible.

Outsourcing as an organisational mechanism. Last but not least, the mechanism of outsourcing should be mentioned as an organisational mechanism to assure coherence. Over the centuries, mathematics has outsourced many (usually applied) sub-domains when they developed their own ways of thinking and working [Lau72]. By considering them not to be a part of mathematics anymore, inconsistencies or conflicts could be removed in an easy way. Even today, there are disciplines of mathematics (like scientific computing and other parts of experimental mathematics) whose standards have removed from the widely accepted mathematical standards. Again, lively discussions have been raised, for example about the computer-generated proof of the Four-Color Theorem, whether these approaches still belong to mathematics or whether the mathematical community must begin to accept such differences [Hei00].

3.4 Conclusion

It turned out to be characteristic for mathematics that its cultural relativity can be hidden more easily than those of other cultural achievements (like language). By means of high coherence and wide consensus, the human dimension and cultural origin of mathematics is obscured more successfully than in other sciences. But for explaining these phenomena, the thesis of the special epistemic status of mathematics is not necessary. In historical investigations instead, various factors have been found within the cultural system of mathematics, which can give accounts for the fact that mathematics is more coherent and less contradictory than other disciplines:

- Mathematics is a highly developed thought-style in which all active parts of knowledge produce a lot of passive linkings.

- Historical investigations show that coherence has been re-established whenever it was questioned. For that, different mechanism have evolved, especially interpractice transitions and consolidations.

- Establishing connectivities is a highly-valued activity for mathematical research, hence, coherence is also a result of a special value system in mathematics. This is the case also with consistency which has even become the inner-mathematical criterion for truth (replacing conformity with reality).

- An important precondition for these decisions lies in the ontological status of mathematical objects, namely in their disconnection from physical reality.

Although the phenomena of consensus and coherence mainly refer to the "front" of mathematics, the most important reasons for their existence concern mathematics in the making, *i.e.*, the "back" of mathematics. This observation agrees with new tendencies in the philosophy of mathematics, namely to focus on mathematical practices and not only on the products (*cf.* [Tym85], Schlimm in this volume). This focus can be deepened by taking a cultural perspective.

4 Outlook: analysing mathematics as a culture

How does mathematics differ from the natural sciences and the arts? We started with this question, went through some epistemological and ontological considerations, and have seen in Section 3 that the most important differences between mathematics and other scientific disciplines lie in the scientific practices and the scientific culture in which practices are embedded. This shift of focus suggests adopting a cultural perspective on the sciences.

The idea to consider sciences as cultures raised in the 1960s in different disciplines, among them anthropology and sociology of knowledge, and has been elaborated in the last decades in higher education research. It is currently adopted in the Austrian project "Science as Culture" [Fis+98], [Arn01] in which several scientific disciplines are compared under a cultural perspective (namely physics, biology, literal arts, and history). The project starts from the assumption that the culture of a discipline "consists of all elements which are characteristic for every culture" [Fis+98, p.5], *i.e.*,

> their traditions, customs and practices, transmitted knowledge, beliefs, morals and rules of conduct, as well as their linguistic and symbolic forms of communication and the meanings they share. To be admitted to membership of a particular sector of the academic profession involves not only sufficient level of technical proficiency in one's intellectual trade but also a proper measure of loyalty to one's collegial group and adherence to its norms. Becher (1984), cited in [Fis+98, p.7]

All these elements together form a scientific culture. The notion comprises "everything a student must acquire in order to become an accepted member of the community" [Arn01, p.2]. Fischer and Arnold show that this wider perspective is very instructive for comparing scientific communities, since it does not only include the body of explicit knowledge and the ontological and epistemological characteristics, but also important parts of implicit knowledge and sociological aspects [ArnFis04]. Although there are already a lot of interesting partial investigations upon the mathematical culture (*e.g.*, [Kit84], [ResvBeFis93], [Hei00]), there is (up to now) no *comprehensive* study on mathematics from the cultural perspective. In [Pre02], this work was started from an educational point of view by collecting a list of central elements of the mathematical culture that affect students when they are engaged with the culture of mathematics in learning processes. But if we really want to understand the differences between mathematics and other sciences, we should put deeper and more systematic investigations on the research agenda.

References.

[Arn01] Markus **Arnold**, Wissenschaftskulturen im Vergleich, Zwischenergebnisse des Projekts *Science as Culture*, Institut für Interdisziplinäre Forschung und Fortbildung 2001

[ArnFis04] Markus **Arnold** and Roland **Fischer**, Disziplinierungen, Kulturen der Wissenschaft im Vergleich, Turia + Kant 2004 [kultur.wissenschaften 11]

[Blo91] David **Bloor**, Knowledge and Social Imagery, University of Chicago Press 1991

[Bur99] Leone **Burton**, The Practices of Mathematicians: What Do They Tell Us About Coming to Know Mathematics?, **Educational Studies in Mathematics** 37 (1999), p.121-143

[Car02] Jessica **Carter**, On the Existence of Mathematical Objects, Ontology and Mathematical Practice, University of Southern Danmark 2002; *PhD Thesis*

[CohSch$_0$86] Robert S. **Cohen** and Thomas **Schnelle** (*eds.*), Cognition and Fact – Materials on Ludwik Fleck, Reidel 1986

[Cro75] Michael J. **Crowe**, Ten "Laws" concerning Patterns of Change in the History of Mathematics, **Historia Mathematica** 2 (1975), p.161-166

[Ern98] Paul **Ernest**, Social Constructivism as a Philosophy of Mathematics, State University of New York Press 1998

[Fis+98] Roland **Fischer** *et al.*, Projektantrag *Science as Culture*, Institut für interdisziplinäre Forschung und Fortbildung 1998

[Fle35] Ludwik **Fleck**, Entstehung und Entwicklung einer wissenschaftlichen Tatsache: Einführung in die Lehre vom Denkstil und vom Denkkollektiv, Schwabe 1935; *english translation in:* [Fle79]

[Fle36] Ludwik **Fleck**, The Problem of Epistemology, *in:* [CohSch$_0$86]

[Fle79] Ludwik **Fleck**, Genesis and Development of a Scientific Fact, University of Chicago Press 1979

[Hei00] Bettina **Heintz**, Die Innenwelt der Mathematik, Zur Kultur und Praxis einer beweisenden Disziplin, Springer 2000

[Her97] Reuben **Hersh**, What is Mathematics, really?, Oxford University Press 1997

[Kit84] Philip **Kitcher**, The Nature of Mathematical Knowledge, Oxford University Press 1984

[Kuh70] Thomas S. **Kuhn**, The Structure of Scientific Revolutions, Chicago University Press 1970

[Lak76] Imre **Lakatos**, Proofs and Refutations: the Logic of Mathematical Discovery, Cambridge University Press 1976

[Lau72] Detlef **Laugwitz**, Anwendbare Mathematik heute, *in:* [Mes72, p.224-252]

[Mes72]	Herbert **Meschkowski** (*ed.*), Grundlagen der modernen Mathematik, Wissenschaftliche Buchgesellschaft 1972
[Pre02]	Susanne **Prediger**, Kommunikationsbarrieren beim Mathematiklernen, Analysen aus kulturalistischer Sicht, *in:* [PreSieLen02, p.91-106]
[PreSieLen02]	Susanne **Prediger**, Franziska **Siebel**, and Katja **Lengnink** (*eds.*), Mathematik und Kommunikation, Verlag Allgemeine Wissenschaft 2002 [Darmstädter Texte zur Allgemeinen Wissenschaft 3]
[ReiHolRot00]	Hans **Reich**, Alfred **Holzbrecher**, and Hans-Joachim **Roth**, (*eds.*), Fachdidaktik interkulturell, Ein Handbuch, Leske + Budrich 2000
[ResvBeFis93]	Sal **Restivo**, Jean P. **van Bendegem**, and Roland **Fischer** (*eds.*), Math Worlds, Philosophical and Social Studies of Mathematics and Mathematics Education, State University of New York Press 1993
[Sch$_1$00]	Joachim **Schroeder**, Mathematik, *in:* [ReiHolRot00, p.451-468]
[Tym85]	Thomas **Tymoczko** (*ed.*), New Directions in the Philosophy of Mathematics, Birkhäuser 1985
[Wel90]	David **Wells**, Are These the Most Beautiful?, **Mathematical Intelligencer** 12 (1990), p.37-41
[Wil81]	Raymond L. **Wilder**, Mathematics as a Cultural System, Pergamon Press 1981
[Whi$_0$93]	Alvin M. **White** (*ed.*), Essays in Humanistic Mathematics, Mathematical Association of America 1993
[Whi$_1$47]	Leslie A. **White**, The Locus of Mathematical Reality: An Anthropological Footnote, **Philosophy of Science** 14 (1947), p.289-303

Received: April 8th, 2003;
In revised version: February 13th, 2004;
Accepted by the editors: April 19th, 2004.

Benedikt **Löwe**, Volker **Peckhaus**, Thoralf **Räsch** (eds.)
Foundations of the Formal Sciences IV
The History of the Concept of the Formal Sciences

Axiomatics and Progress in the Light of 20th Century Philosophy of Science and Mathematics

DIRK SCHLIMM[*]

Department of Philosophy
McGill University
Montréal (QC), H3A 2T7, Canada
E-mail: `dirk.schlimm@mcgill.ca`

> ABSTRACT. This paper addresses the question of how aspects of science have been perceived through history. In particular, I will discuss how the contribution of axiomatics to the development of science and mathematics was viewed in 20th century philosophy of science and philosophy of mathematics. It will turn out that in connection with scientific methodology, in particular regarding its use in the context of discovery, axiomatics has received only very little attention. This is a rather surprising result, since axiomatizations have been employed extensively in mathematics, science, and also by the philosophers themselves.

1 Axiomatics

Euclid's *Elements* and Newton's *Principia* are beyond any doubt among the most widely known theories in mathematics and in science. They are crown jewels in the development of geometry and physics. What both theories have in common is the structure of their presentation, which is *axiomatic* or *deductive*: A number of statements (called *axioms, postulates, hypotheses,* or *laws*, depending on how their status is conceived) are posited, and the central claims of the theory (*e.g.*, Pythagoras's theorem, or Kepler's "laws")

---

[*]I would like to thank Erica Lucast, Uljana Feest, Susanne Prediger, Jeff Speaks, Michael Hallett, and an anonymous referee for helpful comments on earlier drafts of this paper.

are derived as consequences. In addition, all notions of the theory are definable in terms of the primitive notions that occur in the axioms. Henceforth I shall refer to the practice of developing, employing, or studying systems of axioms as *axiomatics*. Notice that in axiomatic presentations it is not necessary for the primitive terms to be considered as uninterpreted symbols, nor must the notion of logical consequence be made explicit. In the former case we speak of a *formal* axiomatization, while the latter distinguishes axiomatizations from *formalizations*, which require a formal language and formal rules of inference. These notions are often conflated in the literature, but they should be kept apart to avoid unwarranted criticism of axiomatics.

To be sure, neither geometry nor physics ended with Euclid's and Newton's theories. Rather, they have inspired a great number of readers, they have been the starting points of various fruitful developments, and they have led to a great many new scientific and mathematical insights. These observations lead directly to the main motivation behind the present paper, namely the question regarding the role that an axiomatic presentation of theories plays in the development of science and mathematics.[1]

The usefulness of axiomatics in theory development is manifold. For example, the formulation of axioms can bring out hidden assumptions, explicate informal concepts, or reveal gaps in the argumentations; once a theory is axiomatized it can be studied through the axioms, and relations to other theories can be established; manipulations of axioms, which can be motivated by empirical findings that contradict some theorem or by attempts to prove the independence of the axioms, can suggest new theories.[2] Furthermore, I believe that axiomatics has a considerable effect on the perception and formulation of analogies, as well as on our capabilities of reasoning about abstract objects.[3]

Although the utility of axiomatic presentations should be of no surprise to working mathematicians or theoretical scientists, I shall show in the present paper that the contribution of axiomatics for the advancement of science and mathematics has not been properly acknowledged in the philosophical literature. To do so I shall present an overview of the main trends in philosophy of science and mathematics with respect to the following two questions:

- How is the change from one theory to another accounted for, *i.e.*, what are the mechanisms underlying theory change?

---

[1] Notice that I am open as to what theories *are*, as long as they can be presented axiomatically.

[2] Non-Euclidean geometries are the most famous outcome of the latter.

[3] A more detailed account of this is planned for the future; *cf.* [Sch06].

- What role is assigned to axiomatics in particular with regard to theory change and discovery?

As it turns out, in philosophy of science very little has been said in this regard. On the contrary, the notions of axiomatics and discovery have often been considered as being opposed to each other. In mathematics there has been a recognition of the creative power of axiomatics, in particular by David Hilbert. However, these views did not catch on in philosophy of mathematics and have been revived only recently.

Before turning our attention to the 20th century, let me briefly mention the major milestones in the history of axiomatics. According to Aristotle, scientific knowledge must be demonstrative, resting on "necessary basic truths" ([*An. post.*, I.vi; 74b], [McK47, p.21]). Euclid's subsequent axiomatization of geometry in the *Elements* was soon considered to be the prototypical presentation of scientific theory. It has inspired works like Newton's *Principia*, Spinoza's *Ethics*, and many many others.

Due to the use of axiomatics in the natural sciences, and to the development and growing acceptance of non-Euclidean geometries, the idea that axioms express necessary truths has been slowly abandoned. Also, starting with the recognition of the point/line duality of projective geometry, the meanings of the primitive terms lost their claim to uniqueness. Frege's invention of predicate logic led to a sharpening of the language of scientific presentations, reducing ambiguities and vagueness, as well as to an increase of rigor in the deductions (see also section 3, on related developments in 19th century mathematics). These developments form the background for the philosophical reflections in the 20th century that are presented next.

2 Philosophy of science

In the following I shall discuss what I consider to be four major families of views in the philosophy of science of the 20th century. Due to space limitations this can only be very sketchy, but I hope to be able to bring out the main positions concerning the questions mentioned above. I look at philosophy of science first, because it has had great impact on the discussions in philosophy of mathematics. Hence, the development in philosophy of mathematics can be better understood when seen in this broader context.

2.1 The received view

By the *received view* in philosophy of science, I will refer to the core of the views that emerged from logical positivism and were dominant from the 1930s until the 1960s.[4] One of the main doctrines of the received view is

---

[4] *Cf.* [Sal+92, p.135]. Nowadays one can also find the label "once received view" [Cra02].

that theories should be considered as linguistic entities, formulated in the language of first-order logic. Empirical meaning is then conferred on the primitive terms by means of coordinative definitions. As a consequence, the distinction between theoretical and observational terms was introduced and the relations between the two have been studied extensively. Specific views on the assessment of scientific theories ranged from *verification*, over *falsification* [Pop34], to *confirmation* [Hem45]. A most important distinction, for our purposes, is made between the *context of discovery* and that of *justification* [Rei38]. In general, the study of activities related to discovery is relegated to psychology, sociology, and history, but is not considered to be of interest for philosophy. Kekulé's dream of a snake biting its own tail, which suggested to him the structure the benzene ring, is seen as the prototypical example of a discovery about which philosophers could have nothing to say.

According to the received view scientific progress is explicated as a succession of theories. It is considered to be cumulative in the sense that old theories are replaced by more inclusive ones (*e.g.*, rigid body mechanics being replaced by classical particle mechanics), or that theories are reduced to others (*e.g.*, thermodynamics being reduced to statistical mechanics). However, more detailed principles providing heuristics for the development of theories are not investigated, since they are thought to lie outside of the context of justification.

Although formal axiomatic presentations of theories were used by philosophers of science adhering to the received view to study properties of theories, particular axiomatizations were not considered to be of philosophical interest, since they are neither unique for a particular set of statements (since different sets of axioms can determine the same set of statements), nor do they determine unique interpretations. Furthermore, syntactic deductions of theorems from axioms yield only tautologies, while scientific discoveries express novel facts. Hempel summarizes these considerations as follows: "[A]xiomatization is basically an expository device," which "can come only after a theory has been developed" [Hem70, p.250]. Thus, axiomatics was employed for presenting and studying scientific theories (and also for explicating philosophical notions like *justification* [Pop34], *explanation* [HemOpp48], and *existence* [Qui48]), but it was not considered in connection with theory development.

2.2 Reactions: Kuhn and Lakatos

In direct opposition to some of the main tenets of the received view, Thomas Kuhn published in 1962 what might well be the most influential book in philosophy of science of the 20th century, *The Structure of Scientific Revo-*

lutions. In what is commonly referred to as the "historic turn" in philosophy of science, emphasis shifted from the internal structure of scientific theories to the actual development of science. Rather than theories, Kuhn considers broader units of scientific progress (*paradigms*), which embody the shared, accepted, and unquestioned views, standards, methods, theories, problems, and goals of the scientists working within a particular tradition. He distinguishes two phases of scientific development: During *normal science* the scientists work on the solution of puzzles guided by the standards and values of the current paradigm. When a considerable number of such puzzles resist a solution a *crisis* emerges, which leads to a proliferation of theories. This crisis is overcome by a *revolution* when a new paradigm is finally accepted that leads to the solution of the anomalies.

Since Kuhn considers different paradigms to be incommensurable, scientific progress, which according to him happens only in the course of scientific revolutions, is not cumulative.[5] Moreover, the scientific changes that are of interest to Kuhn are broader in scope than the move from one theory to another. Thus, it might not surprise us that his notion of theory is rather vague, and that he does not ask where the theories come from. He considers them as "imaginative posits, invented in one piece for application to nature" [Kuh70a, p.12].

Investigation of predictions and determination of values for theoretical constants are regarded by Kuhn as typical problems during normal science. Similarly, he acknowledges that in the process of matching facts with theory scientists work on their theories in order to obtain more statements that can be confirmed or disconfirmed directly and to increase the precision of the predictions. Reformulation of theories in "equivalent but logically and aesthetically more satisfying form", as well as "to exhibit the explicit and implicit lessons" of particular paradigms are also regarded as part of the theoretical work [Kuh70b, p.33]. This part of Kuhn's account of science is very similar to the conception of the received view. However, Kuhn does not consider these developments to be of great value, remarking that "perhaps the most striking feature of the normal research problems [...] is how little they aim to produce major novelties, conceptual or phenomenal" [Kuh70b, p.35]. Kuhn also implies that the process of codification and axiomatization occurs late in the development of a discipline and only in response to a crisis:

> It is, I think, particularly in periods of acknowledged crisis that scientists have turned to philosophical analysis as a device for unlocking the riddles of their field. [...] To the extent that normal research work can be conducted by using the paradigm as a model, rules and assumptions need not be made explicit. [Kuh70b, p.88; p.44-48]

---

[5]The ideas of incommensurability and cumulative progress need some clarification, but this is beyond the purpose of this paper.

Thus, it seems to me that axiomatics is compatible with Kuhn's account of science, but the little he says about it implies that he did not regard it as an important factor for scientific development.

Another very influential reconstruction of science was offered by Imre Lakatos as an advancement of Popper's falsificationism.[6] According to the latter, scientific theories must be empirically falsifiable and should be rejected when such a falsification occurs. One obvious difficulty with this account is that it does not square well with actual scientific practice, where some theories continue to be pursued despite the existence of facts that stand in conflict with them. To overcome this difficulty, Lakatos proposes distinguishing between an irrefutable *hard core* and a *protective belt* of auxiliary hypotheses, which serve to make predictions and can be adjusted when confronted with contradictory empirical evidence [Lak70, p.135]. For example, the hard core of Newton's gravitational theory consists of just his three laws of mechanics and the law of gravitation, while whatever else is needed to apply them is considered to be part of the protective belt.

Lakatos's view is similar to Kuhn's in that it considers broader units of scientific development than theories. These units, called *research programmes*, are successions of theories that share the same hard core. Research programmes are called *progressive* when they allow for novel predictions some of which are confirmed by experience, or *degenerating* when they can only account for empirical evidence in retrospect. Science evolves by replacing degenerating research programmes by progressive ones, *i.e.*, by changes of the theoretical hard core of a programme. But on how these changes come about also Lakatos is silent.

Despite the differences, Kuhn's account of normal science and Lakatos's progressive research programmes are both similar to the characterization of scientific progress of the received view. The notion of theory is more sophisticated in Lakatos than in Kuhn, but, again, the mechanisms of theory change are not explicated and axiomatics is not assigned any particular role in this process.

2.3 Discovery and models

By the mid-20th century most of the tenets of the received view had been challenged. Of particular importance for the present discussion are Norwood R. Hanson and Mary Hesse, who brought the notions of discovery and analogical reasoning back onto the philosophical table.[7]

---

[6]Lakatos's account of science differs in important respects from his views on mathematics, which are discussed in section 3.3 below.

[7]I say "back", because long before modern times, both analogical and deductive reasoning had been discussed in connection with scientific progress (*e.g.*, by Aristotle and Proclus, *cf.* [Pos89, p.148] and [HinRem74]).

Recall, that according to the received view the origin of the formulation of scientific laws was a subject matter for psychology, sociology, or history, but not for philosophy. The fundamental scientific inference was considered to be deduction of data from laws, which served as explanation of the observed phenomena (hypothetico-deductive account, [HemOpp48]). Hanson criticizes this view for not being justified in rejecting the investigation of the origin of scientific laws or hypotheses. He argues that the inference from data to plausible hypotheses is in fact logical, rather than merely psychological [Han58b]. Rather than just being lucky guesswork, Hanson considers the suggestion of new hypotheses to be a reasonable affair that goes beyond inductive generalization, and as such it should be the subject of philosophical reflections.

Peirce's notion of *abduction*, also called *retroduction*, is taken over by Hanson as the logical inference from data to a hypothesis. He explains the origin of scientific laws by the perception of a particular *pattern*, which reveals the conceptual framework within which the data can be systematically organized. Discoveries of scientific laws, according to Hanson, begin with a problem, difficulty, or surprising empirical fact P that the scientist wants to solve or explain. Her reasoning is thereby directed towards developing a hypothesis H, such that if H were true, P would be accounted for [Han58a, p.1086-7]. Such a hypothesis may be obtained, for example, from reasoning by analogy [Han58a, p.1078].

Hanson vehemently rejects the hypothetico-deductive (HD) view of scientific theories, but he also acknowledges that the deduction of consequences from general laws is a crucial ingredient for science. So he writes, for instance, that we can not determine what counts as an *anomaly*, *i.e.*, a deviation from our expectations, "until we have some fairly full theories whose consequences *constitute* our expectations" [Han65, p.52]. Hanson later obscures his own observation by introducing anomalies as conclusions that "although logically 'expected', are psychologically quite unexpected", and the aim of retroduction is to come up with hypotheses that entail the anomaly "as the 'previous' theory may not have done" [Han65]. Presumably, he means that the new hypotheses and consequences are psychologically more satisfactory. However, in the next paragraph Hanson describes the retroductive activity of the scientist as seeking "a novel HD framework within which to reveal the anomaly as logically-to-be-expected" [Han65, p.53].

Thus, although Hanson appears to be quite hostile towards deductive methods and does not give credit to the role of logical deductions in scientific progress, he employs them himself for obtaining consequences of hypotheses. Indeed, it seems to me that both approaches (deductive and retroductive) should be regarded as complementing each other, and that in fact scientists often alternate between them when developing theories. The psychologist Clark Hull, for example, describes theory construction as a process of recurring cycles of hypothesis formulation and testing of consequences. When certain facts can not be accounted for, or certain consequences do not conform to the facts, then the hypotheses have to be amended [Hul52].

Generally, one can interpret Hanson as arguing for widening the scope of philosophy of science by demanding a philosophical investigation of the creative processes behind theory construction. Mary Hesse pursues a very similar goal in her *Models and Analogies in Science* [Hes66]. She distinguishes between *material* and *formal* models; the former are based on pre-theoretic analogies between two observable domains, while the latter are different interpretations of a formal system. Hesse argues that material models surpass formal ones in regard to producing novelties and justifying scientific predictions. Thus, concerning the status of models in science, she maintains, against the received view, the existence of an "essential and objective dependence between an explanatory theory and its model that goes beyond a dispensable and possibly subjective method of discovery" [Hes72, p.356]. To grant that material models are necessary ingredients of scientific theories, however, does not imply that formal models (and the axioms they are models of) do not play any significant role.

Let me point out here what I consider to be an unfortunate pattern in the previous arguments. When new aspects of scientific activity are introduced into the discussion, the new views are often set in contrast to other specific views. This is important for highlighting the values of the new approaches, but it also tends to devalue the insights that have been gained previously. In particular, Hanson and Hesse showed the importance of retroduction and analogical reasoning for theory construction, but in doing so they employed much unnecessary rhetoric against the use deductive methods, which can in fact very easily be seen to complement their own accounts.

Both this pattern of argumentation and the focus on models are also characteristic for the fourth trend in philosophy of science I want to present, namely the semantic view of theories.

2.4 The semantic view of theories

The *semantic view* of theories is a major trend in philosophy of science, which also developed in reaction to the received view. Building on work by Beth and Suppes, its main proponents are van Fraassen [vFr80], Giere [Gie88], Suppe [Sup77], Sneed [Sne71], and Stegmüller [Ste76].

In a series of papers in the 1960s Patrick Suppes argued for an extension of the then still current received view of scientific theories. Regarding theories as an abstract logical calculus in the language of first-order logic augmented by coordinating definitions or empirical interpretations to relate them to the world is too simple a picture, according to Suppes [Sup67]. In particular, he maintains that in practice "formalization [...] in first-order logic is utterly impractical", and suggests including models (understood in the mathematical sense of Tarski [Tar44]) into the philosophical considerations about science. This, he argues, has the advantage of being more natural when complex scientific theories are discussed, and of allowing for a rigorous mathematical (*i.e.*, model theoretic) treatment of various aspects of scientific practice. Moreover, by studying arithmetical models of theories one can also obtain insights into the isomorphic empirical models [Sup67, p.59].

One of Suppes's main points is that actual scientific practice is much more complicated than the simple account of theories suggests: "If someone asks, 'What is a scientific theory?' it seems to me there is no simple response to be given" [Sup67, p.63]. In light of the future developments it should be noted here that Suppes does *not* define theories as a class of models. Rather, he points out that "the explicit consideration of models can lead to a more subtle discussion of the nature of a scientific theory" [Sup67, p.62].

Suppes's considerations have been taken up by van Fraassen, who presents his view, called the *semantic* approach, as being opposed to the "axiomatic and syntactical" analysis of theories [vFr70, p.326]. In contrast to Suppes, who regards semantic and syntactic approaches as complementary, van Fraassen, after initial hesitation, is comfortable of presenting "a view of theories which makes language largely irrelevant to the subject" [vFr87, p.108]. He characterizes the contrast between the syntactic and the semantic view of theories as follows:

> The syntactic picture of a theory identifies it with a body of theorems, stated in one particular language chosen for the expression of that theory. This should be contrasted with the alternative of presenting a theory in the first instance by identifying a class of structures as its models. In this second, semantic, approach the language used to express the theory is neither basic nor unique; the same class of structures could well be described in radically different ways, each with its own limitations. The models occupy center stage. [vFr80, p.44]

The observation that a particular axiomatization of a theory is not unique had been made already by proponents of the received view. However, there the conclusion was to not consider particular axiomatizations as being philosophically illuminating, while van Fraassen draws the conclusion of rejecting a linguistic account of theories altogether.

By pointing to the inadequacies of particular versions of the syntactic approach, van Fraassen argues indirectly for the semantic picture. However, van Fraassen's criticisms may affect particular versions of syntactic approaches, but by no means the syntactic approach in general, as has been noted also by Worrall [Wor84, p.71-73]. The direct argument for the semantic approach is that it is more faithful to the way scientists actually talk and write (*cf.* also [Gie79]). As an example, van Fraassen discusses four "axioms of quantum theory", as they can be found in books on quantum mechanics, and claims that

> they do not look very much like what a logician expects axioms to look like. [...] To think that this theory is here presented axiomatically in the sense that Hilbert presented Euclidean geometry, or Peano arithmetic, in axiomatic form, seems to me simply a mistake. [vFr80, p.65]

It is not clear to me what the distinction is that van Fraassen here alludes to, but it appears to be a result of conflating axiomatization with formalization. When discussing the inadequacy of the syntactic approach he argues against understanding scientific theories as formal deductive systems in the language of first-order logic. In the above quote, however, he contrasts his view with an axiomatization in the sense of Hilbert, which is neither formulated in the language of first-order logic, nor uses explicitly stated rules of inference. Rather, Hilbert presents the primitive terms as uninterpreted, thereby defining a hierarchically structured class of models. Quite similarly, van Fraassen considers the axioms of quantum theory to be "a description of the models of the theory plus a specification of what the empirical substructures are" [vFr80]. Thus, despite van Fraassen's claim to the contrary, it seems to me that the practical differences between axiomatic (*e.g.*, [Hil99]) and semantic approaches are only a matter of emphasis.

In particular, the classes of structures that van Fraassen discusses are all characterized in terms of a system of axioms that they satisfy. So, van Fraassen claims to present an alternative to a linguistic account of theories, but in fact he relies on axioms to determine the class of models that constitute a theory. In other words, his account makes essential use of axioms, but he refuses to regard them as part of what he calls "theories". In addition, he conflates the notion of axiomatization and formalization, and has only very little to say regarding heuristic mechanisms for theory construction and development.

Suppes's suggestion of employing model theoretic techniques in philosophy of science was also put into practice in Joseph Sneed's characterization of the development of scientific theories, in particular of mathematical physics [Sne71]. Sneed attempts to reconstruct the dynamic aspects of theories, *i.e.*, how they grow and change, how they become accepted and rejected. His *nonstatement* view of theories (also referred to as *structuralist* view) rejects the traditional view of theories as sets of sentences formulated in a first-order language, but identifies theories with a class of models (the "core") together with an open set of intended applications. The development of a theory is then characterized by a series of expansions of the core or by changes in the set of intended applications, neither of which need result in more inclusive models or a greater number of applications.

Sneed's account of theory development was put to use by Stegmüller to explicate the theses put forward in Kuhn's *The Structure of Scientific Revolutions*. According to Stegmüller, theories develop in time "through the discovery of new or the rejection of old laws, or the addition of new constraints" [Ste76, p.133]. Notice how claims about core extensions, *i.e.*, about models, are made here in terms of laws or constraints, *i.e.*, in terms of linguistic entities.

Sneed and, following him, Stegmüller give a logical reconstruction of theory development and change, but they do not address (other than in most general terms) how these theory changes come about. In fact, although rejecting the view that theories are best understood as linguistic entities, they do speak of models as being determined by axioms and of changes of models as resulting from changes of axioms. So, Sneed and Stegmüller's account tacitly assumes that axiomatizations affect scientific progress, but, just like Kuhn, Lakatos, and van Fraassen, they do not address this directly.

2.5 Summary

In the received view, theories were understood as sets of sentences, but they were studied in isolation, as if they were static, so to speak. Axiomatic presentations of theories were used in the study of scientific theories, but very little concern was shown for the actual development of theories, nor for the process of discovery in general. Dynamic mechanisms underlying theory development or hypothesis formulation were considered as belonging to the context of discovery and thus as being outside the scope of philosophical investigations. The turn towards the historical and dynamic aspects of science, which include processes of discovery, was accompanied with a move away from theories and linguistic representations. Presumably this was motivated by the need to highlight the contrast to the received view. Thus, we can formulate the two slogans "axiomatic theories without discovery" and "discovery without axiomatic theories" as characterizing the two main directions in 20th century philosophy of science. The relation between axiomatics and discovery has not been the focus of attention in the mainstream and only few philosophers paid careful attention to it, most notably, Patrick Suppes. Unfortunately, it seems that Suppes has either been misinterpreted or neglected.

After this brief recapitulation of 20th century philosophy of science, let us now repeat this exercise, but this time from the point of view of philosophy of mathematics.

3 Philosophy of mathematics

Philosophy of mathematics in the 20th century was highly influenced by late 19th century developments in mathematics. In particular, Frege's invention of the language and calculus of predicate logic [*Begriffsschrift*] began his *logicist* program of reducing mathematical notions to logical ones [*Grundlagen*], which was then carried through (revealing its weaknesses) by Whitehead and Russell in their monumental *Principia Mathematica* [WhiRus10-13]. Closely related are the trend of arithmetizing mathematics [Kle95], *i.e.*, developing mathematics without recourse to geometric intuitions, and the emergence of projective and non-Euclidean geometries, which led to reconsideration and eventual abandonment of the notion of axioms as self-evident truths. Another very influential development was the emergence of set theory in the works of Cantor and Dedekind (*cf.* [Fer99]). Around the turn of the century, however, the paradoxes discovered by Zermelo, Russell, and others, showed that neither the prevailing conception of sets, nor Frege's system of logic provided an ultimate foundation of mathematics.

3.1 Early 20th century

In the wake of the developments just mentioned, but not necessarily causally related to them, two very different approaches to mathematics emerged: On the one hand, L.E.J. Brouwer, building on Kantian views, formulated his philosophy of *intuitionism*, according to which mathematics is an "essentially languageless activity" [Bro52, p.510]. He considered language merely as an aid for communication and memory, and formal logic as restricting mathematical thinking, rather than assisting it. Thus, Brouwer regarded the relation between axiomatics and mathematical creativity as a negative one. On the other hand, David Hilbert worked extensively and very successfully on axiomatizations, in particular in geometry and logic, and he actively promoted axiomatizations in other areas of mathematics and physics (*cf.* [Pec90]). In 1917, he referred to the axiomatic method as a "general method of research" [Hil18, p.405]. For him, axiomatizing a body of knowledge displays the internal conceptual connections and provides a fertile soil for further investigations. He regarded the aim of axiomatically "deepening the foundations" as a fruitful one for all domains of inquiry. Hilbert saw clearly that axiomatics plays an important role in mathematical discovery in a number of ways, only one of which is that it allows rigorous investigations of formal theories themselves, which led to the development of the prosperous discipline of *proof theory*. In the course of the ensuing debate with Brouwer and his followers, the so-called *Grundlagenstreit*, Hilbert's position became known as *formalism*. This is quite unfortunate, since nothing could be more wrong than saying that Hilbert considered mathematics to consist just of formal manipulations of meaningless symbols (*cf.* [Ewa96, p.1106]).

3.2 The received tradition

Despite the great influence Hilbert and his Göttingen school exerted upon mathematics, the mainstream in philosophy of mathematics followed the views of Frege, Russell, and logical positivism, echoing the development in philosophy of science.[8] Accordingly, mathematics was regarded as a purely deductive science, and philosophical discussions revolved around the status of mathematical knowledge (analytic, a priori), mathematical truth (deductivism vs. platonism), the proper foundations of mathematics (logic vs. set theory), and the nature of mathematical objects (platonism, nominalism, neo-logicism, structuralism).

The received tradition considered mathematical discovery as a largely irrational process, just as scientific discovery was seen in the the contemporary reflections on science. For mathematics, the paradigmatic example of a discovery was Poincaré's theorem on Fuchsian functions. According to Poincaré's own account, the theorem popped into his mind quite unexpectedly while he was boarding a bus. Hadamard discusses this and similar examples in his *The Psychology of Invention in the Mathematical Field* [Had45] and especially emphasizes the role of unconscious processes in mathematical creativity. Although formulated over fifty years ago, Hadamard's views are still popular among mathematicians (*cf.* [ChaCon95]).

3.3 New directions

At the time when philosophers of science began formulating alternatives to the received view, a similar turn towards history and practice took place also in philosophy of mathematics, albeit on a much smaller scale. In general, however, the new considerations about science were not carried over to mathematics. Instead, the development in philosophy of science seemed to highlight the fact that science and mathematics are entirely different enterprizes. Of the philosophers who followed the shift towards history and practice and who are thus more likely to reflect on the relation between axiomatics and mathematical progress, I shall discuss Pólya, Lakatos, and Kitcher, and conclude by commenting briefly on some very recent developments in philosophy of mathematics.

In 1945, the mathematician George Pólya initiated almost single-handedly the turn of philosophy of mathematics towards mathematical practice.

---

[8]In the following I use the term *received tradition* for these and related views in philosophy of mathematics.

He distinguishes between two sides of mathematics, which resembles the familiar distinction between the contexts of justification and discovery:

> Yes, mathematics has two faces; it is the rigorous science of Euclid but it is also something else. Mathematics presented in the Euclidean way appears as a systematic, deductive science; but mathematics in the making appears as an experimental, inductive science. [Pól45, p.vii]

By means of numerous examples Pólya investigates the heuristics involved in the invention of mathematics. He is well aware of the novelty of this presentation and writes that "mathematics 'in statu nascendi,' in the process of being invented, has never before been presented in quite this manner" [Pól45]. In his 1954 two volume work *Mathematics and Plausible Inference* [Pól54] Pólya continues the line of inquiry he began in 1945, distinguishing between *demonstrative* reasoning, by which mathematical results are presented, and *plausible* reasoning, which serves "to distinguish a guess from a guess, a more reasonable guess from a less reasonable guess" [Pól54, p.vi]. According to Pólya, the two major forms of plausible reasoning in mathematics are reasoning by induction and by analogy. He explicates the notion of analogy in terms of structure preserving mappings (homomorphisms and isomorphisms) or as being based on "relations that are governed by the same laws." An example that Pólya mentions is the analogy between addition and multiplication of numbers, since they are both commutative, associative, and admit an inverse relation. On similar grounds, subtraction and division are analogous, as are the roles played by 0 and 1.

> In general, *systems of objects subject to the same fundamental laws* (or axioms) may be considered as analogous to each other, and this kind of analogy has a completely clear meaning.
> [Pól54, p.28; emphasis in the original]

Here Pólya points out the importance of axiomatic characterizations of mathematical notions for finding and formulating analogies, *i.e.*, one of the fundamental processes of plausible reasoning by which new mathematics is created.

Imre Lakatos explicitly acknowledges "Pólya's revival of mathematical heuristic and [...] Popper's critical philosophy" as the background of his *Proofs and Refutations*, which is subtitled "The Logic of Mathematical Discovery" [Lak76, p.xii]. Against the received tradition, which he refers to as the *deductivist* view of mathematics Lakatos aims at elaborating the point

> that informal, quasi-empirical, mathematics does not grow through a monotonous increase of the number of indubitable established theorems but through the incessant improvement of guesses by speculation and criticism, by the logic of proofs and refutations. [Lak76, p.5]

Accordingly, Lakatos rejects the attempts of establishing ultimate foundations of mathematics, and also the traditional notion of proof as formal derivations.

Through a careful and detailed analysis of the historical development of the Euler-conjecture on the relation between the number of vertices, edges, and faces of polyhedra, Lakatos's work shows how the content of mathematical concepts is changed in the process of developing proofs. He calls the results of this process *proof-generated concepts* and shows that they completely replace the naive concepts with which the mathematical investigations began.

Since Lakatos considers axiomatic theories to be intimately connected to the view of the received tradition, he does not connect his conclusions to axiomatics, but regards them as being opposed. However, once we admit that axiomatizations do not have to be static, but can evolve, it is only a small step to transfer the strategies that Lakatos identifies to axiomatically characterized notions. Thus, against his own intentions, we can read Lakatos as identifying techniques for reformulating axioms in the light of failures of proof attempts or counterexamples. We can therefore infer from Lakatos's investigations that, by explicitly stating the assumptions made in arguments, axiomatics contributes to the development of mathematics.

The publication of Philip Kitcher's *The Nature of Mathematical Knowledge* has been hailed as another "event of great importance for philosophy of mathematics" [Gro85, p.71]. As the result of a detailed examination of the history of mathematics Kitcher proposes a *naturalist* account, which regards mathematical knowledge as *quasi-empirical* and *fallible* (see the above quotation by Lakatos). He argues for a close connection between science and mathematics and sees himself as standing in what he calls the "maverick tradition" in philosophy of mathematics that originated with Lakatos [AspKit88, p.17].

Kitcher regards the historical development of mathematics as a sequence of *practices*, which are individuated by five distinct, but interrelated, components: The language in use among mathematicians; the set of accepted statements; the questions regarded as important; the reasonings used to justify accepted statements; and methodological views about the character of mathematical proof, and the ordering of mathematical disciplines [Kit83, p.163]. Mathematical progress is characterized by Kitcher as *rational interpractice transitions* that aim to maximize the chances to attain one of the following two epistemological goals: To provide idealized descriptions allowing us to structure our experience, and to attain an intellectual understanding of these descriptions themselves [Kit88]. As particular activities that yield such rational interpractice transitions Kitcher suggests five patterns of mathematical change: Question-answering, question-generation, generalization, rigorization, and systematization [Kit83, p.194].

Although one might be able to find ways in which axiomatics is of use in all five of these patterns, Kitcher discusses axiomatizations only in relation to systematization. Here he mentions the introduction of new terms and principles that provide a unified perspective. He distinguishes between systematization by *axiomatization*, where a small number of principles and definitions are fixed from which previously "scattered" statements are derived, and systematization by *conceptualization*, which "consists in modifying the language to enable statements, questions, and reasonings which were formerly treated separately to be brought together under a common formulation" [Kit83, p.221]. To me both kinds of systematization are aspects of axiomatics, and Kitcher himself seems to conflate the terms of his own distinction by discussing the introduction of the concept of an abstract group as an example of axiomatization. In any case, in contrast to Lakatos, here the usefulness of axiomatics for mathematical progress is explicitly acknowledged.

In addition to its role in rational interpractice transitions, axiomatics can also contribute to the cumulative character of mathematics, which, according to Kitcher, is achieved through *reinterpretation* of previous theories. For example, the discovery of non-Euclidean geometry did not overthrow Euclidean geometry, but rather it led us to change our views about its necessary character and the meanings of the primitive terms. This move can be explicated by the transition from a particular interpretation of an axiom system to another, or a class of other interpretations.

The considerations of Pólya, Lakatos, and Kitcher have recently been taken up by more and more philosophers of mathematics. Regarding the interplay between axiomatics and mathematical discovery I would like to draw attention to the collection edited by Grosholz and Breger, *The Growth of Mathematical Knowledge* [GroBre00]. Herein, many different aspects of the development of mathematics are discussed, the traditional approaches are criticized for not being able to tell an adequate story about the development of mathematics, and the role of abstraction and axiomatization for mathematical progress is emphasized.

3.4 Summary

The development of philosophy of mathematics that I presented can be followed in more detail by considering the following anthologies, each of which contains a number of important contributions reflecting the various trends discussed. Regarding the early views, van Heijenoort's *From Frege to Gödel* [vHe67] and Ewald's *From Kant to Hilbert* [Ewa96] provide many sources;

the received tradition is best represented by the articles in the collection *Philosophy of Mathematics* by Benacerraf and Putnam [BenPut83], while articles pertaining to the newer directions can be found in Tymoczko's *New Directions in Philosophy of Mathematics* [Tym98]. Aspray and Kitcher's *History and Philosophy of Modern Mathematics* [AspKit88] contains an interesting juxtaposition of contributions in the received tradition and also following the newer directions. It also contains an excellent introduction, which presents the development of philosophy of mathematics from a more general perspective than the present paper.

From my, admittedly sketchy, overview about what has been said in philosophy of mathematics regarding the relation between axiomatics and discovery, the parallels to the developments in 20th century philosophy of science should have become obvious. In both areas the received view and received tradition have dominated the discussions for a long time. They were followed by polarized reactions, mainly antagonistic in spirit. Regarding the reflections on the interplay between axiomatics and mathematical progress, we can see a revival of the views first formulated by Hilbert in the early decades of the 20th century; a similar move in philosophy of science has yet to be made.

4 What's next?

Returning to the questions posed at the beginning of this paper, it has now become clear that neither of them has been addressed in a satisfactory manner in 20th century reflections on science and mathematics. In particular, a systematic study of the role that axiomatics plays in theory development is still missing.

I have been deliberately vague regarding the term "axiomatics", because what I consider to be various aspects of it, namely "axiomatic method", "symbolization", "formalization", *etc.*, have been understood in a number of very different ways in the past. For future discussions on methodology in science and mathematics, a better disentanglement of notions and terminology is sorely needed. Moreover, reflections about what scientists say and do seem to profit when the approach is less dogmatic in character, *i.e.*, without the imposition of too strict a priori assumptions. Clearly some focus is necessary, but this should not be gained by completely dismissing alternative aspects and approaches. This is related to what I have found to be an unfortunate recurrent pattern in the discussions, namely that when new points of view are proposed, they are often set in stark contrast to some previous position. This is important for highlighting the novelty of the new approaches, but also tends to devaluate the insights gained by the earlier reflections.

After all, "*Die Mathematik ist ein* BUNTES *Gemisch*" [*Bemerkungen*, p.176], and this should be reflected also in the study of and considerations about the theoretical aspects of science.

Primary Sources.

[*An. post.*]   **Aristotle**, Analytica posteriora, *in:* [TreFor60]

[*Begriffsschrift*]   Gottlob **Frege**, Begriffsschrift, Eine der arithmetischen nachgebildeten Formelsprache des reinen Denkens, Nebert 1879; *english translation in:* [vHe67, p.1-82]

[*Grundlagen*]   Gottlob **Frege**, Grundlagen der Arithmetik, Köbner 1884; *english translation in:* [Aus68]

[*Bemerkungen*]   Ludwig **Wittgenstein**, Bemerkungen über die Grundlagen der Mathematik, *in:* [AnsRhevWr84]

References.

[AnsRhevWr84]   G. Elizabeth M. **Anscombe**, Rush **Rhees**, and Georg Henrik **von Wright** (*eds.*), Ludwig Wittgenstein, Werkausgabe, Band 6: Bemerkungen über die Grundlagen der Mathematik, Suhrkamp 1984 [suhrkamp taschenbuch wissenschaft 506]

[AspKit88]   William **Aspray** and Philip **Kitcher** (*eds.*), History and Philosophy of Modern Mathematics, University of Minnesota Press 1988

[Aus68]   John L. **Austin** (*ed., trans.*), Gottlob Frege, The Foundations of Arithmetic: A Logico-Mathematical Enquiry into the Concept of Number, Northwestern University Press 1968

[BenPut83]   Paul **Benacerraf** and Hilary **Putnam** (*eds.*), Philosophy of Mathematics – Selected readings, 2nd ed.,Cambridge University Press 1983

[Ber65]   Richard J. **Bernstein** (*ed.*), Perspectives on Peirce, Critical Essays on Charles Sanders Peirce, Greenwood Press 1965

[Bor+52]   Edwin G. **Boring**, Herbert S. **Langfeld**, Heinz **Werner**, and Robert M. **Yerkes** (*eds.*), A History of Psychology in Autobiography, vol. IV, Clark University Press 1952

[Bro52]   L. E. J. **Brouwer**, Historical background, principles and methods of intuitionism, **South African Journal of Science** 49 (1952), p.139-146; *reprinted in:* [Hey75, p.508-515]

[ChaCon95]   Jean-Pierre **Changeux** and Alain **Connes**, Conversations on Mind, Matter, and Mathematics, Princeton University Press 1995; *original:* Matière à Pensée, 1989

[Cra02]   Carl F. **Craver**, Structures of scientific theories, *in:* [MacSil02, p.55-79]

[Edw73]   Paul **Edwards** (*ed.*), Encyclopedia of Philosophy, Macmillan 1973

[Ewa96]	William **Ewald** (*ed.*), From Kant to Hilbert: A Source Book in Mathematics, Clarendon Press 1996
[Fer99]	José **Ferreirós**, Labyrinth of Thought, A History of Set Theory and its Role in Modern Mathematics, Birkhäuser 1999
[vFr70]	Bas C. **van Fraassen**, On the extension of Beth's semantics of physical theories, **Philosophy of Science** 37(3) (1970), p.325-339
[vFr80]	Bas C. **van Fraassen**, The Scientific Image, Clarendon Press 1980
[vFr87]	Bas C. **van Fraassen**, The semantic approach to scientific theories, *in:* [Ner87, p.105-124]
[Gie79]	Ronald N. **Giere**, Understanding Scientific Reasoning, Holt, Rinehart and Winston 1979
[Gie88]	Ronald N. **Giere**, Explaining Science: A Cognitive Approach, University of Chicago Press 1988
[Gro85]	Emily **Grosholz**, A new view of mathematical knowledge, Review of The Nature of Mathematical Knowledge, **British Journal for the Philosophy of Science** 36 (1985), p.71-78
[GroBre00]	Emily **Grosholz** and Herbert **Breger** (*eds.*), The Growth of Mathematical Knowledge, Kluwer 2000
[Had45]	Jacques **Hadamard**, The Psychology of Invention in the Mathematical Field, Princeton University Press 1945
[Han58a]	Norwood Russell **Hanson**, The logic of discovery, **The Journal of Philosophy** 55(25) (1958), p.1073-1089
[Han58b]	Norwood Russell **Hanson**, Patterns of Discovery, Cambridge University Press 1958
[Han65]	Norwood Russell **Hanson**, Notes towards a logic of discovery, *in:* [Ber65, p.42-65]
[vHe67]	Jean **van Heijenoort** (*ed.*), From Frege to Gödel: A Sourcebook of Mathematical Logic, Harvard University Press 1967
[Hem45]	Carl G. **Hempel**, Studies in the logic of confirmation, **Mind** 54 (1945), p.1-26, 97-121; *reprinted with a postscript in:* [Hem65, p.3-51]
[Hem65]	Carl G. **Hempel**, Aspects of Scientific Explanation and Other Essays in the Philosophy of Science, Free Press 1965
[Hem70]	Carl G. **Hempel**, Formulation and formalization of scientific theories (a summary-abstract, with discussion), *in:* [Sup77, p.244-265]
[HemOpp48]	Carl G. **Hempel** and Paul **Oppenheim**, Studies in the logic of explanation, **Philosophy of Science** 15 (1948), p.135-175; *reprinted in:* [Hem65, p.245-290]
[Hes66]	Mary B. **Hesse**, Models and Analogies in Science, University of Notre Dame Press 1966
[Hes72]	Mary B. **Hesse**, Models and Analogy in Science, *in:* [Edw73, p.354-359]
[Hey75]	Arend **Heyting** (*ed.*), L. E. J. Brouwer, Collected Works I, Philosophy and Foundations of Mathematics, North-Holland 1975
[Hil99]	David **Hilbert**, Grundlagen der Geometrie, Teubner 1899; *english translation in:* [Hil02]
[Hil02]	David **Hilbert**, The Foundations of Geometry, Open Court 1902

[Hil18]   David **Hilbert**, Axiomatisches Denken, **Mathematische Annalen** 78 (1918) p.405-415; *english translation in:* [Ewa96, p.1105-15]

[HinRem74]   Jaakko **Hintikka** and Unto **Remes**, The Method of Analysis, Its Geometrical Origin and Its General Significance, Kluwer 1974

[Hul52]   Clark L. **Hull**, Autobiography, *in:* [Bor+52, p.143-162]

[Kit83]   Philip **Kitcher**, The Nature of Mathematical Knowledge, Oxford University Press 1983

[Kit88]   Philip **Kitcher**, Mathematical progress, **Revue Internationale de Philosophie** 42 (1988), p.518-540

[Kle95]   Felix **Klein**, Über Arithmetisierung der Mathematik, **Nachrichten der Königlichen Gesellschaft der Wissenschaften Göttingen** 2 (1895); *reprinted in:* [Kle22, p.232-240]

[Kle22]   Felix **Klein**, Gesammelte mathematische Abhandlungen, vol. 2, Springer 1922

[Kuh70a]   Thomas S. **Kuhn**, Logic of discovery or psychology of research?, *in:* [LakMus70, p.1-23]

[Kuh70b]   Thomas S. **Kuhn**, The Structure of Scientific Revolutions, 2nd ed., Chicago University Press 1970

[Lak70]   Imre **Lakatos**, Falsification and the methodology of scientific research programmes, *in:* [LakMus70, p.91-195]

[Lak76]   Imre **Lakatos**, Proofs and Refutations, Cambridge University Press 1976

[LakMus70]   Imre **Lakatos** and Alan **Musgrave** (*eds.*), Criticism and the Growth of Knowledge, Cambridge University Press 1970

[MacSil02]   Peter **Machamer** and Michael **Silberstein** (*eds.*), The Blackwell Guide to the Philosophy of Science, Blackwell Publishers 2002

[McK47]   Richard **McKeon** (*ed.*), Introduction to Aristotle, Chicago University Press 1947

[Mor67]   Sidney **Morgenbesser** (*ed.*), Philosophy of Science Today, Basic Books 1967

[Ner87]   Nancy J. **Nersessian** (*ed.*), The Process of Science, Contemporary Philosophical Approaches to Understanding Scientific Practice, Kluwer 1987

[Pec90]   Volker **Peckhaus**, Hilbertprogramm und Kritische Philosophie, Vandenhoeck und Ruprecht 1990

[Pól45]   George **Pólya**, How to Solve It: A New Aspect of Mathematical Method, Princeton University Press 1945

[Pól54]   George **Pólya**, Mathematics and Plausible Reasoning, 2 vols., Princeton University Press 1954

[Pop34]   Karl R. **Popper**, Logik der Forschung, Springer 1934; *english translation in:* [Pop59]

[Pop59]   Karl R. **Popper**, The Logic of Scientific Discovery, Harper & Row 1959

[Pos89]   Hans **Poser**, Vom Denken in Analogien, **Berichte zur Wissenschaftsgeschichte** 12 (1989), p.145-157

[Qui48]   Willard Van Orman **Quine**, On what there is, **The Review of Metaphysics** 1948; *reprinted in:* [Qui61, p.1-19]

[Qui61]   Willard Van Orman **Quine**, From a logical point of view, Harvard University Press 1961

[Rei38]   Hans **Reichenbach**, Experience and Prediction: An analysis of the foundations and the structure of knowledge, University of Chicago Press 1938

[Sal+92]   Merrilee H. **Salmon**, John **Earman**, Clark **Glymour**, James G. **Lennox**, Peter **Machamer**, J. E. **McGuire**, John D. **Norton**, Wesley C. **Salmon**, and Kenneth F. **Schaffner**, Introduction to the Philosophy of Science, Prentice Hall 1992

[Sch06]   Dirk **Schlimm**, Two ways of analogy, *manuscript* 2006

[Sne71]   Joseph D. **Sneed**, The Logical Structure of Mathematical Physics, Kluwer 1971

[Ste76]   Wolfgang **Stegmüller**, The Structure and Dynamics of Theories, Springer 1976

[Sup67]   Patrick **Suppes**, What is a scientific theory?, *in:* [Mor67, p.55-67]

[Sup77]   Frederick **Suppe** (*ed.*), The Structure of Scientific Theories, 2nd ed., University of Illinois Press 1977

[Tar44]   Alfred **Tarski**, The semantic conception of truth and the foundations of semantics, **Philosophy and Phenomenological Research** 4 (1944)

[TreFor60]   Hugh **Tredennick** and Edward S. **Forster** (*eds., trans.*), Aristotle, Posterior Analytics and Topica, Heinemann 1960 [Loeb Classical Library 391]

[Tym98]   Thomas **Tymoczko** (*ed.*), New Directions in the Philosophy of Mathematics, revised and expanded ed., Princeton University Press 1998

[Wor84]   John **Worrall**, "An Unreal Image", Review of van Fraassen (1980), **British Journal for the Philosophy of Science** 35(1) (1984), p.65-80

[WhiRus10-13]   Alfred N. **Whitehead** and Bertrand **Russell**, Principia Mathematica, 3 vols., Cambridge University Press 1910-13

Received: May 16th, 2003;
In revised version: January 20th, 2004;
Accepted by the editors: February 20th, 2004.

Benedikt **Löwe**, Volker **Peckhaus**, Thoralf **Räsch** (eds.)
Foundations of the Formal Sciences IV
The History of the Concept of the Formal Sciences

Existence, Identity, and the Algebra of Logic

RISTO VILKKO*

Department of Philosophy
P.O. Box 9
00014 University of Helsinki, Finland
E-mail: risto.vilkko@helsinki.fi

> ABSTRACT. During the 20th century it became commonplace to assume that verbs for being are multiply ambiguous between the *is* of predication, the *is* of existence, the *is* of identity, and the *is* of subsumption. This assumption is also known as the Frege-Russell ambiguity thesis, for its currency is due largely to Gottlob Frege and Bertrand Russell. However, after the Middle Ages no philosopher assumed such multiple ambiguity before the 19th century. This paper approaches the development of the Frege-Russell ambiguity thesis from a historical point of view. Special attention is paid to the ideas of George Boole and Augustus De Morgan.

One of the most interesting open problems in the history of formal sciences concerns the rise of modern logic epitomized by the Frege-Russell theory of quantifiers. One of the cornerstones of this theory is the distinction between the allegedly different meanings of verbs for being. According to received wisdom, such verbs are multiply ambiguous between the *is* of predication, the *is* of existence, the *is* of identity, and the *is* of subsumption. This view, also known as the Frege-Russell ambiguity thesis, is built into the notations that have been used in logic since the turn of the 20th century, in that the allegedly different meanings are expressed differently in the usual logical notations.[1] The *is* of identity is expressed by the identity sign

---

*This paper is based on a research project which I share with Professor Jaakko Hintikka (Boston University). I want to thank Professor Hintikka for invaluable inspiration and collaboration. Also, I take this occasion to thank my anonymous referee for insightful constructive comments.

[1] Stanisław Leśniewski's notation provides an important exception here.

$a = b$, the *is* of predication by a singular term's filling the argument slot of a predicative expression $P(a)$, the *is* of existence by the existential quantifier $(\exists x)P(x)$, and the *is* of subsumption by a general conditional of the form $(\forall x)(x\epsilon S \supset x\epsilon P)$. Both Gottlob Frege and Bertrand Russell attached great importance to the ambiguity of the verb *is*. During the 20th century it became commonplace to subscribe to this thesis even though it is not necessary or even fully obvious (*cf.* [Hin79]; [Mat79]). But then again, it turns out that after the Middle Ages no philosopher assumed such multiple ambiguity before the 19th century. What happened? How did the Frege-Russell thesis come about? In what follows, I approach these questions from a historical point of view. I first say a few introductory words about the treatment of existence in Aristotle and Kant, and thereafter focus on the 19th century English developments in the field of the algebra of logic and on the ideas of George Boole and Augustus De Morgan in particular.

Aristotle considered the Frege-Russell distinction but rejected it. His treatment of existence in the context of a syllogistically constructed science was in rough agreement with the ancient Greek language, in which there were no separate verbs for existence. Existence was expressed by the absolute construction with ἐστίν which looks like a special case of predication, *e.g.*, "Zeus is" as a limiting case of such statements as "Zeus is a god" or "Zeus is powerful". In effect, Aristotle treated the different Frege-Russell senses of different components in the force of ἐστίν. *Each of these components could be absent or present on any one occasion of the use of ἐστίν.* In syllogistic reasoning, existence was sometimes present as part of the force of the predicate term, sometimes absent. The existential force trickled down from the most general terms of the sense in question along a sequence of syllogistic conclusions. Hence, in any one particular science existential force had to be assured, according to Aristotle, only for the widest generic term defining the field of that science:

> Thus we assume the meaning alike of unity, straight, and triangular; but while as regards unity and magnitude we assume also the fact of their existence, in the case of the remainder proof is required.
> [*An. post.*, A 10, 76a, 34-36]

Existence could not serve alone as a predicate term because it would have been too broad a term, not restricted to any one category and thus not an essence of anything [*An. post.*, B 7, 92b, 13-15]. In this sense, according to Aristotle, existence was not a predicate. However, it could be a part of the force of a predicate term.

It is often said that Kant's discussion of existence includes a criticism of the idea that existence is a predicate. In fact, it includes a stronger criticism, namely the rejection of the idea that existence could be even a part of the force of a predicate term. According to Kant, existence adds nothing to a concept of a thing:

> '*Being*' is obviously not a real predicate; that is, it is not a concept of something which could be added to the concept of a thing. [...] The small word 'is' adds no new predicate. [*Kr. d. r. V.*, B 625]

This does not mean that Kant embraced the Frege-Russell thesis. It means that at the turn of the 19th century the notion of existence became homeless, as far as the logical representation of different propositions was concerned. It must be admitted, though, that Kant's criticism served to disassociate the predicative and the existential uses of *is* from one another. According to Leila Haaparanta, Kant seems to have inspired the Frege-Russell distinction [Haa86].

After Kant the next major development in logical theory was the algebra of logic that originated in England around the mid-19th century.[2] The following two ideas came to the forefront: (1) the operators corresponding to the syllogistical standard forms of universal and particular judgments were treated as duals; (2) universal judgments were taken to be relative to some universe of discourse and were inevitably taken as the non-existence of exceptions in that domain. Because of the duality, existential quantifier expressions came to express existence. The homeless notion of existence thus found a new home, no longer in the predicative *is* but in the existential quantifier.

Before moving on to take a closer look upon the ideas of Augustus De Morgan and George Boole, it is important to acknowledge that neither of them introduced existential or universal quantifiers, and therefore they must not be regarded as early pioneers of the predicate calculus. However, at least De Morgan seems to have been aware of at least some of the difficulties that arose in the absence of quantifiers [Smi82, p.24].

If the situation into which Kant had thrust all thinkers was thus felt outside philosophy, it is only natural that it was perceived independently and more or less simultaneously by several philosophers. One way of trying to deal with it was to make the Frege-Russell distinction, or some part of it. Therefore it is not surprising to find parts of the Frege-Russell thesis put forward by, *e.g.*, De Morgan, Mitchell, and Peirce. Frege's new logic, which he introduced in his *Begriffsschrift* (1879), was therefore not in all respects a unique discovery that could have been made by someone like Frege at any time. His groundbreaking results –including the distinctions between

---

[2] Strictly speaking it was upon Leibniz's initiative that the idea of an algebraic structure of logic began to grow — even though it was Boole who really started its systematic development.

allegedly different senses of being– were achieved very much in a particular historical situation.

One cannot really speak of coincidences in this connection. This becomes even more evident when it is noted that for logicians there existed an obvious way of finding a new home to the orphaned existential force, even in the case of syllogistic premises. This was to assign it only to the particular quantifier expression, which was later turned into our now familiar existential quantifier. This transfer was encouraged by other facts. For mathematicians like De Morgan and Boole, universality came to mean universality in some universe of discourse which is the ultimate subject of that discourse. In virtue of the duality of universally and particularly quantified statements reflected in their interdefinability, this meant that the universal statements express universality and the existential ones existence with respect to the given domain. The key notion in Boole's algebra of logic was the notion of the universe of discourse. It did not necessarily refer to the actual universe but whatever system of objects we choose to speak about. In his 1847 mathematical analysis of logic Boole understood the universe as comprehending

> every conceivable class of objects whether actually existing or not, it being premised that the same individual may be found in more than one class, inasmuch as it may possess more than one quality in common with other individuals. [Boole, MAL, p.15]

In other words, Boole's universe is the only class which contains all the individuals that exist in any class. This is in perfect agreement with De Morgan's 1847 characterization of his universe of discourse as a range of ideas which is either expressed or understood as containing the whole matter under consideration [De Morgan, FL, p.38].

In his late manuscript *Logic and Reasoning*, Boole says succinctly that the limiting conceptions of universe and nothing express simply the ideas of existence and non-existence [Boole, L & R, p.218]. Rush Rhees has claimed that there may be important reasons for this late change of mind [Boole, L & R, p.30]. This indeed was an important change but it was not quite as late as Rhees suggests. Boole expressed the is of existence by $x = 1$, *e.g.*, "Something exists" and, respectively, $x = 0$ for "Something does not exist" already in his greatest work, *An Investigation of the Laws of Thought* [Boole, ILT, p.189f]. In any case, it is interesting to compare this late definition of the universe as expressing simply the idea of existence to the earlier definition of 1847, where the universe covered every class of objects whether actually existing or not. With regard to this comparison it is important to pay attention to the fact that the phrase "actually existing or not", in the earlier definition, refers to classes and not to objects.

In the beginning of his professional career, around the turn of the 1830s, Boole's friend and colleague Augustus De Morgan considered the *is* of predication and the *is* of identity as fundamental non-reducible copulas, and gave a relational analysis of both of them.³ In the early 1920s, Bertrand Russell also considered it "a disgrace to the human race" that it has chosen to employ the same verb for so very different ideas as the *is* of predication and the *is* of identity [Rus19, p.172]. It was Russell's intention to remedy this deplorable state of affairs by means of a new symbolic logical language. De Morgan's early treatise *On the Study and Difficulty of Mathematics* (1831) also takes into account formulations which seem to involve the is of subsumption, *i.e.*, expressions like "All the □ is (contained) in the ○". However, at that time De Morgan came to the conclusion that in the last analysis to say that "All the circle is in the square" is merely to say that "Every point of the circle is a point of the square". Hence, expressions of class inclusion reduced to expressions of simple identity [Mer90, p.29-33].

In his first significant contribution in the field of logic, the 1839 paper *First Notions of Logic*,⁴ De Morgan made the rather bold and controversial claim about all propositions being either assertions or denials. According to him, "X is Y" and "X is not Y" were "the two [most simple] forms to which all propositions may be reduced" [*De Morgan, FL*, p.2]. The task of formal logic was to handle, organize, and arrange expressions of simple identity and non-identity. In other words, during that time the man who a little bit later became one of the most important architects of the logic of relations considered seriously the idea of only one basic relation, *i.e.*, identity. Under these circumstances, for example, expressions of the form $(\forall x)(P(x) \supset Q(x))$ were taken to be equivalent to those of the form $(\forall x)(\exists y)(P(x) \supset (Q(y) \,\&\, x = y))$. However, as Daniel Merrill has pointed out, De Morgan did not make it quite clear why the latter form should be regarded as more basic than the first one [Mer90, p.38].

De Morgan was very sensitive with regard to semantical issues, perhaps even more sensitive than Boole. In the third chapter of his *Formal Logic* De Morgan discusses first the general characteristics of the terms of a proposi-

---

³Thomas Reid did not want to construe the "A is equal to B" as a categorical proposition. In his opinion, the subject, the predicate and the relation in this phrase made three terms, whereas categorical propositions of the form "A is B" consisted of two terms only. Sir William Hamilton, in his turn, subscribed to the more traditional view of only one affirmative copula. De Morgan disagreed with both Reid and Hamilton. Against Reid, he held the opinion that "A is B" and "A is equal to B had the same logical form. It was just that according to De Morgan, there were two copulas, *is* and *is equal to*. On the inferential level De Morgan agreed with Reid and contradicted with Hamilton: in his view the logical rules for *is* and *is equal to* were equally fundamental. [Mer90, p.29f]

⁴This essay is included in De Morgan's *Formal Logic* as the first chapter of the book [*De Morgan, FL*, p.1-25].

tion, and concentrates thereafter on those of the connecting copulae *is* and *is not*. He summarizes the most common uses of the verb *is* as follows:

> (1) Absolute identity, as in "The thing he sold you *is* the one I sold him";
>
> (2) Agreement in a certain particular or particulars understood, as in "He *is* a Caucasian" said of a European in reference to the colour of his skin;
>
> (3) Possession of a quality, as in "The rose *is* red";
>
> (4) Reference of a species to its genus, as in "Man *is* an animal"; and
>
> (5) Existential use, as in "Man *is*, i.e., exists".[5]   [*De Morgan, FL,* p.53]

De Morgan considered the fifth use, *i.e.* "the use of the verb alone", to be independent of all the other uses which he wanted to bring down to the *is* of identity.[6] As a result, he held on to the *is* of identity as the most fundamental use of the verb and promoted the *is* of existence as another independent and irreducible use. A few years later, in his second pamphlet on the syllogism [*De Morgan, Syll.*], De Morgan still asserted his basic view of the reducibility of (almost) all uses of *is* to the *is* of identity. However, he admitted that "is" also had another basic meaning. He did not make this another meaning explicit but evidently it must have been the *is* of existence. In *Formal Logic* we have, in any case, at least a part of the Frege-Russell distinction before Frege and Russell. However, De Morgan clearly did not take it the way we do now. On a few pages of his *Formal Logic* he writes about the different *senses,* the different *meanings,* and the different *uses* of the verb *is* [*De Morgan, FL,* p.49-54]. It seems as if he did not have a clear opinion about whether the differences in the use of the verb *is* are due to the multiple ambiguity of a single word or to differences in the context in which it occurs. His treatment is problematic in several ways. For example, as Merrill has shown, his comments about the mathematician's *is* in the sense of "is equal to" are somewhat confusing [Mer90, p.55]. I shall not, however, dig any deeper into this ground here.

Boole and other 19th century algebraic logicians applied the theory and practice of setting up equations and solving them algebraically. As is well known, in algebraic theory and practice the identity sign not only asserts numerical identities but functional dependencies between mathematical objects. Jaakko Hintikka has called this use of the notion of identity –which occurs between variables or variables and constants– the equational use of identity [Hin04]. With respect to this kind of identity it is only known that

---

[5]Daniel Merrill provides an exhaustive scrutiny of De Morgan's five main senses of "is" in his book *Augustus De Morgan and the Logic of Relations* [Mer90, p.51-55].

[6]De Morgan considered all such statements as (1) doubly singular ("This one A is this one B"), (2) singly singular ("This one A is one of the Bs"), or (3) quantified identity statements ("Every A is one of the Bs", "Some A is not one of the Bs") [*De Morgan, FL,* p.53f]; [Mer90, p.54].

the unknown factors in an equation have their numerical identities. It is the task of the computer (man or machine) to establish these identities for the purpose of finding out their predicates. In accordance with the algebraic technique of Boole and others, one seeks first for dependencies between different factors in a configuration. These dependencies are then expressed in the form of equations, and the unknown elements identified by solving the equations. This technique, with its equational use of identity, is a descendant of a method of analysis that goes back to ancient Greece [Hin04]. Frege's treatment of identity, in turn, did not capture the role of identity in this algebraic equational sense. As Hintikka has observed, the quite common understanding about Frege having taken his notion of identity from mathematics "must be taken with qualifications. For it has been seen that the most important use of this notion in mathematics was the equational one which is not captured by Frege's notation" [Hin04].

Hintikka has pointed out that Frege apparently did not introduce the notion of identity into his formal language in order to facilitate the expressing of dependence relations. For Frege identity ($a = b$) expressed simply the identity of the references of two singular constants. To put it at its crudest, for Frege identities merely asserted that two specific entities are one and the same. He did not regard function symbols as expressions of dependencies in the first place, but as means of forming complex names. For instance, as far as Frege was concerned, the equation $3^2 = 9$ did not express a dependence between numbers 3, 2 and 9. In his view, it expressed the sameness of the references of the two names "3^2" and "9". As Hintikka has correctly remarked in a forthcoming contribution, the form of this equation is not $f(a,b) = c$ but $a = b$. In this connection it is also important to remember that, unlike, *e.g.*, Giuseppe Peano, Frege did not count functions among primitive non-logical symbols of his formula language [Hin04].

Indeed, during the late 1890s Frege and Peano had a lively debate about primitives and the use of identity in formal languages (*cf.* [*Ueber die Begr.*]; [*Frege, Briefw.*, p.181-198]; see also [Gra00, p.231, 248f]). Unlike Frege, Peano included functions among his non-logical primitives and therefore had no problem about the cognitive value or informativeness of identities. Consequently he did not develop anything like Frege's sense-reference distinction. In his voluminous recent work, *The Search for Mathematical Roots 1870-1940,* Ivor Grattan-Guinness has correctly summarized this exchange of thoughts by stating that "Frege showed himself to be the sharper logician and philosopher; but on mathematical matters he was less strong, for he puzzled over Peano's use of classes in a way which revealed his own misunderstandings" [Gra00, p.249].

In spite of the efforts of a few pioneers in the field of the algebra of logic, algebra was not generally considered, around the mid-nineteenth century, as the model example for demonstrative sciences. In comparison with the perfectly deductive geometry, it was regarded a mere toolkit for the purposes of computation. The surprisingly poor reception of Boole's groundbreaking results was a consequence of this mathematical climate: Even Boole's own countrymen showed but little interest towards his results before William Stanley Jevons revised his theory in the 1860s. On the other side of the English Channel his work remained practically unknown for decades (*cf.* [Pec97, p.214f]). However, there is good reason to believe that Frege's new approach to logic, which departed from standard algebraic theory and practise, was one of the most important reasons why it took so long before the great significance of his results was properly recognized. Even at the turn of the 20th century, twenty years after the *Begriffsschrift,* there were only a few experts in the foundations of mathematics who knew Frege's work well enough to respect its significance. For the average working mathematician of that time, his results had nothing to say. On the one hand, every working mathematician was familiar with setting up equations and solving them algebraically. On the other hand, Frege's formula language was not only visually but also in many theoretical respects like nothing they had ever seen before. Moreover, it considered foundational issues which had no influence on everyday mathematical practise.

Almost all of those who wrote essay reviews on Frege's first presentation of his new symbolic logic agreed that his original two-dimensional notation appeared to be cumbersome, inconvenient and excessively complex. What is more, most of his critics were astonished by the fact that Frege almost completely ignored what had been going on, during the first three quarters of the 19th century, in the field of the algebra of logic both in Britain and in Germany. Only one of his reviewers cared to pay attention to the fundamental difference between Frege's approach and that of the algebraists [Las79, p.248]. This was no surprise to Frege. He understood from the very beginning that the majority of both mathematicians and philosophers might shrink back from his novelties.[7]

The identity sense of verbs for being gained importance when relations and functions were included among the notions studied in logic. In Aristotelian syllogistic, it did not make much difference whether a phrase like "some teachers are wise" was in effect parsed as "some teachers are identical with members of the class of wise people" or as "some teachers have the predicate of wisdom". However, relational expressions like "the teacher of Alexander the Great" could not be handled like this. The situation in logic was perceived to be like the situation in algebra, where identities and

---

[7] *Cf.* [Vil98]; [Vil02, p.129-140].

predications under the sense of satisfying certain conditions had to be distinguished from one another. This development was connected with the gradual change of the notion of relation from a relational predicate to a genuine entity linking its two terms.[8]

In one respect, Frege went to the other extreme away from Aristotle. For Aristotle, existence was attributed to particular objects in the sense that it was part of the force of predicates of objects. Frege denied that existence could be attributed to individuals and sought to construe it as a higher-order predicate expressing the non-emptiness of a lower-order one (cf. [Haa85, Haa86]). In this new sense, existence was a predicate for Frege, too.

Primary Sources.

[An. post.]	**Aristotle**, Analytica posteriora, in: [Ros28]
[Boole, MAL]	George **Boole**, The Mathematical Analysis of Logic, Being an Essay Towards a Calculus of Deductive Reasoning, Macmillan 1847
[Boole, ILT]	George **Boole**, An Investigation of the Laws of Thought, on which are Founded the Mathematical Theories of Logic and Probabilities, Macmillan 1854
[Boole, L & R]	George **Boole**, Logic and Reasoning, in: [Rhe52, p.211-229]
[De Morgan, SDM]	Augustus **De Morgan**, On the Study and Difficulty of Mathematics, London 1831
[De Morgan, FL]	Augustus **De Morgan**, Formal Logic, or, The Calculus of Inference, Necessary and Probable, Taylor and Walton 1847
[De Morgan, Syll.]	Augustus **De Morgan**, On the Syllogism II, On the Symbols of Logic, the Theory of the Syllogism, and in Particular of the Copula, **Transactions of the Cambridge Philosophical Society** 9 (1850), p.79-127
[Begriffsschrift]	Gottlob **Frege**, Begriffsschrift, eine der arithmetischen nachgebildete Formelsprache des reinen Denkens, Nebert 1879
[Ueber die Begr.]	Gottlob **Frege**, Ueber die Begriffsschrift der Herrn Peano und meine eigene, **Berichte über die Verhandlungen der Königlich Sächsischen Gesellschaft der Wissenschaften zu Leipzig, Mathematisch-Physische Klasse** XLVIII (1897), p.361-378
[Frege, Briefw.]	Gottlob **Frege**, Wissenschaftlicher Briefwechsel, in: [Gab+76]
[Kr. d. r. V.]	Immanuel **Kant**, Critik der reinen Vernunft, Hartknoch 1787

---

[8] Cf. [Hin04].

References.

[Flø04] Guttorm **Fløistad** (*ed.*), Language, Meaning, Interpretation, Springer 2004 [Philosophical Problems Today 2]

[Gab+76] Gottfried **Gabriel**, Hans **Hermes**, Friedrich **Kambartel**, Christian **Thiel**, and Albert **Veraart**, Gottlob Frege, Wissenschaftlicher Briefwechsel, Meiner 1976 [Nachgelassene Schriften und wiss. Briefwechsel 2]

[Gra00] Ivor **Grattan-Guinness**, The Search for Mathematical Roots 1870-1940, Logics, Set Theories and the Foundations of Mathematics from Cantor Through Russell to Gödel, Princeton University Press 2000

[Haa85] Leila **Haaparanta**, Frege's Doctrine of Being, University of Helsinki 1985 [Acta Philosophica Fennica 39]

[Haa86] Leila **Haaparanta**, On Frege's Concept of Being, *in:* [KnuHin86, p.269-289]

[Hin79] Jaakko **Hintikka**, " Is", Semantical Games, and Semantical Ambiguity, **Journal of Philosophical Logic** 8 (1979), p.433-468

[Hin04] Jaakko **Hintikka**, On the Different Identities of Identity: a Historical and Critical Essay, *in:* [Flø04, p.117-139]

[KnuHin86] Simo **Knuuttila** and Jaakko **Hintikka** (*eds.*), The Logic of Being: Historical Studies, Kluwer 1986 [Synthese Historical Library 28]

[Las79] Kurd **Lasswitz**, Rezension von Freges Begriffsschrift, **Jenaer Literaturzeitung** 6 (1879), p.248-249

[Mat79] Benson **Mates**, Identity and Predication in Plato, **Phronesis** 24 (1979), p.211-229

[Mer90] Daniel D. **Merrill**, Augustus De Morgan and the Logic of Relations, Kluwer 1990 [The New Synthese Historical Library 38]

[Pec97] Volker **Peckhaus**, Logik, Mathesis universalis und allgemeine Wissenschaft, Leibniz und die Wiederentdeckung der formalen Logik im 19. Jahrhundert, Akademie Verlag Berlin 1997 [Logica nova]

[Rhe52] Rush **Rheese** (*ed.*), George Boole, Collected logical works, Volume I, Studies in Logic and Probability, Open Court 1952

[Ros28] William David **Ross** (*ed.*), The Works of Aristotle, Volume I, *Categoriae* and *de Interpretatione* and *Analytica Priora* and *Analytica Posteriora* and *Topica* and *de Sophisticis Elenchis*, Oxford University Press 1928

[Rus19] Bertrand **Russell**, Introduction to Mathematical Philosophy, Allen + Unwin 1919

[Smi82] G.C. **Smith**, The Boole-De Morgan Correspondence 1842-1864, Clarendon Press 1982 [Oxford logic guides]

[Vil98] Risto **Vilkko**, The Reception of Frege's Begriffsschrift, **Historia Mathematica** 25 (1998), p.412-422

[Vil02] Risto **Vilkko**, A Hundred Years of Logical Investigations, Reform Efforts of Logic in Germany, 1781-1879, Mentis 2002

[VilHin∞]   Risto **Vilkko** and Jaakko **Hintikka**, Existence and Predication from Aristotle to Frege, Proceedings of the Entretiens of Institute International de Philosophie, Madrid, September 17-21, 2002, *forthcoming*

Received: March 27th, 2003;
In revised version: January 18th, 2004;
Accepted by the editors: February 20th, 2004.

Benedikt **Löwe**, Volker **Peckhaus**, Thoralf **Räsch** (eds.)
Foundations of the Formal Sciences IV
The History of the Concept of the Formal Sciences

The Formal Aspect of the Fourteenth Century Concept of Consequence

STEPHANIE WEBER-SCHROTH

Philosophisches Seminar
Georg-August-Universität Göttingen
Humboldtallee 19
37073 Göttingen, Germany
Email: stephanie.weber@phil.uni-goettingen.de

The theory of consequences is one of the most important developments in logic during the Middle Ages. One of the first treatises on this theory is Walter Burley's *De consequentiis* (On Consequences), written in 1302 which is followed by numerous treatises in the 14th century. The best known of them are the treatises on consequences in William of Ockham's *Summa logicae* and Burley's *De puritate artis logicae*. Apart from these very elaborated works, we also find shorter academic treatises in the Oxford tradition,[1] *e.g.*, compilatory tracts of rules to be learned as well as commentaries of different lengths with more or less detailed explanations on them. It is assumed that in the 14th century the theory of consequences was an integral part of the logic curricula at universities.

The focus here will be on a typical treatise on consequences (*De consequentiis*) by the English author Richard Billingham. He was well known for his *Speculum puerorum*, a famous and influential text in the 14th and 15th century logic curriculum, which discusses rules for proving all kinds of propositions.

In order to expound the formal aspect of the 14th century theory of consequences, it is helpful to look at the definitions of consequences in these trea-

---

[1]The predominant division of consequences in the middle of the 14th century, however, is found in the treatises of authors like John Buridan and Pseudo-Scotus who are representatives of the Paris tradition ([J. Buridan DC, I-3 and I-4, p.20-24], [Pseudo-Scotus, Quaestio X, p.287*sq*]; *cf.* [Sch91, p.XXIIIf]; [Mor71, p.135]; *cf.* also [Wol95]).

tises. Billingham defines a consequence as a certain connection (*quoddam aggregatum*) consisting of an antecedent and a consequent with a sign of the consequence (*nota consequentie*) like *ergo, igitur* or *quia*.² The antecedent precedes the sign of the consequence whereas the consequent follows it:

> *Consequentia est quoddam aggregatum ex antecedente et consequente cum nota consequentie. Sunt note consequentie: "ergo", "igitur", "quia" et consimilia. Antecedens preterquam in conditionalibus et causalibus est illud quod precedit notam consequentie, et consequens est illud quod subsequitur notam consequentie.*
> [S, I, 1]³

He distinguishes between a formal consequence (*consequentia formalis*) and a material consequence (*consequentia materialis*). In a formal consequence the consequent is understood in the antecedent (*consequens est de intellectu antecedentis*), e.g., "Petrus is a man; therefore Petrus is an animal":

> *Consequentia formalis est illa, quando consequens est de intellectu antecedentis. Exemplum: "Petrus est homo; ergo Petrus est animal."*
> [S, I, 2.2]⁴

This definition of a formal consequence is quite common among Billingham's contemporaries and can be found in many treatises on the theory of consequences, e.g., in the treatise by Richard Lavenham (*consequentia formalis est quando consequens necessario est de intellectu antecedentis* [R. Lavenham DC, p.99, 2]). William of Osma and his contemporary, here called "Anonymus_P", define this kind of consequence in the same way, using a similar formulation: *consequens intelligitur formaliter in antecedente* [W. of Osma DC, I, 1ª regula]; *cf.* [Anonymus_P DC, f. 1vb]. Likewise, Robert Fland gives the following definition of a formal consequence:

> *Prima [regula] est ista: Ubi consequens intelligitur in antecedente formaliter. Verbi gratia, ista consequentia est formalis "Homo est; igitur animal est" quia hoc consequens "animal" formaliter intelligitur in antecedente, scilicet, "homo".*
> [R. Fland DC, p.57, 1]

We know from an account of the stoic conception of implication by Sextus Empiricus that, presumably, this definition was also used in megaric-stoic

---

²Like other medieval logicians Billingham does not distinguish between consequences, conditional and causal propositions. *Cf.*, *e.g.*, [R. Fland DC, p.62, 20] and [J. Buridan DC, I-3, p.21, 8-18 and p.22, 60-64].

³*Cf.* [O, I, 1] and [T, I, 1]. The different versions of Billingham's treatise on consequences are quoted corresponding to the critical edition of Richard Billingham's *De consequentiis* as follows: Salamanca [S]; Oxford [O]; Casanatense [C] and the Toledo commentary on Billingham's treatise [T]. [R] is the abbreviation for the version of Ripoll that is not edited. For bibliographical details see the references at the end of this paper.

⁴*Cf.* [T, I, 7].

logic, though in a general way, without distinguishing between formal and material consequences:

> And those who judge by "implication" declare that a hypothetical syllogism is true when its consequent is potentially included in its antecedent; [...].
> [Sextus Empiricus, II, 112].

Even though there existed a Latin translation of the *Outlines of Pyrrhonism* in Saint Victor near Paris in the second half of the 13th century (the text was ascribed to Aristotle),[5] it is still not established whether it had influenced medieval logic. In any case, we cannot assume that Billingham or his contemporaries knew this text.

In the 12th century the *intelligitur in* appears in connection with the definition of a true conditional proposition in the *Summa dialectice artis* by William of Lucca, a successor of Peter Abaelard:

> *Veluti hec propositio "homo est" ita quando antecedit ad istam "animal est" quod non solum ista ad illam necessario sequitur, sed etiam in illa intelligitur, unde hec ypothetica bona et in conditionali sensu vera est "si est homo, est animal."* [W. of Lucca, XII, no. 12.04, p.206sq]

However, it remains unclear how William of Lucca came to know this formulation, since we do not find it in Abaelard, and it cannot be assumed that he was acquainted with the *Outlines of Pyrrhonism*. Neither can we assume a connection between his work and the treatises of Billingham and his contemporaries, as William of Lucca's *Summa dialectice artis* was not widely known.

Billingham's formulation *consequens est de intellectu antecedentis* is also found in Burley's treatise on consequences of 1302. There, Burley defines a natural consequence (*consequentia naturalis*) as a consequence whose consequent is understood in the antecedent:

> *Consequentia naturalis est quando consequens est de intellectu antecedentis, nec antecedens potest esse verum nisi consequens sit verum; ut "si homo est, animal est."* [W. Burley DC, V, 70, p.128]

However, in spite of the fact that the works of Burley (and Ockham) were well known in the middle of the 14th century, we cannot determine whether Billingham knew them. Thus, the exact origin of Billingham's requirement for a formal consequence, *viz.*, *consequens est de intellectu antecedentis* or *consequens intelligitur in antecedente* remains unclear.

The requirement that the consequent has to be understood in the antecedent is a necessary and sufficient condition for a formal consequence in

---

[5] *Cf.* [Bäu91].

the treatises of Billingham and his contemporaries. According to this condition it is impossible for the antecedent to be true without the consequent, or, in other words: it is impossible that the antecedent is true while the consequent occurs and is false.[6] But what does "understanding the consequent in the antecedent" mean, and what is the formal aspect of this kind of consequence in relation to the material consequence?

It is assumed that in Billingham's distinction between formal and material consequences the term "formal" is chosen intentionally, whereas the term "material" is used only as a complementary term which does not indicate anything as to the nature of the material consequence. In a formal consequence the authors presuppose some relation between the antecedent and the consequent. In the above mentioned example "Petrus is a man; therefore Petrus is an animal" the relation is a topical one between the predicate terms "man" and "animal". In the antecedent "man" is predicated of Petrus, and since "man" is the species term to "animal", the latter can be predicated of Petrus in the consequent. We can therefore say that the predicate term "animal" in the consequent is understood in the predicate term "man" in the antecedent. This consequence holds by means of the *locus* "from a species to its genus" and the corresponding maxim: "Of which the species is predicated, the genus is predicated."[7] Thus, the consequence "Petrus is a man; therefore Petrus is an animal" is valid (*valet/est bona*) in virtue of a relation between two terms in the antecedent and the consequent. This is an example of Ockham's second group of formal consequences, which are immediately valid by means of an intrinsic medium (*medium intrinsecum*) and mediately valid by means of an extrinsic medium (*medium extrinsecum*):

> *Quaedam [consequentia formalis] tenet per medium intrinsecum immediate, et mediate per medium extrinsecum ⟨non⟩ respiciens generales condiciones propositionum, ut veritatem vel falsitatem, necessitatem vel impossibilitatem [...].*
> [W. of Ockham SL, III-3, 1, p.589, 50-53][8]

The consequence in Billingham's example holds (*tenet*) immediately by means of an intrinsic medium, *i.e.*, a proposition which can be formulated with the two terms "man" and "animal", in this case "A man is an animal."[9] The example holds mediately by means of an extrinsic medium, *i.e.*,

---

[6] *Cf.*, e.g., [T, I, 6]: *Consequentia que valet est in qua antecedens non potest esse verum sine consequente, ut: "tu curris; ergo tu moveris."*

[7] *Cf.* [W. of Sherwood, p.90, 145] and [Petrus Hispanus, V, p.64, 10].

[8] I agree with Schupp's correction [Sch93, especially p.215], who points out that the extrinsic medium of a formal consequence does not consider the general conditions of propositions like truth or falsity, necessity or impossibility.

[9] *Cf.* [W. of Ockham SL, III-3, 1, p.588, 24-28].

a general rule, here concerning the topical relation between the two terms "man" and "animal". In this case, the extrinsic medium corresponds to the maxim above: "Of which the species is predicated, the genus is predicated."[10] Thus, the consequence "Petrus is a man; therefore Petrus is an animal" holds because of two media.[11]

In the *Tractatus longior* of his treatise *De puritate artis logicae*, Burley uses a different terminology than Ockham. Instead of a formal consequence he speaks of a "natural consequence" (*consequentia naturalis*), which he defines as a consequence whose consequent is included in the antecedent (*antecedens includit consequens*). A natural consequence holds by means of an intrinsic *locus*:

> *Quaedam naturalis, et est quando antecedens includit consequens; et talis consequentia tenet per locum intrinsecum.*
> [W. Burley TL, II-1, 1, p.61, 6-8].

In his early treatise on consequences (*De consequentiis*), Burley gives the following example of a natural consequence (quoted above): "If a man is, an animal is." It seems reasonable to suppose that the intrinsic *locus* can be understood as a general rule about the topical relation between the antecedent and the consequent. Therefore, Burley's natural consequence corresponds to Ockham's second group of formal consequences, and his intrinsic *locus* to the extrinsic medium in Ockham's text.[12] With the term "natural" (versus "accidental", see below) Burley stresses the importance of this kind of consequence.

Instructive are the examples of formal consequences given in the texts of Billingham's contemporaries, as well as in the Salamanca copy of Billingham's treatise itself. In those texts which define a formal consequence as one where the consequent is understood in the antecedent, the authors usually add enthymematic consequences whose consequent is understood in the antecedent in virtue of a topical relation of terms, *e.g.*, "Petrus is a man; therefore Petrus is an animal" [S, I, 2.2], (see above), "a man runs; therefore an animal runs"[13], "a man is; therefore an animal is" [R. Fland DC, p.57, 1], "if a man is, an animal is" [W. Burley DC, V, 70, p.128], for a natural consequence, see above.[14] Most of these texts in the Oxford tradition give the impression that in a proper formal (natural) consequence the consequent

---

[10] *Cf.* [W. of Ockham SL, III-3, 1, p.588, 36-44].
[11] *Cf.* [Sch88, p.77 *sq*].
[12] *Cf.* [Sch88, p.79] and [Sch93, p.219].
[13] *Cf.* [W. of Osma DC, I, 1ᵃ regula]; [R. Lavenham DC, p.99, 2]; [T, I, 7]; [Anonymus_P DC, f. 2ʳᵃ].
[14] *Cf.* Thomas Manlevelt's example of a simple conditional proposition (*conditionalis simplex*); [T. Manlevelt DC, f. 60ʳ].

is understood (*intelligitur in/est de intellectu*) in the antecedent due to a topical relation of terms. This view is confirmed by the Toledo commentary on Billingham's treatise, which points out that a formal consequence can be either an enthymematic or a syllogistic consequence:

> Nota quod consequentia formalis est in qua consequens est de formali intellectu antecedentis, ut: "tu curris; ergo tu moveris"; ista est duplex, quia quedam est syllogistica, alia est entymematica. [T, I, 15]

The example the commentator first lists for a formal consequence is valid in virtue of the topical relation between the predicate terms: "You run; therefore you move." The predicate term of the consequent is related to the predicate term of the antecedent as a genus term is to the species term. It is therefore understood in the antecedent, *i.e.*, the consequence is validated by the *locus* "from a species to its genus".

However, in these texts various examples that do not hold in virtue of a topical relation are added to the listed rules for valid consequences. Let us consider a rule for copulative propositions: "A consequence from the whole copulative proposition to one of its (principal) parts is valid" (*A tota copulativa ad alteram eius partem (principalem) valet consequentia*). In two copies of Billingham's treatise the following example is given: "Socrates runs and Plato disputes; therefore Plato disputes."[15] Here, the requirement for a formal consequence is not fulfilled by a topical relation. Nevertheless, it makes sense to say that the consequent is understood in the antecedent because the consequent is part of the antecedent.

Another rule to consider in this context is a rule about exclusive propositions: "A consequence from an affirmative exclusive proposition to its universal proposition with transposed and direct terms is valid" (*Ab exclusiva affirmativa ad eius universalem de terminis transpositis in terminis rectis valet consequentia*). The example "Only a man runs; therefore everything running is a man"[16] does not hold in virtue of a topical relation between the antecedent and the consequent either. Here, the consequent follows from the antecedent by a syntactical transformation and is therefore understood in the antecedent, *i.e.*, the example is also a formal consequence.

To explain the requirement of the *intelligitur in* it is not sufficient to refer to the topics. The validity of consequences like the two just mentioned cannot be demonstrated by reasons pertaining to a relation of terms in the antecedent and the consequent. Ockham mentions these two possibilities in relation to his first group of formal consequences. This group includes all consequences that hold by means of an extrinsic medium, *i.e.*, by means of

---

[15] [S, III, 2] and [C, III, 2].
[16] [S, V, 4]; [O, V, 3]; [C, V, 2.b]; *cf.* [R, f. 2ᵛ]; *cf.* also [T, IV, 7].

a rule concerning the form of the propositions and not the categorematic terms:

> *Consequentia autem quae tenet per medium extrinsecum est quando tenet per aliquam regulam generalem quae non plus respicit illos terminos quam alios. Sicut ista consequentia "tantum homo est asinus, igitur omnis asinus est homo" non tenet per aliquam propositionem veram formatam ex istis terminis "homo" et "asinus", sed per istam regulam generalem "exclusiva et universalis de terminis transpositis idem significant et convertuntur." Et per talia media tenent omnes syllogismi.*
> [W. of Ockham SL, III-3, 1, p.588, 28-35]

In these cases the consequences hold in virtue of a syntactical relation (*ratione complexorum*)[17] between the antecedent and the consequent. For the same reason all syllogistic consequences belong to Ockham's first group of formal consequences. But formal consequences like "Socrates runs and Plato disputes; therefore Plato disputes", "Only a man runs; therefore everything running is a man" and syllogistic consequences cannot be explained by a unified concept of consequences. This means that the extrinsic medium which defines Ockham's first group of formal consequences does not describe a unified group. While rules for valid consequences, like "A consequence from the whole copulative proposition to one of its (principal) parts is valid" belong to the logic of propositions, rules like "A consequence from an affirmative exclusive proposition to its universal proposition with transposed and direct terms is valid", and rules for syllogistic consequences belong to the logic of terms. What we see here is that the modern division into term logic and logic of propositions is not an appropriate instrument to analyze medieval logic.

Let us take another look at the definition of a formal consequence and consider the different rules for disjunctive propositions. Difficulties arise with respect to the following rule: "A consequence from one part of a disjunctive proposition to the whole disjunctive proposition is valid" (*A parte disiunctive ad totam disiunctivam est bona consequentia*), e.g., "Socrates runs; therefore Socrates runs or Plato disputes" [T, III, 3].[18] The question of whether the requirement for a formal consequence, *i.e.*, the *intelligitur in*, is fulfilled or not, and if it is fulfilled, how it could be explained, is answered neither in Billingham's treatise nor in the Toledo commentary on Billingham. In the example, the antecedent and the consequent are not related due to their categorematic terms. But in this case it might also be difficult to explain the *intelligitur in* by a syntactical relation between the antecedent and the consequent. However, it is generally supposed that this consequence and similar ones are formal consequences because neither of

---
[17] *Cf.* [W. of Ockham, I-4, 1, p.15, 1-8].
[18] *Cf.* [S, IV, 3]; [O, IV, 3]; [C, IV, 3]; [R, f. 2ʳ].

the two rules for material consequences (mentioned below) can be applied to them.[19]

There is a close connection between the formal consequence in the treatises of Billingham and his contemporaries, and Ockham's and Burley's definition of a formal (natural) consequence. According to this definition, the consequent can be inferred from the antecedent either in virtue of a semantical or a syntactical relation. Neither in Billingham's treatise nor in those of his contemporaries do we find a division of formal consequences like the one in Ockham's (and Burley's)[20] treatises. Rather, their requirement for a formal consequence, *i.e.*, the *intelligitur in*, comprises the two groups without explicitly distinguishing between them.

With regard to the examples added to the definition of a formal consequence, we have just seen that a proper formal consequence is valid due to a topical relation between the terms in the antecedent and the consequent. The importance of these consequences, which hold by means of the dialectical *loci*, can also be confirmed by a further passage in the Toledo commentary on Billingham's treatise. The commentator states that in a valid consequence the antecedent cannot be true without the consequence. Although this is true for all formal and material consequences, the example he gives is again a formal consequence whose consequent is understood in the antecedent due to a topical relation of terms:

> *Consequentia que valet est in qua antecedens non potest esse verum sine consequente, ut:* "*tu curris; ergo tu moveris.*" [T, I, 6]

We do not conclusively know whether Billingham required the *intelligitur in* only for proper formal consequences or whether all formal consequences had to fulfil it. It is possible that Billingham and his contemporaries considered the *intelligitur in* as decisive for proper formal consequences but did not require it of consequences that are valid in virtue of a syntactical relation between the antecedent and the consequent. However, if they required the *intelligitur in* of all consequences which are valid but not material, then they might have had difficulties in explaining how it is fulfilled with regard to some of their rules. But probably this fact did not worry medieval logicians.

The concept of the formal consequence according to the Oxford tradition obviously differs from the Paris tradition. John Buridan, *e.g.*, regards consequences that are valid in virtue of a topical relation as material consequences:

---

[19] For more detailed information *cf.* my commentary in: [Web03, I.1.2, IV.3, IV.4, and VII.5].
[20] [W. Burley TL, II-1, 3, p.86, 9-15].

Sed consequentia materialis est cui non omnis propositio consimilis in forma ⟨quae formaretur⟩ esset bona consequentia, uel, sicut communiter dicitur, quae non tenet in omnibus terminis forma consimili retenta; uerbi gratia, "homo currit; ergo animal currit", quia in his terminis non ualet "equus ambulat; ergo lignum ambulat." [J. Buridan DC, I-4, p.23, 10-14]

But in the Oxford tradition as well as in the Paris tradition material consequences are of less importance than formal consequences, *i.e.*, while for English authors like Billingham and his contemporaries proper formal consequences are based upon a dialectical *locus*, authors of the Paris tradition regard these consequences as material consequences and therefore as less important.

How much the topics influenced the development of the theory of consequences is a matter of debate. It is in particular Eleonore Stump who argues, in her article *Topics: Their Development and Absorption into Consequences* [Stu82b], that the theory of consequences developed out of the doctrine of the topics. Green-Pedersen, however, points out that the textual situation does not support the widely shared view which ascribes a leading role to the topics. He, instead, regards the treatises on syncategorematic words, and the sophism-collections arranged after syncategoremes, as the places where the greatest number of similar or even identical discussions can be found. Furthermore, he mentions Boethius' *De hypotheticis syllogismis* and Aristotle's *Prior analytics* as possible ancient source-books.[21] Regardless of how much influence on the theory of consequences the doctrine of the topics actually had, it might plausibly be assumed that it was not the only source.

Let us finally have a brief look at the concept of a material consequence (*consequentia materialis*), which Billingham and his contemporaries (as well as Ockham and Burley) distinguish from the concept of a formal consequence.

Billingham defines the material consequence as follows: A consequence is material if the antecedent is impossible or the consequent necessary, and if the consequent is not understood in the antecedent (*consequens non intelligitur in antecedente*):

---

[21][Gre84, especially p.291*sq* and p.294*sq*]. It should be mentioned here that Green-Pedersen focusses on the division of consequences corresponding to the contintental tradition of, for instance, Buridan, that later became predominant. However, the theory of consequences developed in England around 1300. There, as seen above, the authors regarded consequences which are valid by means of a dialectic *locus* not only as formal consequences but even as proper formal consequences. Thus, in the Oxford tradition these consequences were of much more importance than in the Paris tradition where they were regarded as material consequences that were of little interest. On material consequences and the topics in Buridan see also [Zup03, ch.6].

> *Consequentia materialis est illa, ubi antecedens est propositio impossibilis et consequens non intelligitur in antecedente; vel consequentia materialis est, ubi consequens est necessarium et non intelligitur in antecedente. Exemplum de primo: "homo est asinus; ergo homo est capra"; ista consequentia est bona de materia, quia antecedens est propositio impossibilis, et consequens non intelligitur in antecedente. Exemplum de secundo: "Sortes disputat; ergo deus est"; ista consequentia est bona de materia, quia consequens est necessarium et non intelligitur in antecedente.*
>
> [S, I, 2.1][22]

It is remarkable that Billingham uses the criterion of the *intelligitur in* to distinguish the material consequence from the formal one, but in order to define the material consequence this is not sufficient. In addition to this negative definition he lists two cases:

1. The antecedent is impossible. In this case it is not possible that the antecedent is true without the consequent, *i.e.*, a consequence with an impossible antecedent is valid in any case, *e.g.*: "A man is an ass; therefore a man is a goat." The corresponding rule can be found in the commentary of Toledo: "From the impossible anything follows" (*Ex impossibili sequitur quodlibet*) [T, I, 9.1].

2. The consequent is necessary. In this case again it is impossible that the antecedent is true, and the consequent occurs and is false, *i.e.*, a consequence with a necessary consequent is also valid in any case, *e.g.*: "Socrates disputes; therefore God exists." The corresponding rule is: "The necessary follows from anything" (*Necessarium sequitur a quolibet*) [T, I, 9.2].

Billingham's text suggests that these two cases (and all consequences where these well-known rules are applied in some way) cover all possibilities for material consequences. These consequences are valid only because of the truth values of their antecedent and their consequent. Thus, the minimal condition for the validity of a consequence becomes the sufficient condition for a material consequence.

Unlike Billingham, Ockham defines in his *Summa* the material consequence independently from the formal consequence. A material consequence holds by means of an extrinsic medium (*medium extrinsecum*) concerning the truth, falsity, necessity or impossibility of both the antecedent and the consequent. Truth, falsity, *etc.*, depend on the specific contents of the terms, *i.e.*, a material consequence holds in virtue of its terms:

> *Consequentia materialis est quando tenet praecise ratione terminorum et ratione alicuius medii extrinseci respicientis praecise generales condiciones*

---

[22] *Cf.* [O, I, 2.1]; [T, I, 7].

> *propositionum; cuiusmodi sunt tales "si homo currit, Deus est; homo est asinus, igitur Deus non est" et huiusmodi.*
> [W. of Ockham SL, III-3, 1, p.589, 55-58][23]

The proposition "A man is an ass", *e.g.*, is impossible because the predicate term "ass" is incompatible with the subject term "man". The extrinsic medium refers to the above mentioned rules "From the impossible anything follows" and "The necessary follows from anything." Ockham does not regard consequences that are valid according to these rules to be as important as formal consequences, and he remarks that their rules are not used frequently:

> *Aliae regulae dantur, quod "ex impossibili sequitur quodlibet" et quod "necessarium sequitur ad quodlibet"; [...]. Sed tales consequentiae non sunt formales, et ideo istae regulae non sunt multum usitatae.*
> [W. of Ockham SL, III-3, 38, p.730, 88-p.731, 92]

This evaluative attitude concerning the material consequence is also found in Burley. In the *Tractatus longior* of his treatise *De puritate artis logicae* and in his early treatise *De consequentiis* he distinguishes the "natural" consequence from the accidental consequence (*consequentia accidentalis*):

> *Consequentia accidentalis est, quae tenet per locum extrinsecum, et est quando antecedens non includit consequens, sed tenet per quandam regulam extrinsecam, ut: "Si homo est asinus, tu sedes"; haec consequentia est bona et tenet per hanc regulam: "Ex impossibili sequitur quodlibet."*
> [W. Burley TL, II-1, 1, p.61, 8-12]

The accidental consequence corresponds to Ockham's material consequence. In using the terms "natural" versus "accidental" Burley also indicates that the accidental consequence is of less interest than the formal consequence.

These evaluative remarks about the material or accidental consequence express a general view in the 14th century. The reason for it might be that a consequence which holds only in virtue of the truth values of the antecedent and the consequent is not as useful in disputations as a formal consequence.[24] Billingham accepts the material consequences as valid and lists two different cases with examples at the beginning of his treatise, but then moves on to the formal consequences.[25] His distinction between formal

---

[23] I agree with Schupp's correction [Sch93, p.215], who points out that the *non* before *ratione alicuius medii extrinseci respicientis praecise generales condiciones propositionum* has to be deleted because a material consequence holds precisely in virtue of these general conditions of propositions.

[24] *Cf.* in this context the tradition of the treatises on obligations (*De obligationibus*). The aim of obligations disputations was to demonstrate the skills in applying logical rules in the practice of argumentation. *Cf., e.g.,* [Spa82] and [Stu82a].

[25] [S, I, 2.2]; *cf.* [O, I, 2.1]; [R, f. 1^r].

and material consequences is not one where the two types are deemed equal either.

To conclude, the concept of the *consequentia formalis* in the 14th century theory of consequences according to the Oxford tradition highlights the difference between the medieval English and the modern concept of what it is to be formal. For English logicians in the 14th century a consequence is formal if the consequent follows from the antecedent either in virtue of a semantical or a syntactical relation between them. But a proper formal consequence holds in virtue of a topical relation between two terms of the antecedent and the consequent. Due to this relation the criterion of the *intelligitur in* is fulfilled, and it is impossible for the antecedent to be true without the consequent. The definition at the beginning of Billingham's treatise is based on this concept of a formal consequence. It was modified without further explanation, according to the listed rules in his treatise. The question of how to explain the *intelligitur in* in consequences that are valid because of a syntactical relation, like in the above mentioned example: "Socrates runs; therefore Socrates runs or Plato disputes", is not addressed in the treatises of Billingham and his contemporaries. Neither do we get information about whether all formal consequences, or only those that are based on a dialectical *locus*, had to fulfil the criterion of the *intelligitur in*.

However, medieval logicians in the 14th century probably did not have problems with their "incoherent" concept of a formal consequence, nor did the lack of a structural criterion to distinguish a formal consequence from a material one worry them, because their division of consequences into formal and material was motivated by practical use in their disputations.

The concept of the formal consequence in the Oxford tradition stays closer to ordinary language than the concept in the Paris tradition. As we have seen above, Buridan, *e.g.*, rules out consequences that are valid in virtue of a topical relation of terms. For him, these consequences are material and therefore of less interest. Formal consequences hold because of their forms, regardless of any semantical relations between the antecedent and the consequent.[26] Thus, we might say that the concept of the formal consequence according to the Paris tradition is nearer the modern concept of what it is to be formal than the corresponding concept within the Oxford tradition is.[27]

---

[26] *Cf.* [J. Buridan DC, I-4, p.22*sq*, 5-9].

[27] For the differences between English and continental logic in the early 14th century *cf.* also A. de Libera's investigations of the differences between these traditions in the 13th century [dLi82].

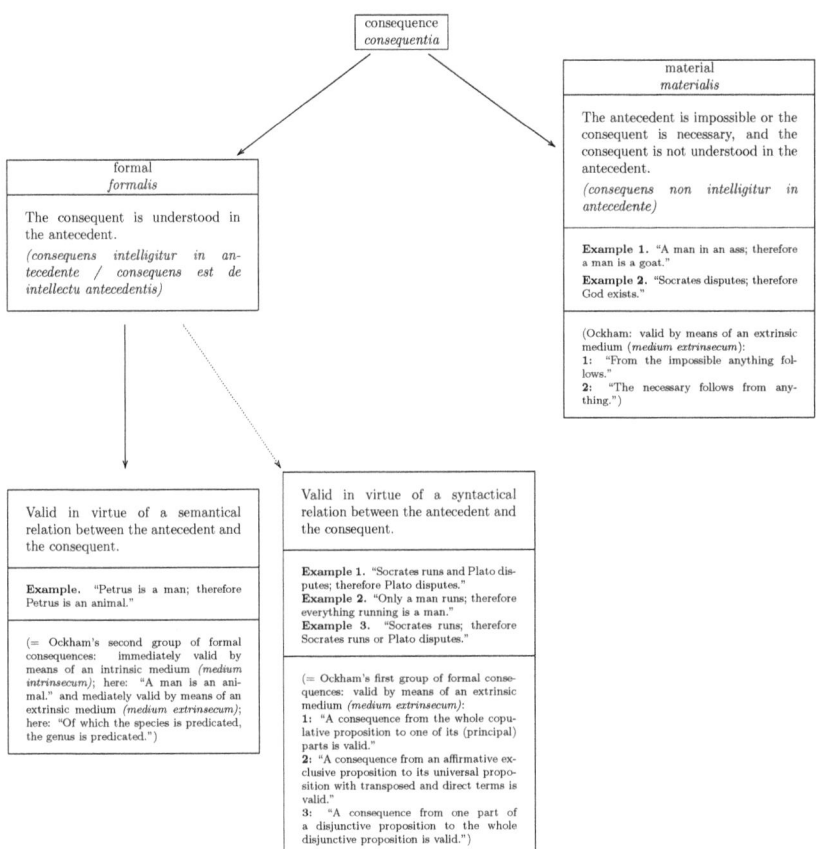

Figure 1. The 14th century concept of consequence in the Oxford tradition, based on Billingham's treatise on consequences.

Primary Sources.

[Anonymus$_P$ DC]	**Anonymus**$_P$, De consequentiis, cod. Padua, Biblioteca Universitaria 1123, ff. 1vb-3va; *manuscript*
[J. Buridan DC]	**John Buridan**, Tractatus de consequentiis, *in:* [Hub76]
[Petrus Hispanus]	**Petrus Hispanus**, Tractatus called afterwards Summule logicales, *in:* [dRi72]
[Pseudo-Scotus]	**Pseudo-Scotus**, In Librum primum Priorum Analyticorum Aristotelis Quaestiones, *in:* [Wad68a]
[C]	**Richard Billingham**, De consequentiis, Casanatense version, *in:* [Web03, p.63-76]
[O]	**Richard Billingham**, De consequentiis, Oxford version, *in:* [Web03, p.45-61]
[R]	**Richard Billingham**, De consequentiis, cod. Barcelona, Archivo de la Corona de Aragón, Ms. Ripoll 166, ff. 1^r-5^r; *manuscript, version not edited/manuscript damaged*
[S]	**Richard Billingham**, De consequentiis, Salamanca version, *in:* [Web03, p.27-43]
[T]	**Richard Billingham**, De consequentiis, Toledo commentary, *in:* [Web03, p.77-125]
[R. Lavenham DC]	**Richard Lavenham**, Consequentiae, *in:* [Spa74]
[R. Fland DC]	**Robert Fland**, Consequentiae, *in:* [Spa76]
[Sextus Empiricus]	**Sextus Empiricus**, Outlines of Pyrrhonism, *in:* [Bur55]
[T. Manlevelt DC]	**Thomas Manlevelt**, De consequentiis, cod. Erfurt Q271, ff. 55v-62r; *manuscript*
[W. Burley DC]	**Walter Burley**, De consequentiis, *in:* [Gre80]
[W. Burley TL]	**Walter Burley**, De puritate artis logicae, Tractatus Longior with a Revised Edition of the Tractatus Brevior, *in:* [Boe55]
[W. of Lucca]	**William of Lucca**, Summa dialectice artis, *in:* [Poz75]
[W. of Ockham]	**William of Ockham**, Scriptum in librum primum sententiarum ordinatio, distinctiones IV-XVIII, *in:* [Etz77]
[W. of Ockham SL]	**William of Ockham**, Summa logicae, *in:* [BoeGálBro74]
[W. of Osma DC]	**William of Osma**, De consequentiis, *in:* [Sch91]
[W. of Sherwood]	**William of Sherwood**, Introductiones in logicam, *in:* [BraKan95]

References.

[Bäu91]	Clemens **Bäumker**, Eine bisher unbekannte mittelalterliche lateinische Übersetzung der " Pyrroneioi Hypotyposeis" des Sextus Empiricus, **Archiv für Geschichte der Philosophie** 4 (1891), p.574-577
[Boe55]	Philotheus **Boehner** (*ed.*), Walter Burley: De puritate artis logicae, Tractatus Longior with a Revised Edition of the Tractatus Brevior, Franciscan Institute 1955

[BoeGálBro74]	Philotheus **Boehner**, Gedeon **Gál**, and Stephen F. Brown (*eds.*), William of Ockham: Summa logicae, Franciscan Institute 1974 [Opera Philosophica I]
[BraKan95]	Hartmut **Brands** and Christoph **Kann** (*eds.*) William of Sherwood: Introductiones in logicam, textkritisch hrsg., übersetzt, eingeleitet und mit Anmerkungen versehen, Meiner 1995
[Bur55]	Robert G. **Bury** (*ed.*), Sextus Empiricus: Outlines of Pyrrhonism, Heinemann 1955 [Loeb Classical Library]
[Etz77]	Girard I. **Etzkorn** (*ed.*), William of Ockham: Scriptum in librum primum sententiarum ordinatio, distinctiones IV-XVIII, Franciscan Institute 1977 [Opera Theologica III]
[Gre80]	Niels Jørgen **Green-Pedersen** (*ed.*), Walter Burley: De consequentiis, **Franciscan Studies** 40 (1980), p.102-166
[Gre84]	Niels Jørgen **Green-Pedersen**, The Tradition of the Topics in the Middle Ages, The Commentaries on Aristotle's and Boethius' "Topics", Philosophia-Verlag 1984
[Hub76]	Hubert **Hubien** (*ed.*), John Buridan: Tractatus de consequentiis, Publications universitaires, Louvain 1976 [Philosophes médiévaux XVI]
[Jac93]	Klaus **Jacobi** (*ed.*), Argumentationstheorie, Scholastische Forschungen zu den logischen und semantischen Regeln korrekten Folgerns, Brill 1993
[KreKenPin82]	Norman **Kretzmann**, Anthony **Kenny**, and Jan **Pinborg** (*eds.*), The Cambridge History of Later Medieval Philosophy, From the Rediscovery of Aristotle to the Disintegration of Scholasticism 1100–1600, Cambridge University Press 1982
[dLi82]	Alain **de Libera**, The Oxford and Paris traditions in logic, *in:* [KreKenPin82, p.174-187]
[Mit95]	Jürgen **Mittelstraß** (*ed.*), Enzyklopädie Philosophie und Wissenschaftstheorie, volume I, Metzler 1995
[Mor71]	Edgar **Morscher**, Der Begriff "*consequentia*" in der mittelalterlichen Logik, **Archiv für Begriffsgeschichte** 15 (1971), p.133-139
[ODo74]	J. Reginald **O'Donnell** (*ed.*), Essays in Honour of Anton Charles Pegis, Pontifical institute of mediaeval studies, Toronto 1974
[Poz75]	Lorenzo **Pozzi** (*ed.*), William of Lucca: Summa dialectice artis, Liviana 1975
[dRi72]	Lambertus Marie **de Rijk** (*ed.*), Petrus Hispanus: Tractatus called afterwards Summule logicales, Van Gorcum & Co. 1972
[Sch88]	Franz **Schupp**, Logical Problems of the Medieval Theory of Consequences, with the Edition of the "Liber consequentiarum", Bibliopolis – Edizioni di filosofia e scienze 1988
[Sch91]	Franz **Schupp**, William of Osma: De consequentiis, Über die Folgerungen, textkritisch hrsg., übersetzt, eingeleitet und kommentiert, Meiner 1991
[Sch93]	Franz **Schupp**, Zur Textrekonstruktion der formalen und materialen Folgerung in der kritischen Ockham-Ausgabe, *in:* [Jac93, p.213-221]
[Spa74]	Paul Vincent **Spade** (*ed.*), Richard Lavenham: Consequentiae, *in:* [ODo74, p.99-112]

[Spa76]	Paul Vincent **Spade** (*ed.*), Robert Fland: Consequentiae, **Mediaeval Studies** 38 (1976), p.54-84
[Spa82]	Paul Vincent **Spade**, Obligations, Developments in the Fourteenth Century, *in:* [KreKenPin82, p.335-341]
[Stu82a]	Eleonore **Stump**, Obligations: From the Beginning to the Early Fourteenth Century, *in:* [KreKenPin82, p.315-334]
[Stu82b]	Eleonore **Stump**, Topics: Their Development and Absorption into Consequences, *in:* [KreKenPin82, p.273-299]
[Wad68a]	Lucas **Wadding** (*ed.*), Pseudo-Scotus: In Librum primum Priorum Analyticorum Aristotelis Quaestiones, *in:* [Wad68b, p.273-330]
[Wad68b]	Lucas **Wadding** (*ed.*), John Duns Scotus: Opera omnia, volume I, Lyon 1639; *reprint:* Georg Olms 1968
[Web03]	Stephanie **Weber** (*ed.*), Richard Billingham: *"De consequentiis"* mit Toledo-Kommentar, Kritisch herausgegeben, eingeleitet und kommentiert, Grüner 2003 [Bochumer Studien zur Philosophie 38]
[Wol95]	Gereon **Wolters**, *Consequentiae*, *in:* [Mit95, p.415f]
[Zup03]	Jack **Zupko**, John Buridan, Portrait of a Fourteenth-Century Arts Master, University of Notre Dame Press 2003

Received: May 16th, 2003;
In revised version: February 12th, 2004;
Accepted by the editors: March 1st, 2004.

www.ingramcontent.com/pod-product-compliance
Ingram Content Group UK Ltd.
Pitfield, Milton Keynes, MK11 3LW, UK
UKHW021317180426
11947UKWH00015B/1284